BIRKHÄUSER

A Lost Mathematician, Takeo Nakasawa

The Forgotten Father of Matroid Theory

Hirokazu Nishimura and Susumu Kuroda, Editors

Birkhäuser
Basel · Boston · Berlin

Editors:

Hirokazu Nishimura
University Tsukuba
Dept. Mathematics
Tennodai 1-1-1
305-0006 Tsukuba
Ibaraki-ken
Japan
e-mail: logic@math.tsukuba.ac.jp

Susumu Kuroda
Ludwigstraße 4
94032 Passau
Germany

Library of Congress Control Number: 2008939300

Bibliographic information published by Die Deutsche Bibliothek.
Die Deutsche Bibliothek lists this publication in the Deutsche Nationalbibliografie; detailed bibliographic data is available in the Internet at http://dnb.ddb.de

ISBN 978-3-7643-8572-9 Birkhäuser Verlag AG, Basel - Boston - Berlin

Basel · Boston · Berlin
P.O. Box 133, CH-4010 Basel, Switzerland
Part of Springer Science+Business Media
Printed on acid-free paper produced from chlorine-free pulp. TCF ∞
Printed in Germany

ISBN 978-3-7643-8572-9 e-ISBN 978-3-7643-8573-6

9 8 7 6 5 4 3 2 1 www.birkhauser.ch

Dedicated to Mie Nakasawa, who is a sister-in-law of Takeo Nakasawa. Fortunalely she is still alive, and Takeo Nakasawa is still alive in her memory.

Preface

Takeo Nakasawa was a very unfortunate mathematician. He was born in Kochi prefecture of Japan in 1913, and he died of dystrophia at the age of 33 in Khabarovsk in the Soviet Union in 1946. He was forced to live his youth in Japan in an era of a fanatical government, which entered upon the Fifteen Years War (1931–1945) without any clear purpose or any efficient leadership. His discovery of matroid theory, made independently of the work of American mathematician Hassler Whitney, remained unnoticed or forgotten for a long time. As far as I know, it was Joseph P. S. Kung who discovered Nakasawa in his thorough investigation on the history of matroid theory in the 1980s. Nakasawa wrote four papers, the first three of which were concerned with what is now called matroid theory. The remaining paper was concerned with general topology. All four papers were written in German in the 1930s, and they were published in the same university bulletin, namely, the bulletin of the Tokyo University of Arts and Sciences, from which Nakasawa graduated in 1935 and where he stayed as an assistant until August 1938. His four papers were published during 1935–1938. His career as a mathematician ended in 1938 or so. He left Japan for Manchuria to seek a new life in a new frontier. In Manchuria he worked as a bureaucrat and married a Japanese woman in July 1939. He had three children there. At the end of the Second World War, the Soviet Union invaded Manchuria and took away many Japanese soldiers to such harsh places as Siberia, where they were forced to perform hard labor in camps without sufficient food or shelter. Nakasawa was one of these unfortunate Japanese soldiers to suffer such a brutal death in Siberia.

In summer 2006 Akira Saito, a professor of the Department of Computer Science and System Analysis of Nihon University, asked members of the Institute of Mathematics of our university, via Professor Akito Tsuboi of our institute, how to write "Takeo Nakasawa" in Chinese characters. This was the first time I ever heard the name. By consulting Zentralblatt Math, I could easily identify his four papers, all of which were published in the bulletin of the Tokyo University of Arts and Sciences (which became the Tokyo University of Education after the Second World War). In the 1970s our university, called University of Tsukuba, replaced the Tokyo University of Education. For this historical reason, these three universities have the same association of alumni, and the central library of our university has a special corner for materials related to the above association of alumni. Therefore it was not difficult for me to find that Nakasawa was an assistant of the Tokyo University of Arts and Sciences when his four papers were published, and that Nakasawa had died by 1953. Then a question arose in my mind of when and how Nakasawa died and why at such a relatively young age. Although a complete answer to the question could not be determined, I could easily find how to write Takeo Nakasawa in Chinese characters, and I conveyed the result to Professor Saito.

Hiroyuki Tachikawa, who is presently an emeritus professor of our university and was once a student of Kiiti Morita, came to our university two days a week after his retirement. Kiiti Morita is famous for Morita equivalence and Morita duality. I consulted with Tachikawa about this question, and he was so kind as to get in touch with Tokunosuke Yoshida, who was once a lecturer of Tokyo high school for teachers and had connections with some people surrounding Nakasawa. According to Yoshida, Nakasawa presumably went to Manchuria after he was dismissed from the Tokyo University of Arts and Sciences. Then an idea came to my mind that if he went to Manchuria in the 1930s and had stayed there until the end of the Second World War, it was very probable that he was detained in Siberia or some other places in the Soviet Union by the invading Soviet forces and was forced to labor there. I know well that many Japanese soldiers had the same or similar experiences. Then another idea came to my mind that Nakasawa might have died in the Soviet Union, because I know well that many former Japanese soldiers detained in Siberia or other places in the Soviet Union at the end of the Second World War had died there for unbalanced nutrition or some other diseases. In December 2006 I asked Yasukuni Shrine, where the war dead of modern Japan was apotheosized, about Takeo Nakasawa. Yasukuni Shrine was kind enough to investigate its records, and it got seven hits on the name "Takeo Nakasawa". Only one of the seven Takeo Nakasawas had the same birthday as the one I was looking for. To confirm my discovery, I asked Yasukuni Shrine where Takeo Nakasawa was born. A spokesperson there replied "Kochi prefecture (of Japan)", which gave me confidence that I had really found the mathematician I had been seeking.

A fortunate discovery was that Nakasawa's sister in law was still alive. Her name is Mie Nakasawa. Even if Nakasawa were alive, he would at the time of my investigation have been 94 years old. This meant that it would not be easy to find someone who knew him personally. This is particularly so for a person who died almost unnoticed in the Soviet Union at the age of 33 in 1946.

Thanks to M. Nakasawa, I learned that Nakasawa had married in Manchuria and had three children. Thanks to her generous permission, I was able to look at a copy of Takeo Nakasawa's family register. Although M. Nakasawa was only 7 years old or so when Takeo Nakasawa left for Manchuria, and although she had not met him since then, she was of much help in my preparation of this book. In particular, she had finally found Takeo Nakasawa's photos, which we once believed to have been lost under the utter confusion of the Second World War. I am very happy to be able to include these photos in the book. There are more than ten photos, ranging from his childhood to one in 1944 (two years before his death), in which he stood with his first son and his first daughter.

If Nakasawa's four papers in German were available only in the bulletin of the Tokyo University of Arts and Sciences, they would remain almost unnoticed forever. It is never easy to access the publications of an extinct university, and German is not so popular as it once was in mathematics. This is the reason why I have decided to translate his four German papers into English and to publish them in a book form. The translation was not a simple matter. Since I am not familiar with German, I asked Professor Kuroda, who is an expert in German, to cooperate with me in translation as the coauthor. Professor Kuroda kindly accepted my offer. Frankly speaking, Nakasawa's German was not necessarily sophisticated, and so my English translation is not verbatim. I tried to make Nakasawa's works as accessible as possible to our contemporary mathematicians, not to mathematicians in the 1930s. Nakasawa's notation was considerably antic. By way of example, he wrote $A \leftarrow\ \rightarrow B$ in place of $A \leftrightarrow B$ (two propositions A and B are equivalent) and ${}^*a_2, \ldots a_n$ in place of $\hat{a}_1, a_2, \ldots a_n$ (the sequence obtained from the sequence $a_1, a_2, \ldots a_n$ by elimination of the first element a_1). In this sense I am not only the translator but the strict referee of his papers. Since Professor Kuroda was not familiar with mathematics, I asked him to translate Nakasawa's four German papers into English without taking notice of mathematical expressions or mathematical jargon at all. Professor Kuroda was so efficient as to give me the English translation of Nakasawa's four papers, one a month for four consecutive months. My work began upon these first English drafts. Strictly speaking, the fourth paper of Nakasawa, which was concerned with general topology, was dealt with principally by Professor Kazuhiro Kawamura of our institute, who is so reserved as to decline to post his name as a coauthor. Nevertheless, I gladly acknowledge that Kawamura's contribution to this book was essential and crucial.

The 1930s, when Nakasawa did his own research on linear dependence and general topology, were an age of storm and stress (Sturm und Drang) both in the world and in Japan. In Germany, the Weimar Constitution was replaced by the Nazi party in 1933, which caused a great exodus of European intellectuals into the U.S. In the early days of the 1930s, Japan occupied the whole of Manchuria to establish its puppet state called Manchukuo, while the military became a powerful political power and Japan became more and more fanatical at home. To understand Nakasawa's life, the reader should understand this situation, in particular, why Japan was led to the Fifteen Years War (1931–1945) internationally at a time when it was a terribly fanatical state domestically. Since European and American readers are not necessarily familiar with Japanese history, I have tried to tell this story in the first four chapters. Those chapters are presented without references, which does not mean that I would like to claim originality there. On the contrary, I have exploited the common knowledge of history, and I have consulted many Japanese books written by true historians. It would be absurd to believe that a mathematician could present original results in history, and I make no such claim. The fifth chapter is devoted to a brief review of the mathematical landscape surrounding Nakasawa in the 1930s. To understand someone in the past, it is necessary to understand him or her historically, i.e., in the historical context where the person at issue was destined to live.

Acknowledgements In preparing this book, I owe much to many people. First of all, I must express my deep thanks to Professor Akira Saito, who let me know that there was a great mathematician called Takeo Nakasawa. Without him, I would never have know even his name. I would like to express my sincere gratitude to Yasukuni Shrine, which has apotheosized Takeo Nakasawa among the war dead for decades. I am thankful to Professor Kikai, a professor in philosophy of our university, who introduced me to Professor Kuroda. I am deeply indebted to Tachikawa and Yoshida, who have turned my attention to Manchuria. I am also indebted to Professor Kawamura, who did most of the preparation of the complete English translation of Nakasawa's fourth paper based on Kuroda's tentative translation into English. I am very glad to acknowledge my indebtedness to M. Nakasawa, without whom it would have been impossible for me to trace the personal trail of Takeo Nakasawa. I am also glad to acknowledge my indebtedness to Hamada, who worked in the bureau of welfare in Kochi Prefecture, sent me Takeo Nakasawa's military record, and introduced me to M. Nakasawa. I owe much to Professor Günter M. Ziegler who, amidst his demanding responsibilities as president of the Mathematical Society of Germany, took the time to help make this conjectured book become a reality by recommending it to Birkhäuser. I owe much also to Thomas Hempfling and Karin Neidhart of Birkhäuser for their endeavors in publishing this book. Last but not least, I am deeply indebted for financial support to the Institute of Mathematics of the University of Tsukuba.

Hirokazu Nishimura
Institute of Mathematics, University of Tsukuba
Tsukuba, Ibaraki, 305–8571
Japan

Contents

The Life of Takeo Nakasawa

When I speak of Takeo Nakasawa, I cannot help but recall Sharaku Toshusai of the Edo period. He was an artist who specialized in ukiyo-e (Japanese woodblock prints). His artistic career was exceedingly short, only about ten months during the years 1794 to 1795. It appears to be unknown when he was born or when he died. The only certain fact is that he left approximately 150 ukiyo-e, modeled mainly on Kabuki actors. Among them, "Ebizo Ichikawa" is the masterpiece of masterpieces. It is modeled on the Kabuki actor Danjuro Ichikawa V after this name passed to the sixth in the succession. The name Ebizo Ichikawa is used by the fifth after the succession. Just as a good performance of a talented actor or actress inspires the audience, this ukiyo-e undoubtedly inspired its contemporaries. It is clear that Sharaku Toshusai opened up a new field of ukiyo-e by drawing the physical features of actors with stark exaggeration and by omitting the subject's softer aspects in order to more vividly describe the human personality. He is documented in *Ukiyo-e Ruiko* [Considerations on Currents of Ukiyo-e], which is a book published for the first time in 1800. Appearance in this book led to high acclaim for him, since of the many artists represented, only 37 were ukiyo-e artists. However, it seems that we will never know anything more about the man, but only about his works: It has been said by some that Sharaku Toshusai is the same person as the Noh-actor Jurobei Saito in Awa, Shikoku, but this is now considered to be doubtful at best.

Mathematician Takeo Nakasawa is reminiscent of artist Sharaku Toshusai in that his time of activity was also extremely short. His life, if not to the extent of Sharaku Toshusai's, is nevertheless veiled in mystery. His work did not attract attention when he was alive and, after he finished his mathematical work he was swiftly forgotten, receiving high recognition only a long time after he died.

Now let us look at his profile. He was born on 5 February, 1913, in Kochi Prefecture, which lies in the Shikoku island of Japan. In the following year the First World War broke out; in the previous year the Meiji Emperor had passed away, and the renowned army general and war hero Maresuke Nogi and his wife died the death of martyrs. Thus an era of Japanese history had ended.

It can be said that Nakasawa's life started at that time with many vicissitudes. His family genealogy might well confuse the reader. He was born not as Takeo Nakasawa but as Takeo Sogabe and became Takeo Nakasawa only at the age of 11 on 18 March 1924, when he was adopted as a son by Morinao Nakasawa and his wife Shigeki Nakasawa. Takeo Nakasawa's real father was Tomoshiro Sogabe. His real mother Morie Sogabe was an elder sister of Morinao Nakasawa. She was born as Morie Nakasawa, and became Morie Sogabe after her marriage with Tomoshiro Sogabe. Morinao Nakasawa and Shigeki Nakasawa had three children before they adopted Takeo, after all of their three children had died. Morinao Nakasawa and Shigeki Nakasawa had two daughters after they had adopted Takeo, but one of them died of dysentery at the age of 5. Fortunately, the other daughter survives and was kind enough to share with me much information about Takeo Nakasawa.

In 1935 (the tenth year of the Showa period), he graduated from the department of mathematics, Tokyo University of Arts and Sciences, which is the predecessor of Tokyo University of Education to be succeeded by University of Tsukuba in the 1970s. His supervisor was Koshiro Nakamura, born in 1901 and 12 years senior to Nakasawa. In passing, Kiichi Morita, a Japanese mathematician very famous for the concepts of Morita equivalence and Morita duality, was born on February 11, 1915 (the fourth year of the Taisho period) and so was two years junior to Nakasawa. Morita also studied at Tokyo University of Arts and Sciences. Shortly after his graduation, Nakasawa was hired as an assistant at that University where he stayed until being discharged on 22 August, 1938. With the first three of his four papers written in German, all of which were published in the proceedings of the Tokyo University of Arts and Sciences during the period 1935–1938, he shares with Hassler Whitney in the U.S the distinction of co-founder of the theory of matroids.

While we will not explore the details of Nakasawa's mathematical work in this chapter, it should be noted in particular that those papers were published when he was in his early 20s. He was undoubtedly an early maturing genius: He wrote superb papers in his twenties. It is likely that when asked for their papers written in their twenties, many mathematicians, even those who are called experts in their fields, will break out into a cold sweat. At the age of 45, Teiji Takagi, very famous in number theory, published his epoch-making paper on class field theory, which made his name everlasting. At the age of 43, Kiichi Morita, an algebraist as well as a topologist, released what is now called Morita theory. In the mid-1960s, Minoru Tomita, very famous in the theory of operator algebras, wrote a paper on the prototype of the theory to be called later the Tomita-Takesaki theory. As he was born in 1924, this paper was written when he was over the age of 40. In fact, this paper was not published, because it was beyond the referee's understanding.

Finally we note that it was not easy in the 1930s and 1940s for Japanese mathematicians to have their ideas acknowledged by European or American mathematical circles. I will give a well-known example. The Japanese mathematician Kiyoshi Ito arrived at the idea of stochastic differential equations in 1942, when he wrote a paper in Japanese and published it in a Japanese journal. The theory of stochastic differential equations was undoubtedly so important as to have yielded Black-Scholes equations in financial engineering in the 1970s as a by-product. Nevertheless, Ito published his idea on stochastic differential equations in English in 1951. Imagine what would have happened if Ito had died in the 1940s.

Let us return to the profile of Nakasawa. He seems to have gone to Manchuria to look for a new start after he was discharged from his job on August 22, 1938. I have no clear idea of what he did in the following period. In July 2006, however, I found out that he is listed as deceased on the 1953 membership list of Meikei-kai, the alumni association of Tokyo University of Arts and Science, Tokyo University of Education, and the University of Tsukuba. In December 2006, by chance, it occurred to me that I might ask the Yasukuni Shrine, a national shrine for the spirits of the war dead of modern Japan, if he is enshrined there. I found seven Takeo Nakasawa's on the list, but after restricting them by birthdays, found the unique one. There is little doubt that he is really the Takeo Nakasawa who I have been looking for, since not only the birthday but the registered prefecture coincided with those of the desired one. On June 20, 1946 (Showa 21), the announcement of his death was sent to the Yasukuni Shrine by the Ministry of Health and Welfare. According to this report, he died of disease during the war at a hospital in Khabarovsk in Russia. Based on this fact, we can conjecture his probable story as being a familiar one for Japanese soldiers who met the end of the war in Manchuria: Maybe he was conscripted there and then seized as a captive by the USSR Forces which advanced southward at a furious pace just before the end of the war. Presumably he was subjected to slave labor in various places in Siberia, without sufficient food and shelter. At last his body screamed out in agony and he was brought to the hospital in Khabarovsk, but it was too late. He died at age 33. Here I will tentatively finish my personal odyssey on the trail of Takeo Nakasawa. It started at the resource center of the University of Tsukuba Library and ended at the Yasukuni Shrine, leaving me in a deep depression, not knowing what forces of nature or man to blame for Nakasawa's tragedy.

To further support the probability of this conjectured story, you should know that Manchuria when Takeo Nakasawa went there was not just the name of a region but a formally independent state called Manchukuo. In March 1932, Manchukuo was established with the last Emperor of the Qing Dynasty (one of the former names of China) as its head. The emperor at that time in China, by the name of Puyi, reigned as sovereign of Manchukuo. However, the new state was actually built as a puppet state, indeed a colony, of Imperial Japan. This colonization was effected by the Japanese Forces on its own volition without any negotiation with the Japanese government. At that time, the Japanese army seemed to have been terribly afraid of the tremendous progress of the First Five-Year Plan of the USSR and to have wanted to secure the occupation of Manchuria and Nei Mongol (Inner Mongolia) because of the anti-Japanese movement in China by Zhang Xueliang and his advocates.

The Manchurian incident, which prefaced the foundation of Manchukuo, started in September 1931, which was also the beginning of Japan's notorious 15-years war. The motion picture "The Last Emperor" starred Ryuichi Sakamoto, who played the role of Masahiko Amakasu, the emperor in the dark in Manchuria in contrast to Puyi as the emperor in daytime in Manchuria, and its soundtracks were also produced by him. The movie clarifies the story of Manchukuo, though it was dramatized with the usual exaggeration and transmogrifications of Hollywood productions. The Japanese government sent many Japanese citizens to Manchukuo as teams of immigrants. Because Japan at that time was economically destitute as a result of the world-wide depression, this immigration was intended partly to reduce the population inside Japan, but it was also partly intended to secure Manchukuo as a food supply station for Japan. In 1936, Kooki Hirota's cabinet put forward a plan to immigrate 5 million Japanese to Manchuria in 20 years from 1936 to 1956 by announcing the, so-called, Promoting Program for Reclaiming of Manchuria and Nei Mongolia Immigration. What a grandiose program it was! In fact, following this program, 200,000 people were sent to Manchuria or Nei Mongolia as farmers. There is no doubt that Takeo Nakasawa went willingly to Manchuria against such a background. The young Japanese as farmers in Manchukuo were called "warriors of sickles", an alias that provides unmistakable evidence of this exodus being a national policy. This immigration had actually been stopped as a formal program when Japan entered a period during which it lost command of the air and the sea on the Huang Hai (Yellow Sea) and the Sea of Japan, and Manchukuo had been brought to inevitable collapse. However, this possibility must not have crossed the mind of young Nakasawa, who apparently had been to Manchukuo in the late 1930s, in his mid 20s: Not a few people, as Nakasawa must have, dreamed of Manchukuo as a new frontier at that time.

Manchukuo consisted of the three eastern provinces Liaoning, Jilin, and Heilongjiang, in addition to Rehe and Nei Mongolia. It had five ethnic groups: the Hans, Manchus, Mongolians, Koreans, and Japanese. Its slogan was "Five ethnics in harmony." This phrase would certainly have won the grand prize in a "buzzword contest" if it had been presented in contemporary Japan. This slogan, as well as the name *Oodoo raku do* [the righteous realm of peace and prosperity; Arcadia], was invented by a dentist named Kaisaku Ozawa. Many readers have not heard this name, but the fact that he is the father of the world-famous conductor Seiji Ozawa will certainly ring a bell. The name *Seiji* is written as 征爾 in Kanji characters, and these characters were taken after Seishiro Itagaki and Kanji Ishiwara, who were the actual executors of the Manchurian Incident. If the origin is traced in this way, it should be no surprise that I felt something making waves in my mind when I first saw the name.

Let us get back to the point. As I mentioned previously, the five ethnic groups had lived in Manchukuo. It is said that quite a few Russians (about sixty thousand) lived there, especially in Harbin or along the Chinese Eastern Railway. Taking this fact into consideration, Kaisaku Ozawa should have said "Six Ethnics in Harmony". Since Japan won the First Sino-Japanese War in the end of the nineteenth century, Manchuria had been the place where Japan and Russia (at that time, the Romanov Dynasty) had fought over the territory and influence on it. After the Tripartite Intervention (France, Germany and Russia), the Quing granted the right of construction to Russia, and the Chinese Eastern Railway was fully opened to traffic in 1901. Due to this circumstance, many Russians originally lived along the railway, and more Russians came there as refugees because of the Russian Revolution in March 1917. Although the slogan "Five ethnics in harmony" was nothing but a slogan, it is indeed undeniable that Manchukuo was the crossroads where different cultures met.

It is generally said that Japanese people who were born in Manchukuo and grew up there got plenty of culture shock when they came back to the homeland after the end of the war. Li Xianglan (Yoshiko Yamaguchi, under her real name) is symbolic of the multicultural characteristic of Manchukuo. She was a native Japanese who had a good command of both Japanese and Chinese. She came to public notice as a Chinese actress from the Manchurian Motion Picture Association in 1938, her Japanese origin being veiled in secrecy. She was intended as a symbol of sham harmony between Japan and China. In the progress of the Fifteen Years' War, both the mainland of Japan and the Manchukuo partook of the dreadful experiences of wartime. It was, however, a quintessential "blue rose" to place Manchukuo under the control of mainland Japan. It was surprising to see at that time pictures of a woman in Manchukuo modeling a refined set of bathing suits. It seems to me that people in the mainland of Japan at that time would not even have imagined such pictures. I have no doubt that Harbin in Manchukuo was an international city at that time.

What we are interested in, however, is what Takeo Nakasawa did in the Manchukuo described above. After he gave up the world of mathematics (for reasons we do not know), he passed the examination for elite bureaucrats. There is evidence that he first worked at the State Council in Hsinking (or Xinjing), the capital of Manchukuo. In passing we note that Nobusuke Kishi, who is the maternal grandfather of the former Prime Minister Shinzo Abe, was on the State Council at that time. Nakasawa married a lady from Kyushu in 1939 (Showa 14). According to the 1940 membership list of Meikei-kai, his working place was the Department of General Affairs, the Heiho Public Institution of Secretariats. I must admit to not understanding his personal address: the list says "Koa Juku, Kokuga Town, Kokuga [the private school for Asia development, Heiho Town, Heiho]." Nowadays, Koa Juku reminds me of the groups of people who drive trucks on Sundays with martial songs at full blast and yell out irrational slogans from the extreme right. I do not understand why such a word appears on his personal address. It would be much appreciated if someone could enlighten me on this matter.

To return to our historical discourse, since the Battle of Midway in June 1942, the tactical situation for the Japanese Forces was deteriorating day by day, and it was far beyond consideration for the Japanese government to manage immigration to Manchuria. From around the latter half of the year 1943, the Kanto Army had to commit a major part of its military forces, and a significant number of weapons, to the Southern battle line. Moreover, in April 1945, the USSR notified Japan that it would not renew the Soviet-Japanese Neutrality Pact, which would expire after a year. This was based on the so-called Yalta Agreement, which was sealed among the US, the Great Britain and the USSR in February 1945, in the Crimean Peninsula. After receiving this notification, the Kanto Army decided to give up three-quarters of Manchukuo and to wage a war of attrition for the remaining part if and when the USSR invaded it. However, these facts were not revealed to Japanese settlers in Manchukuo and even today a significant number of people might believe that the USSR suddenly assailed Manchukuo by unilaterally abrogating the Soviet-Japanese Neutrality Pact; but this is a vulgar view of the matter.

The fact is that the Japanese government at that time disowned the Japanese settlers in Manchukuo. The origin of the detention of Japanese soldiers in Siberian labor camps and the problem of the war-displaced children, which has not completely been solved, lies here. After the USSR notified Japan that they would not renew the pact, there were a few months up to their actual invasion of Manchukuo, and it seems that the Japanese government or the imperial army had enough time to relocate the people there to safe areas. However, it was almost impossible at that time to expect the Supreme Council of War-Leadership in Japan to do such a thing, because the possibility of defeat in the war loomed large, and the council was too occupied mentally and physically to think of anything beyond daily routines.

On the other hand, in Europe, the USSR Forces had surrounded Berlin, and took control of it in April 1945, the date when the USSR notified Japan of non-renewal of the pact. Desperate at the certainty of the fall of Berlin, Hitler killed himself. In May 1945, Germany finally surrendered unconditionally, as Italy had already done in September 1943. Because of these developments, the USSR had no more reason to bottle up its huge army in the European battle line, so transferred it to the Far East of the Eurasian continent in a surreptitious fashion. After the transfer was complete, the USSR attacked Manchukuo. An old saying is that a lion makes an all-out effort to catch only one rabbit, and this apparently was true of the USSR attack against Manchukuo. The USSR greatly over-prepared for the attack, since the power of the USSR Forces which were already in the Far East was surely enough to destroy the terribly weakened Manchukuo. Perhaps the bitter experience of the Russo-Japanese War some decades ago haunted the thoughts and dreams of the generals of the USSR Forces.

Takeo Nakasawa was drafted into the army on May 7, 1945 as a second-class soldier, a rank that implies he was a first-time draftee. This was, generally speaking, the rank at which a novice in the army started his military career. The Kanto Army had succeeded in gathering a military unit of 780,000 soldiers. This military unit was quite large in number in, but the reality of its strength was not so reassuring: The Kanto Army inducted any and all men who could be of any possible use to make up for the shortage of soldiers. In the words of one of the cadre of the Kanto Army, "even a dead man is better than nothing." However, no air unit or tank regiment existed, and not all soldiers had even a basic rifle. On the other side, the USSR Forces had more than 5,000 aircraft, more than 5,000 tanks and about 1,750,000 soldiers. According to military experts, this was equivalent to 20 times the size of the Kanto Army.

When this huge Army of the Soviet furiously came from all three directions in the north of Manchukuo at the signal of the time tone of 0 o'clock of August 9, 1945, the abandoned Japanese settlers could do nothing but shriek in agony. Prior to the invasion of the USSR Forces, a rumor was spread in Manchukuo that vicious criminals had been released from USSR prisons and were providing the first line of invasion. When the USSR Forces actually came, the rumor appeared true to the people in Manchukuo and they were terrified. Pillages, assaults, rapes ... all the horrors of war were heaped upon them. It is said that only 100,000 people escaped to the mainland of Japan safely among an estimated 300,000 settlers in Manchukuo. The 600,000 abandoned soldiers were detained in Siberia, forced to work in severe conditions without enough food, and ten percent of them, 60,000, lost their lives. The detention of soldiers started at the end of the summer or at the beginning of the autumn, and the winter in Siberia of minus 40 degrees Celsius divided those who had enough physical strength from those who did not. Many people who lost their lives in Siberia died in the first winter or not long after.

Takeo Nakasawa probably fell mortally ill and died before the next winter came. According to the notification from the Ministry of Health and Welfare, he died of malnutrition. From the formal viewpoint of the USSR at that time, malnutrition was not a disease, and it did not prevent him from being forced to labor. He had three children but they died of disease during the forced march returning to the mainland of Japan from Manchuria. Only his beloved wife survived to come home but a little while later, for reasons unknown, she lost contact with Nakasawa's family.

There is a museum called *Mugon Kan* (Museum of Silent Artists) in Ueda city, Nagano prefecture. You can go there by car in 30 minutes from Ueda-Sugadaira Interchange on the Joshinetsu expressway. With a backdrop of Mt. Asama, its building reminds us of monasteries in medieval Europe. It shows more than 300 works. When I came into the building, I felt the strange sense of tension which overwhelmingly filled the museum. The works presented in this museum were drawn by young art students who died in the Asia-Pacific War. Some drew their own portraits, and others drew their wives or girl friends. Some drew their families, and others drew landscapes. One of these pictures is reminiscent of boys and girls in the contemporary style of that time, who were called modern boys and modern girls. The artist of this picture had changed his paintbrush into a rifle and set it in a wartime background. How ironic this was! These young artists did not have enough time to make their talent in art effloresce, because they were born in an outrageous and irrational time. There are some unfinished drawings. A story is told of one young art student who was drawing the naked body of his girl friend. He promised her to finish it when he came back alive from the battlefield. Even the background of the picture was not finished. The girl friend wept inconsolably and the artist was unable to utter even a single further word. They spent their last moments together in silence, feeling the intensity of each other's grief. Against a background of numerous Japanese rising sun flags the soldiers were waved off to the battlefield. The young artist died at age 27 at Luzon in the Philippines. The picture remained unfinished

When I contemplate the life of Takeo Nakasawa, I recall the Old Testament of Job. Unlike his demeanor in the New Testament, the God of the Old Testament is fierce, and commits genocide when he is angry. The cities of Sodom and Gomorrah in Genesis were destroyed completely for incurring the displeasure of God. Especially in Job, God assigns ordeals to righteous Job one after another. Those who have faith in sublunary principles or in the law of Karma will not understand this. Milton, the seventeenth century British poet, stated that he did not want to believe in a god like this. Sima Qian, the author of the Records of the Grand Historian in ancient China, wondered whether there was any divine justice in the world. Here, it is the justice of God that is in question: Does God really have righteousness? Reading the Old Testament of Job faithfully, we will arrive at the predetermination of Calvin, but the road to this can be very hard. There are a lot of religionists in the world, but few reach the level of Calvin. Some religionists of newly-risen religions tend to teach sublunary principles by attributing a kind of mysterious power to that religion, and they themselves tend to make full use of the benefits to be derived. On the one hand, in the Old Testament, Job gave up asking questions of God, and happily left his fate to God's decisions. On the other hand, Takeo Nakasawa apparently died in the middle of trials by God. As he did not die on the battlefield but in a hospital, he might presumably have had some time to reflect upon and evaluate his life. Had he survived, I and the world might have learned something of his thoughts. That is impossible. We must be content with knowing that he was promoted to first-class soldier on October 10, 1945, two months after the end of the war. When he died on June 20, 1946, he was elevated to the superior class and into the top rank of soldiers. In the normal case, a promotion like this would not have been admitted, but clearly Nakasawa was recognized by some people as exceptional.

I would like to finish this chapter by mentioning that Nakasawa was included in the Iwanami Dictionary of Mathematics, fourth edition published in March 2007. His work had not been mentioned at all up to the third edition. We finish this chapter by reporting this fact to the spirit of Nakasawa that is left to us even after he has gone.

Figure 1: *Top left:* 13 July 1922 – From the right: Takeo Nakasawa's real brother (15 years old), his real sister (18 years old), and Takeo Nakasawa (9 years old) *Top right:* 25 January 1928 – Takeo Nakasawa was 14 years old. *Bottom left:* March 1929 – Takeo Nakasawa was 16 years old. *Bottom right:* Date: unknown – Takeo Nakasawa's real sister. She got married and had one son and three daughters.

Figure 2: *Top left:* 6 July 1930 – Takeo Nakasawa was 17 years old. *Top right:* 26 June 1932 – Takeo Nakasawa was 19 years old. He entered the University of Arts and Sciences this year. *Bottom left:* 17 October 1932 *Bottom right:* 5 February 1936 – Takeo Nakasawa was an assistant of the University of Arts and Sciences. The photo was taken as the memory of his birthday.

Figure 3: *Top:* Date: unknown – The photo was received by Takeo Nakasawa's real sister in Japan on 4 August 1941. Takeo Nakasawa was with his wife and his first daughter in Manchuria. *Bottom:* 5 May 1936 – Takeo Nakasawa was in Nihonbashi with his real sister.

Figure 4: *Top:* 29 March 1942 – Takeo Nakasawa with his wife, his first daughter (a year and 8 months old) and his first son (4 months old) in Manchuria. *Bottom:* 8 January 1944 – Takeo Nakasawa's last photo.

Figure 5: *Top:* 25 June 1939 – Takeo Nakasawa's marriage. Since it was held in Dalian, Nakasawa's parents and other relatives could not attend the ceremony. *Bottom:* 25 June 1939 – At the marriage.

South Manchurian Railway Company (1906–1945)

Manchuria is the native home of the Jurchens, who built the Qing Dynasty (1612–1912) in China. In a narrow sense, Manchuria means the North-East district in modern China and usually means the three east provinces, namely, Liaoning, Jilin and Heilongjiang. Across the Amur River and its tributary Ussuri, the Northeastern area meets the East Siberia district. Across the Yalu River, the Southern area meets the Korean Peninsula. The Western area is separated from the Mongolian Plateau by the Greater Khingan Range. In a broad sense, the East district of the Inner-Mongolia Autonomous Region was added to the three provinces mentioned above. In this regard, the present Chinese government insists that the name "Northeast district of China" should be used instead of "Manchuria." One reason for this is that they have the intention of getting rid of the abhorrent past connection to Manchukuo, which was undoubtedly a Japanese puppet state. Another is that they want to claim that this land has always been inalienably possessed by China since ancient times. I am afraid to say that this allegation by the Chinese government is, however, pretty impossible.

Qin Shi Huangdi (BC 361–338), who unified China for the first time, had built the famous Great Wall of China to prepare for possible invasions by other races than the Han race. Manchuria lies completely outside the Great Wall of China, which was intended to protect China's cultural orbit against such invasions. The successive dynasties of China have repeatedly repaired and reconstructed the Great Wall of China. In fact, the area around Manchuria is called "Kangai no chi [the land outside Shanhaiguan]" because it is placed outside the Shanhaiguan. Shanhaiguan is the eastern end of the Great Wall of China. The name "*Kan*to" in Kanto Army means the eastern district of Shanhaiguan in China, and there is no connection with the Kanto district including the Tokyo metropolitan area in Japan. In passing, we note that the "Kanto" district in Japan once meant the Eastern part of "Oosaka-no Seki [Barrier of Oosaka]" in Shiga Prefecture in ancient times. Manchuria was the land of rise and fall of races living to the north of China. Even before the time of the Qing Dynasty, the Jurchens living in Manchuria invaded China to establish the Dynasty Jin (1115–1234). Jin ruled the northern half of China, while the Chinese-origin Song Dynasty was driven south. Gaozong (reign: 1127–1162), the first emperor of the Southern Song Dynasty, concluded a peace pact with the Jin Dynasty in 1142, and was forced to send silver and silk to the Jin Dynasty every year to pay liege homage. The perception by the Han people that they were the center of the world is considerably comparable to the same attitude held by the ancient Greeks: It is said that the ancient Greeks called themselves Hellenes and other races barbaroi, which means people who speak languages offensive to civilized ears. To the Chinese boasting of their culture and country, this kind of humiliating story belongs to the dreadful past which they want to forget. It is really embarrassing that being culturally superior does not necessarily mean being militarily superior.

The Qing dynasty, whose origin lay in Manchuria, gave special treatment to Manchuria. It forbade Han people to enter the area around Manchuria. However at a later time of the Qing Dynasty, it reluctantly gave silent approval to Han people's entry into the area, because the Qing dynasty had lost its power and desperately needed to produce food for soldiers facing the Russian menace. After the collapse of the Qing Dynasty due to the Chinese Revolution in 1911 and the establishment of the Republic of China in the following year, armed factions behaved willfully in various parts of the country. Because of the social unrest in the central part of China, many Han people settled in Manchuria.

In the thirteenth century, Moscow fell under the control of the confederation of Kipchak khan (1243–1502), a descendant of the Mongol Empire, which was built by Genghis Khan (reign: 1206–1227) across the Eurasian Continent. In the fifteenth century, Russia had for all practical purposes been emancipated from the rule by Kipchak khan and expanded to the east, and to the south. In 1689, under Peter the Great of the Romanov imperial dynasty, Russia concluded the Nerchinsk treaty with the fourth emperor of the Qing dynasty, the Kangxi emperor, to draw up a borderline between the two countries. This treaty was largely advantageous to China: China had obtained the south of Stanovoy Mountains, the north of the Heilongjiang River and the east of Ussuri, which constituted the area called Outer Manchuria. If Outer Manchuria is included, the area of Manchuria reaches 1,550,000 square kilometers. Regrettably for China, Outer Manchuria was ceded to Russia by the two unequal treaties, the Treaty of Aigun and the Treaty of Beijing in the mid-nineteenth century, when the invasion of the West into the East was rampant.

In the mid-nineteenth century, the modern era of Japan started with the Meiji Restoration, whose slogan was restoration of the regal government. Till the end of the pacific war in 1945, however, the national goal was wealth and military strength. This reflected nineteenth century trends in Europe. At that time in Europe, imperialism was at its best, and economic wealth and a large military capacity were considered to be essential for first-class nations. In the First Sino-Japanese War (1894–1895) at the end of the nineteenth century, Japan defeated the Qing Dynasty, which had been considered a sleeping lion. This victory greatly surprised the European nations. For the first time in history, the Japanese saw the land of Manchuria when the Japanese Forces set foot on the soil of Liaodong Bandao. Although Japan won Liaodong Bandao as a result of its victory in the First Sino-Japanese War, Germany, France and Russia forced its return to China. The German emperor, Wilhelm the second, who was a famous "yellow peril" agitator, led these three countries. This was the so-called Tripartite Intervention. Japan was of course reluctant to admit the restitution in such a humiliating manner.

Since Russia took over Liaodong Bandao and subsequently established a fortress in Lüshun, defeating Russia had become Japan's earnest desire. This desire was fulfilled in the Russo-Japanese war (1904–1905). As a result of Japan's victory in the war, the Kanto Territory Area Command was moved to Lüshun in 1906. The Kanto supreme viceroy, the chief of the Kanto Territory Area Command, was also the General Officer or the Lieutenant General of the army or the navy. The purpose of establishing the Kanto Territory Area Command was to withdraw the Japanese Forces, which wanted to stay in south Manchuria even after the treaty of Portsmouth, and to allay the suspicions of the United States and Great Britain against Japan. The elder statesman Hirobumi Ito persuaded the Japanese Forces to accept this pullout. During the Russo-Japanese war, the United States and Great Britain helped Japan, not because they wanted to see Japan defeat Russia decisively, but because they wanted to prevent Russia from invading East Asia further. Japan's victory in the Russo-Japanese war gave them no joy but began a new worry.

Colonization outside one's country was a worldwide trend at that time, not unique to Japan. However, Japan's aggressive military goals were shown very clearly without any apology from the beginning, and this had hurt the occupied countries more severely than some other colonizations. In cases of European countries and the United States, the army was not the first entity to appear in a country to be colonized. Firstly, missionary priests visited the country as messengers of God,. They were followed by merchants, who were only then followed by men in khaki. The Kanto Territory Area Command took a role in the governance of Kanto, which comprised Lüshun and Dalian in the south of Liaodong, and their attached land. This governance was the beginning of the real relationship between Japan and Manchuria. Around the same time, the South Manchuria Railway Company was established to manage the East Qing Railway between Changchun and Lüshun, its branch lines and their railway zone, which were transferred from China to Japan in the treaty of Portsmouth. In addition to the Kanto Territory Area Command and the South Manchuria Railway Company, Japan's Foreign Ministry, which had consular offices in various locations in Manchuria under the governance of the Qing Dynasty, claimed that the management of Manchuria was one of its diplomatic responsibilities. As a result, the management of Manchuria by Japan was conducted by triumvirate. They competed for their own turf. In this regard, Takashi Hara, who became Prime Minister after the first world war, and was called "*heimin saiso*" [a commoner premier], had written some comments in his diary, which became an extremely valuable document for the history of politics during the Meiji and Taisho period. It says that diplomats such as consuls do not know business but only diplomacy, the South Manchuria Railway Company does not know diplomacy but only business, and the territory area commands know neither business nor diplomacy. Because Japan did not have a consistent and coherent plan for the management of Manchuria, it created a really sinuous political structure.

Just as the relationship between the United States and the Soviet Union during the cold war, the situation of two giants keeping each other in check is generally stable, unless one of them goes mad. This holds true also for the Japanese-Russia relationship after the Russo-Japanese war: Japan and Russia kept each other in check on both sides of Manchuria. They made four agreements after the war. Only the third one was concluded as a completely secret treaty, but the others were partly open to the public and partly remained secret. In the first Russo-Japanese agreement in 1907, Russia admitted Japan's predominance in Korea, and Japan admitted Russia's predominance in Outer Mongolia. The borderline between Japan and Russia in Manchuria was also determined. Roughly speaking, it was in the middle of Harbin and Jilin. The reason why Japan and Russia determined these things was that the United States seemed much interested in Manchuria. By way of example, the rail baron of America, Harriman, had proposed to establish a Japanese-American joint venture, and the Secretary of State, Knox, proposed to neutralize the authority of the South Manchurian Railway Company. The second agreement of Japan and Russia in 1910 was nothing but a merely additional one. Further Russo-Japanese agreements were not planned at the conclusion of the second Russo-Japanese agreement, but world affairs moved rapidly in some preceding years of the First World War. In 1908, the Young Turk Revolution erupted in Ottoman Turkey. In November 1911, Outer Mongolia became independent from China due to the Chinese Revolution, which occurred in October 1911. Taking these circumstances into consideration, the third Russo-Japanese agreement was concluded in July 1912. The division of Nei Mongolia was discussed and decided. As a result, Japan obtained the eastern part of Nei Mongolia in addition to South Manchuria, which Japan had already possessed. Moreover, the First World War made Russia run short in military supply due to the severe fight with Germany. Because Russia hoped for help from Japan, and because the elder statesman Arimoto Yamagata and the Japanese Forces, who wanted to expand its influence in Manchuria, were willing to have close connections with Russia, the fourth agreement was concluded in 1916. This might be called a military alliance rather than a mere pact. It claimed that Japan and Russia would cooperate together in order for China not to be governed by any other country. This was what was to be called the Treaty of Mutual Corporation and Security between Russia and Japan, preceding the Treaty of Mutual Corporation and Security between the United States and Japan concluded after the Second World War. In March 1917, a socialist revolution took place in Russia. But for the Russian Revolution, world history might have been completely different. This was the first socialism revolution in the world. Because of this, the far-east area, including Manchuria, became really unstable. Moreover, the Soviet Union made every secret of the Romanov imperial dynasty open to the public, and suggested that the other countries mentioned in the fourth agreement were the US and Great Britain. This outspoken behavior of the Soviet Union broke down the well-established relationship between Japan and the Romanov dynasty.

In modern times, we cannot help saying that Japan has not been skillful in getting crucial information and analyzing it. The Russo-Japanese war was no exception. In fact, Japan finally won the Russo-Japanese war, but it was even until right before the Portsmouth Peace Conference that Japan did not know whether the East Qing Railway Company was managed by the Russian government or by a private concern. Needless to say, Japan did not know at all whether the management of the railway paid well or not. At that time, the principal purpose in having that railway was military, that is, to prevent Russia from going south. The economic point of view was lacking completely. Considering the severe financial condition in Japan due to the Russo-Japanese war, the opinion that the branch line of the East Qing Railway was not needed was quite strong in the government at that time. Because of this, in October 1905, the Japanese government tied up the preparatory contract on the co-management of the south branch line by Japan and America with the American rail baron Harriman. Nevertheless, this contract itself was withdrawn, because the Foreign Minister Jutaro Komura, who was abroad during the negotiation, opposed it fiercely. As we see the situation, it is easy to understand the pressed government at that time. To win the Russo-Japanese war, Japan had paid about 17 million Yen by excessive issuing of war bonds. The number of soldiers in Japan during those years was usually 20 million, but 1.09 million soldiers went to the war and 380 thousand died. Because the South Manchurian Railway Company was acquired at such huge costs, it was politically difficult to give it up simply for financial reasons.

The first president of the South Manchurian Railway Company was Shimpei Goto. He had been the president of the company for about two years, from November 1906 to December 1908. Nevertheless, his name and his deeds became a legend in the company. Japan had already won Taiwan from the Qing Dynasty as a consequence of Japan's victory in the first Sino-Japanese war, long before it won Manchuria. Goto, as the president of Taiwan and responsible since 1898 for people's livelihood, had achieved considerable successes in various fields, such as the improvement of public safety by punitive expeditions against marauding bands, a decrease in the number of opium addicts created by the opium monopoly, stabilization of tax income by land reallocation, which streamlined the entangled web of land possession, countermeasures against contagious diseases, which showed Goto's true value as a doctor, and improvement of transportation, which was achieved mainly by building railways. This established the foundations of modern Taiwan. In fact, the population of Taiwan was 2.6 million people in 1896 but increased up to 6.6 million people by 1943. Goto was hired as the president of the South Manchurian Railway Company because of this remarkable success in Taiwan. It is said that he drew upon the English East India Company in his management of the South Manchuria Railway Company.

His achievement in Taiwan unification was much better than his record as president of the South Manchurian Railway Company. This was because the management of Manchuria was in the hands of the Tripartite, namely, the South Manchurian Railway Company, the Foreign Ministry, and the Kanto Territory Area Command, which seems to have interfered with his ability to manage the railway. In addition, his time as president of the South Manchurian Railway Company was very short: He had to leave Manchuria to become a member of the Cabinet in mainland Japan. It was Gentaro Kodama who recommended Goto to be the President of the South Manchurian Railway Company. Kodama was a legendary military man who led Japan to victory in the Russo-Japanese war (1904–1905). During those years, Russia had usually 15 times as much military power as Japan did, and on the basis of their national budgets, Russia was eight times as big as Japan. Goto was the chief of the general staff of the Manchurian Army and was indeed a genius in military affairs. Kodama was the person who had recommended Goto for administrator of the Taiwan peoples' livelihood. Kodama, communicating with Shimpei Goto, proposed a plan to manage Manchuria, right before the end of the Russo-Japanese war. The plan claimed that the only formula to manage postwar Manchuria was that the management of hundreds of facilities should be done in disguise of the management of the railway. This plan was, I would rather say, an unrealized one.

It is surely said that the management of the South Manchuria Railway Company had diversified aspects of business such as the coal pit development in Fushun, the steel industry in Anshan, and the establishment of such high quality European style hotels as the Yamato Hotel in the main cities along the railway. Nowadays many railway companies have hotels along the railways, but at that time, it was exceptional and highly epochal. Needless to say, management of terminal ports by a railway company is very convenient for transportation. Imperial Russia made the East Qing Railway build the Dalian port and manage it. Imperial Japan also made the South Manchurian Railway Company expand and manage the Dalian port. In August 1908, the South Manchurian Railway Company started operating a regular shipping route from

Shanghai to Dalian. Thanks to this initiative, the international route from Shanghai via the South Manchurian Railway, the East Qing Railway, and the Siberian Railway to Europe had been completed, and Dalian had been transmogrified into a European-like city, which had wider roads, as was favored by Goto.

In the same year, the South Manchurian Railway Company had established the Dalian Electric Railway as a subsidiary company. The company had established a streetcar system in Dalian city and managed it. Dalian was the place where the headquarters of the South Manchurian Railway Company lay, and it made every effort to make Dalian an international city as large as those in Europe, just as Russia had made Harbin an international city by building the East Qing Railway. The second president of the South Manchurian Railway Company was Korekimi Nakamura from the Finance Ministry. He and the well-known literary magnate of Japan, Soseki Natsume, had been good friends in the same schools, namely, the Meiji English Language School, and the Preparatory Course to University of Tokyo (to be called the First High School later). Nakamura invited Natsume to Manchuria in the summer of 1909. After Natsume came back from Manchuria, he wrote a book on his travels, entitled "Mankan Tokoro Dokoro [From place to place in Manchuria and Korea]" in a newspaper serial form. He wrote that, in Manchuria, people expressed their opinions frankly in various places, new ideas were being realized successfully every day, and people were twice as well rewarded as their counterparts in mainland Japan. They stated also that the cultural revolution in Manchuria progressed much faster than the one in mainland Japan, and the culture in Manchuria was closer to Europe than to mainland Japan. According to the travel writings by Natsume, while people in mainland Japan did not know Manchuria at all, the Japanese in Manchuria seemed to enjoy modern lives such as night clubs, bars, golf clubs, yacht clubs, and amusement parks. When Natsume arrived at Dalian, Nakamura was in the middle of watching a baseball game. As you can infer from this, at that time in Dalian, watching baseball games had become usual in daily life, and a sort of urban culture had grown up.

In October 1913, regular transportation service was started, connecting Dalian and cities around the sea coast of China such as Hong Kong and Guangzhou. It seems that they wanted to make the Dalian port a large trade hub between Europe and Asia. This story was very romantic, deserving to be called a truly continental and Eurasian idea. The power of action without concern for profits by the South Manchurian Railway Company was deemed marvelous beyond mere compliments. From the insular point of view of an ordinary Japanese in mainland Japan, regular travel services between Dalian and some main cities in Japan seemed enough, but easy access to Manchuria caused people to begin thinking and behaving in broader terms. Some may regard the South Manchurian Railway Company as an embryonic form of the third sector, which comprises companies managed by both the government and by private sectors in contemporary Japan. But the representatives of the present third sectors are former government officials without creativity at all. It would be too impolite to compare the third sector to the South Manchurian Railway Company.

Japan had also acquired some attached areas along the East Qing Railway in the Treaty of Portsmouth after the Russo-Japanese war. In these areas, the Kanto Territory Area Command was responsible for general governance and police activity, and the South Manchurian Railway Company was engaged in civil engineering, education, and medical and sanitary affairs. As a result, the company had the face of a local government. The attached areas along the railway were quite akin to foreign settlements in China of Great Britain and other European countries, and Japan enjoyed extraterritoriality in these areas. Because the attached areas along the railway did not find themselves in already established cities of China, building cities in the attached areas along the railway was like building cities in the heath or in the moor. Roads were built in a reticular pattern. Some circular parks were laid out from place to place, and these parks were connected by roads. This way of building cities was highly inspired by the remodeling plan of Paris by Napoleon III in the middle of the nineteenth century. In this regard, Goto himself seems to have pictured the Champs d'Elysees in Paris or Unter den Linden in Berlin.

Following the rapid spread of urban areas in the attached areas along the railway, supplying water to these new cities became exigent. The South Manchurian Railway Company built water supply systems here and there. After streets and water supply systems, people needed medical care and education, that is, hospitals and elementary schools. Elementary Schools were built from place to place in the attached areas along the railway, but it was not true that the attached area of every station found an elementary school. Therefore the South Manchurian Railway Company decided to bear the fares for children coming to school by using the railway. How wonderful it was! In 1925, hopeful children began to learn Chinese in elementary schools. Even for the Chinese, schools called *Kokaido* had been built. I do not say that there was no prejudice or discrimination against the Chinese in Manchuria at all, but at least the Japanese living in China were aware of the very transparent fact that they lived in an atmosphere replete with Chinese.

In addition to these facilities, libraries and parks were built along the railway. The South Manchurian Railway Company paid all costs to acquire land and to build such new facilities. The South Manchurian Railway Company intended that the management for these facilities should be paid off by public charges collected from the residents and from rents. The fact is, however, that the company stayed mired in the red, as far as these enterprises related to the attached area were concerned. The South Manchurian Railway Company had paid 400 million yen for these enterprises from 1906 to 1936, but it earned only 100 million yen during the same period, so that the company was in the red by 300 million yen. It had a great surplus, as far as the management of the railway was concerned, and it had never been in the red as a whole. The great surplus of the railway management compensated for the great deficit in the management of the attached area along the railway.

Because of bad results in budgetary balance, the company had closed the marine transportation and the steel industry, but it had not even tried to close the management of the attached area along the railway or to build urban spaces in bleak plains. Shimpei Goto stated that he would like to realize earthly supremacy by pursuing heavenly supremacy, and the management of the attached areas along the railway so as to improve the lives of their residents was no other than pursuing heavenly supremacy in Goto's sense. Giving up the management of the attached areas along the railway was politically out of the question, because the South Manchurian Railway Company was a state policy cooperation. When Manchukuo was established and the South Manchuria Railway Company, which enjoyed extraterritoriality within Manchukuo, became a pain in the neck to the Kanto Army in the middle of 1930, the company was forced to give up its management of the attached areas along the railway. The formal end of the company came with the end of the war in 1945, but by surrendering the management of the attached areas along the railway, the company had become a living corpse. As the fifteen years war progressed, pursuing heavenly supremacy was forgotten and Japan tried to establish worldly supremacy directly.

Goto, taking advantage of his experiences in Taiwan, started a new business of investigation, similar to contemporary think-tanks. This initiative developed successfully into the information and research division of the South Manchuria Railway Company. Nevertheless, the huge amount of information collected in such a manner had never affected the real diplomacy or military strategies of Imperial Japan. It had been left on the shelf. Even now Japanese researchers on China rely heavily on information that was collected at random by people utilizing odd moments. I do not understand why Japanese politicians and military men did not make use of it. This compares poorly with American politicians and military men who let such social scientists as Ruth Benedict, the famous author of the Chrysanthemum and the Sword, do research on Japan, and applied the results to diplomacy and military strategies against Japan in practice.

Around 1910, Japan finally entered a period of placing party politics over clan politics. A semi-governmental corporation such as the South Manchuria Railway Company was liable to be a place to hunt for concessions. In fact, in 1918, a famous bribery case was revealed. Seiichi Nakanishi, the vice president of the company and a member of the Seiyukai political group at that time, was prosecuted, but found to be innocent. This is a usual pattern for such corrupt acts in Japan. Aside from the truth of the case, it can be said that the management of the South Manchuria Railway Company was affected greatly by the movements of party politics. Hara, who was the president of Seiyukai, was murdered at the Tokyo station in November 1921, when he was in the middle of being criticized for his apparent cult of money. In 1919, when the Prime Minister was Hara, the Kanto Territorial Area Command was divided into two sections, namely, the Kanto Agency for civilian affairs, and the Kanto Army for military affairs. By means of this division, the triumvirate of the Kanto Territory Area Command, the South Manchuria Railway Company, and the Foreign Ministry was replaced by the quadrumvirate comprising the Kanto Army, the Kanto Agency, the South Manchuria Railway Company, and the Foreign Ministry. This system was by no means efficient.

Speaking about political activity in Manchuria, military policemen from the Kanto Army, the common police of the Kanto Agency, and the consular police of the Foreign Ministry insisted on their own turfs. At that time, the Kanto Army had only ten thousand men, while the number of members of the South Manchuria Railway Company was more than thirty thousand. But ten years later, after the establishment of Manchukuo, the power balance between the Kanto Army and the South Manchuria Railway Company had reversed completely. Moreover, the Ministry of Colonial Affairs, built by Prime Minister Giichi Tanaka in 1929, made the situation more complicated. Although the ministry was very small, it had the right to appoint or dismiss the chief of the Kanto Army and the president and the vice president of the South Manchuria Railway Company, which made the situation tangled This Ministry initiated the Manchuria Immigration Policy and in 1932, when Saito was Prime Minister, it sent the first armed immigrants to Manchuria. The complex situation of the quadrumvirate and the new Ministry of Colonial Affairs was somehow simplified by the establishment of Manchukuo. The Commander of the Kanto Army automatically became the ambassador to Manchukuo and the chief of the Kanto Agency. In December 1934, the authorities possessed by the Ministry of Colonial Affairs were transferred to the newly established Manchuria Bureau, and the Minister of the Army automatically became the chief of the Bureau. In this way the Ministry of Colonial Affairs was completely defeated. Moreover, in June 1936, Japan decided to abolish its extraterritoriality in Manchukuo. At the end of 1937, the governance of land and health affairs was transferred to Manchukuo, and the management of education was transferred to the related Japanese authorities. As a result, the management of the attached areas along the railway by the South Manchuria Railway Company was completely over. This meant that the company became a mere railway company both in a good sense and in a bad sense. This was the beginning of the end of the story about the South Manchuria Railway Company.

The Road to the Fifteen Years War (1931–1945)

The establishment of Manchukuo, its prosperity and its collapse happened from the 1930s to the 1940s. At that time, the potential problems which the government of the Meiji period had happened to take onto itself and which had continued to be skillfully contained finally became actual. In the middle of the nineteenth century, Japan put an end to the Edo period which was the time of the Samurai and started to walk on the road to a modern state. Because of this circumstance, I have to begin this chapter with the Meiji restoration.

The principal goal for the leaders in the Meiji period was to build a prosperous country and to have a strong army. This meant that the government had to modernize its economy and its military forces. In this regard, other Asian countries were far behind Japan at that time. If modernization had failed or had been delayed, Japan would have become a satellite nation or a colony of Great Britain, France, the United States or some other European country, as had been seen in a number of other Asian countries. Because of this situation, Japan had to be modernized swiftly. To purchase up-to-date weapons was not enough to modernize the military. In the Edo period, there was a class system; the warriors, the top of the caste, used to be engaged in military affairs. However, in November 1872, the Meiji government emphasized, by announcing a national service charge, that to be a soldier was a national obligation. Beginning in January 1873, all Japanese men had by the call-up law to be soldiers for a certain period of time. There were many exceptions to this system. For example, householders, government clerks, and people who paid a "substitution fee" could escape this obligation. As a result, the military obligation weighed heavily on the second and the third sons of farmers. Some of these exceptions were abolished by an amendment to the law in 1879, and they were completely abolished in 1889.

To modernize the country, the government had not only to rebuild the system that governs people but also to modernize people themselves. In this respect, military services provided good opportunities. Modern people were not produced spontaneously. They were not bamboo shoots, which grow naturally. Even time perception based on watches, which became common in modern times, was not common at all in the Edo period. People had not started wearing watches until the Showa period. Thus a radical change of time perception was seen in the twentieth century, a century in which relatively local battles became all-out wars. In pre-modern societies, perception of time was very ambiguous at best. Had this condition prevailed, neither our modern society nor its modern army would have worked properly. Prompted by the military's adoption of fatigues and combat boots, ordinary people started to wear Western style clothes and shoes instead of traditional Japanese kimonos and yukata or footwear such as tabi and waraji. Even without military obligation, this change might have taken place, but the change would have been much slower. The same thing can be said about the growing popularity of European foods and the use of beds and chairs.

In the Edo period, each district had its own indigenous authority. That is to say, each district had its own political system, economic system, and language. The language differed from one district to another, and communication between distinct districts was not easy. Not only compulsory education but also military obligation had contributed to the spread of standard Japanese. Without a standard language in a country, orders in military forces would not be understood. In addition, the idea of democracy, which is based on equality among people, was well propagated in the army. The army indeed has its own class system within its confines, but it does not matter at all if one is rich or poor, of high or low origin. The same phenomenon is seen in prisons. Both models represent democracy in a vacuum bottle. Moreover, the army gave lower-class youngsters plenty of opportunities to rise to higher classes. The tiny, but real, hope of drafted soldiers was to become high-class soldiers, a distinction that was introduced in 1877. After the completion of obligatory military services, high-class soldiers were welcomed with admiration in their local societies. In the latter half of the 1880s, an advancement system from high-class soldiers to corporals, sergeants and sergeant majors was established. Noncommissioned officers were a good option for the second or third sons of farmers, who had little chance of getting a good job after competing their service. In later years the Japanese Army gradually became a principal obstacle against the modernization of Japan, but we should note that it contributed a lot to build modern Japan at its grass roots in earlier times of modernization.

It was Masujiro Ohmura from the Choshu domain that built the foundation of the Imperial Japanese Army. He was born in 1824, and was once a doctor. As a doctor, his reputation was not very good. He became secretary of defense of the new Meiji government due to his good performance for directing strategies in the second war by the Edo shogunate against the Choshu domain and in the Boshin War between the former shogunate army and the army of the new Meiji government from 1868 to 1869. In September 1869, he was stabbed by an assassin. Because of the injury, he died in November 1869. Aritomo Yamagata, who came back from Europe and followed Ohmura, rose to the top of the army. During establishment of the Imperial Japanese Army, Yamagata's lust for power played a decisive role. A similar thing happened in the Soviet Union. In the establishment of the Soviet system, Stalin's lust for power was so strong that the Soviet Empire had suffered from it until its collapse and even after Stalin's death in 1953. In 1922, Yamagata died. Because he was an elder statesman in the days of the Meiji Restoration, a national funeral was held. Regrettably, almost all attendants were members of the military or police. On the other hand, when Shigenobu Ohkuma died, at almost the same time, the funeral was totally private, but many famous people from various fields and even many of the general public were present at the funeral. The Meiji Emperor did not like Yamagata and called him a grasshopper, because the Japanese figure representing Yamagata's name looked like one. The Taisho Emperor, who followed the Meiji Emperor, drove him out of the Imperial Palace while asking his attendants to give him something to take with him. Yamagata was disliked by many people. He was famous for loving medals so much that Takashi Hara, who was known as a "people's prime minister", turned him away bluntly by simply saying that his love of medals was due to his being only a son of a common foot soldier.

The independence of the supreme command, which was characteristic of the Japanese military up to the Second World War, prepared the way for the Fifteen Years War. The supreme command held the exclusive right to lead the Japanese forces. The independence of the supreme command meant that, in effect, it was held solely by the Emperor, and the cabinet or the diet held no power over it. The Meiji Constitution, which had been valid until the Second World War ended, claimed that the prerogatives of the Emperor, such as the legislative authority or the foreign diplomatic authority including the supreme command, were to be exercised subject to advice by appropriate ministers. Nevertheless, on the ground of the eleventh article of the Meiji Constitution, which claimed that the Emperor had command of the military, the supreme command became a sacred right belonging to the Emperor so that it could not even be touched by the cabinet or the diet. Around 1930, the politics of Japan came to be dominated by the military. At that time, the phrase "Do not interfere in the sacred supreme command" was chanted again and again. This phrase was highly threatening to politicians and the general public, who were almost forbidden to say a word on military affairs. In April 1930, the London Naval Treaty was concluded and Japan ratified it in October 1930, which offended many right-wingers and military men. This ratification was regarded as interference with the supreme command, and the angry chief commander Hiroharu Kato submitted to the Emperor a letter of resignation. The

Prime Minister Osachi Hamaguchi was attacked by a right-wing assassin at the Tokyo Station and died in August of the following year. It has been said generally that independence of the supreme command was triggered by the rebellion in the 1870s of the complaining warrior class such as Shimpei Eto and Takamori Saigo. However, we should note that they led a completely private army, not the national army. The truth seems to have been that the Meiji Government, because only a few years had passed since its establishment, and because its foundations were very fragile, wanted to keep the army away from severe political strife. If political strife and rebellion by the army had taken place cooperatively and simultaneously, the fragile Meiji Government would not have survived. As a result, in December 5, 1877, the staff section, which previously was a part of the Ministry of War, was separated from that ministry and became an independent Staff Headquarters, freeing them from the hands of the government and coming under the direct control of the Emperor. Although this was a very important decision, the matter was not discussed thoroughly within the government, and the real starting point of the Fifteen Years War, which took place fifty years later, was that decision. Aritomo Yamagata had a deep personal feeling for Staff Headquarters. When another staff office within the army was about to be established, following the Prussian model, he objected violently and the plan was finally abolished. From the standpoint of Yamagata, who wanted to keep his authority over the personnel issues of the army as chief of the Staff Headquarters, this plan was nothing but a nightmare. The devouring ambition of Yamagata was satisfied steadily on the basis of his control of Staff Headquarters, not that of the Emperor. He quickly rose in rank and became Prime Minister in 1889, only ten years later.

When the supreme command became independent of the government, there arose the serious problem that the government and the military might lose communication and cooperation with each other in case a general national defense policy was demanded, or in case a war was actually waged. Generally speaking, foreign policies, financial policies and military policies cannot be separated either in peace or in war. When these are truly integrated, effective diplomacy will be achieved when it can and wars can be waged efficiently when it cannot. A good example of unification of these three functions was seen in Prussia, or the second Reich, in the latter half of the nineteenth century, when the ruler was Wilhelm the first, Bismark was Chancellor and von Mortke was Chief of the Staff Headquarters. Prussia waged war after war in order to establish a unified Germany: The war against Denmark in 1864, the Austro-Prussian War in 1866, and the Franco-Prussian War in 1870. After these wars ended and a unified Germany was established, Germany had no wars for the succeeding twenty years. On the contrary, Prussia, or a unified Germany, made a great effort to keep the peace of the whole of Europe. In particular, Bismark played a decisive role in the Congress of Berlin concluding the Russo-Turkish war (1877–1878). This goal was achieved because they tactically established good diplomatic relationships with other European countries. The path towards war was clear, as was the path towards avoidance of war. The purpose of the Franco-Prussian War was not to annex France to Germany but to make the southern states such as Bayern swear allegiance to the flag of the German Empire. If Japan had been in the place of Prussia at that time, it would have tried to swallow France desperately, only to lose its way in the mud.

In summary, Prussia carefully fashioned its reason for the war and declared it with marvelous clarity: In January 1871, the establishment of the German Empire was declared in the Hall of Mirrors at the Palace of Versailles. The scenario was very clear. On the other hand, Japan during the Fifteen Years War had nothing positive to boast of. It did not know why it started the war, was not interested much in cooperation between diplomacy and the military, started a war without any serious plan, and had no idea of how they might end it – completely opposite to the Prussian case. An obvious question arises. Why did the leaders of the Meiji Government allow the independence of the supreme command? Did they not even consider the probabilities and consequences if the military went out of control?

Such leaders as the brilliant Hirobumi Ito, who established the independence of the supreme command, were bound together by a strong feeling of solidarity. They shared the same fundamental principles in politics, because at that time, those who were at the center of the government belonged to the same political faction – and they were all born and raised in the southwest of Japan. With naive confidence that they could unify politics and military affairs if necessary, and without any serious discussions, they made this crucial determination to separate the supreme command from the government.

Hirobumi Ito, who was Prime Minister during the First Sino-Japanese War (1894–1895), was a good example of real experience in unifying politics and military affairs in crucial times. In time of war, operations were conducted by the Emperor-centered Headquarters, which civil officers were not allowed to enter. Formally, everybody in he Headquarters except for the Emperor was a military officer. Nevertheless, during the First Sino-Japanese War, the integration of politics and military affairs was realized by Ito's strong leadership in the imperial staff conference and the conference of the headquarters of the military. During the Russo-Japanese War, Aritomo Yamagata and the Prime Minister Taro Katsura shared the same point of view. Moreover, since people like Ito, Yamagata, and Katsura grew up in the Edo period, they were educated as sons of samurai. The principles of Confucianism came first in the education of samurai. They were forced to read such Confucian sacred books as the Analects of Confucius, again and again, so as to completely and accurately memorize them. There was no specific ability to be acquired by doing so, but this was a kind of liberal arts program that enabled them to prepare themselves to live with other people in the existing social structure. And this ability was very important in thinking about the strategies of war in a broader perspective. On the other hand, education in the school of the army, established in the Meiji period so as to cultivate commissioned officers, emphasized the strategies of war exclusively. This was based on the simple idea that the principal aim of the military was to wage battles. Political science, economics and history, which were needed to think about real strategies, were not taught at all in the army's school. People with this kind of education were nothing but military technocrats with knowledge only of war strategies. Of course people with this kind of knowledge were necessary and could be staff officers in army divisions or less, but were not suitable to be officers in command, who were supposed to plan national defense policies while harmonizing military affairs, politics and economics.

A typical military technocrat of this kind was Hideki Tojo, who was Prime Minister when Japan attacked Pearl Harbor, which began the war against the United States of America in December 1941. He was still Prime Minister when the Japanese Forces were defeated in Saipan in July 1944 and was apparently destined from the beginning to lose the war. Naoki Hoshino became Chief Secretary of the Cabinet right after establishment of the Tojo Cabinet, and he was one of the few closest persons to Tojo. He stated that Tojo could do anything he was told to do, but he could not determine what to do by himself. Kanji Ishihara, who was one of the few leading figures of the Manchurian Incident (1931–1932), criticized Tojo more cynically. When he was questioned by the presiding justice of the Tokyo Judgment (the trial of war criminals at the end of the Second World War) whether there was a serious disagreement between him and Tojo, he replied bluntly that he had some opinions while Tojo had no serious opinion at all so that there could be no disagreement between him and Tojo. German sociologist Max Weber has claimed that the best bureaucrat is the worst politician. This statement holds literally for Tojo. The tragedy of the Fifteen Years War showed what would happen when a system such as the independent supreme command started working relentlessly under people with a myopic viewpoint who had a firm grip on command.

The exact time when the military was established as an independent political power was probably when Japan had won the First Sino-Japanese War (1894–1895) and the Russo-Japanese War (1904–1905). As early as in 1900, under the second Yamagata Cabinet, the reformation of the military system was conducted, and it was decided that military ministers should be military men in active service. Strictly speaking, this meant that the Minister of the Army and the Minister of the Navy always had to be an active general or an active lieutenant general, and if the military wanted to break up the Cabinet, it would suffice to withdraw one or both of the ministers. This was the most powerful weapon of the military. This was no other than military control over politics, in sharp contrast to civilian control over military affairs. The military, equipped with an independent supreme command and under the rule of military ministers who were active generals or lieutenant generals, were afraid of nothing. In fact, in 1912, when the second Saionji Cabinet tried to reduce the military budget that had seen a tremendous increase due to the First Sino-

Japanese War and the Russo-Japanese War, it was forced to resign en bloc, stalemated by the sole resignation of the Minister of the Army and the army's refusal to appoint a new minister. In 1913, under the Gombei Yamamoto Cabinet, this rule had been weakened so that retired generals or retired lieutenant generals could, in principle, also be military ministers. As it turned out, no retired officers were ever actually appointed as ministers. However, in 1936, just after the military rebellion called the February 26 Incident, the Koki Hirota Cabinet cancelled this weakening to restore the original rule that the Minister of the Army and the Minister of the Navy always had to be an active general or an active lieutenant general, which destined Imperial Japan to see its final days.

The Staff Headquarters, which was set up in 1877, was intended for the army. For the navy, a Military Department was set up in February 1884. A Unified Staff Headquarters to lead both the army and the navy at the same time was planned by Kaoru Inoue and Hirobumi Ito to be set up in 1886. But in the final analysis, because of strong opposition by Aritomo Yamagata and Iwao Oyama, the Unified Staff Headquarters was divided into two departments, the Staff Headquarters of the Army and the Staff Headquarters of the Navy. The name of the Staff Headquarters of the Navy was changed several times thereafter. In February 1889, it became the Navy Staff Headquarters, and later in May 1893, it became the Navy Command Department. Because of this situation, the military had four captains, namely, Minister of the Army, Minister of the Navy, General of the Staff Headquarters, and Chief of the Navy Command Department. The Ministry of the Army and the Ministry of the Navy were concerned with management of the military, such as its personnel and budgets. Their Ministers attended the cabinet council meetings and they advised the Emperor. On the other hand, the Staff Headquarters and the Military Command Department were concerned with planning of national defenses, strategies of wars, and actual operations of military forces. The General of the Staff Headquarters and the Chief of the Command Department worked as the chief commanders to support the Emperor. They were the supreme leaders of the military. Because ministers could not be engaged in the affairs of military commands, the former two Ministries were completely separated from the Staff Headquarters and the Military Command Department. In February 1944, Prime Minister Hideki Tojo, who at that time was also Minister of the Army and Home Minister, held concurrently the position of General of Staff Headquarters. In concordance with this, the Minister of the Navy, Shigetaro Shimada, who was called a minion of Tojo, concurrently held the position of General of the Command Department, who would have been called the Chief of the Command Department before 1932

In May 1893, in order to prepare for the threatening First Sino-Japanese War (the actual war occurred in 1894–1895), the Wartime Imperial General Headquarters Measure was ratified, as a result of which the Imperial General Headquarters would be set up as the supreme department of command to be led directly by the Emperor in wartime. All the staff of the Imperial General Headquarters had to belong to the military, so that it consisted of the Minister of the Army, the Minister of the Navy, the General of Staff Headquarters, the Chief of Command Department, and so on. Civilians were completely shut out. Just as the judicial authority is usually separated from the legislative body or the government in modern countries, the leaders of the Meiji period separated the supreme command from the government. This decision helped keep the military far away from the strife of pork-barrel politics in peacetime, but it kept the military from cooperating with politics in wartime. When it came to the First Sino-Japanese War or the Japanese-Russo War, politically experienced elder statesmen such as Hirobumi Ito played an extraordinary and unlawful role. However if they had retired, it was clear that such a military system without politics would not function well. Moreover, since the military affairs had to be dealt with exclusively by the Imperial General Headquarters in wartime, the Staff Headquarters and the Command Department were not allowed to be concerned with more than strategies of homeland defense in peacetime, so long as they abided by the law. This gives a good example of double-talk or the discrepancy between superficial policies and real practices, which has been seen without interruption in Japanese history since the formation of Japan as a country in mythical times. Such a dual control is not, frankly speaking, effective in decision making, which was demonstrated dramatically in the Second Sino-Japanese War (1937–1945) and the Pacific War (1941–1945). Because the army and the navy were completely independent organizations, which were only under direct control of the Emperor himself, they had no unified military strategies and no unified military expansion plans. They came to make desperate efforts to expand and refine their turfs only for their own departments. This deep-rooted tendency towards a disharmonious duet of the Japanese Forces had not improved by the end of the war. Some readers might think about the air force, but at that time there was no air force as an independent organization in Japan. Both the army and the navy had their own air force squadrons.

During the early days of the imperial army and the imperial navy, many military bureaucrats were lacking any regular military education. After the end of the First Sino-Japanese War (1894–1895), staff officers who graduated from the military academy and the military college came to dominate the main positions in the Ministry of the Army and Staff Headquarters so that the military bureaucratic system of the Japanese Imperial Army had been established. The rules called the General Law were the compiled basic rules to integrate and standardize the education of the army. In four or five years around 1910, the rules were reformulated drastically. The former version of the General Law looked like a literal copy of the corresponding laws in Germany and France, but after the reformulation, it emphasized spiritualism without rationality, a stubborn approval of offense-oriented credos celebrating the omnipotence of assaults with guns and swords by infantrymen, and extreme uniformity of the entire army. The close relationship between the army and the ideology of the imperial system was to be clearly recognized. During the Fifteen Years War, the General Law became the sacred book like the Bible to Christians. In the Pacific War, the principal enemy was, of course, the United States, but even during this time the General Law presupposed Russia or the Soviet Union as the chief foe. When traditional thinking went so far that it fell overboard, it was really ridiculous and very sad.

During the First World War, the United States and the European countries realized that they could not win without waging an all-out war. Because of this, those countries modernized their forces very rapidly. The Japanese forces, however, lagged behind in modernization, because it had enjoyed the fruits of the First World War while it was far away from that war's battleground and free from its disasters. In European countries, battlefield cooking cars were introduced. Firstly, they were horse carts, and later automobiles. Because of this, European warriors did not have to cook their food individually, so they could concentrate on fighting rather than on housekeeping chores. The Japanese Imperial Army, by contrast, relied on camping pots, which each warrior had to carry until the very end of the Second World War. This forced Japanese warriors who went on the battle line to look for water and fuel when they ran out of them. When they could not find any, they had to make forays into villages and towns and wreck private houses. Chinese people used to call the Japanese Army the *inferior* army, not the *imperial* army. This was not a failing of individual Japanese warriors but the structural problem of the Japanese Imperial Army as a whole. Marches by the Japanese infantry force had used every possible technique such as pushchairs, rickshaws, and carts to help warriors carry their personal luggage. In the Second Sino-Japanese War, the Japanese Army used captured Chinese people. The number of Chinese people forced to accompany Japanese warriors was almost equal to the number of the Japanese warriors. It was usual that each warrior carried baggage which weighed more than thirty kilograms. Here we saw terrible pictures of an army that had fallen behind the international trend of mechanization. In terms of the massive use of automobiles, the First World War had started the new era. But in this regard, the Japanese Army was far behind European countries. The Japanese Army used horses until the very end of the Second World War. In 1936, the United States produced more than four million cars a year but Japan only produced about ten thousand. Moreover, cars produced in Japan at that time were poor in performance, and they were in constant need of repairs. Due to this, the Minister of the Army used imported foreign cars instead of home-made cars. When they accompanied the Emperor, the car had to be in working order so that they were forced to use imported foreign cars for the sake of reliability. As a result, transportation had to rely heavily on horses. In 1958, a statue of dead horses was enshrined in the Yasukuni Shrine, a famous shrine to the war dead of imperial Japan. In 1994, the program of the tenth memorial service for the war-horses that died in the war included a prayer for the souls of the seven hundred thousand dead horses in all battlegrounds, who failed to return to their motherland. We express sincere condolences to these dead horses of imperial Japan.

The Fifteen Years War (1931–1945)

In September 1931, the Manchurian Incident took place. The man who wrote out its scenario was Kanji Ishihara. He was a military man. He was in a sense exceptional among military men because he could speak of strategies of war and actually did speak of them. At that time, almost all military men could not even talk about strategies, because they were terribly myopic. He was born on 18 January 1889. He entered the Central Army Cadet School in September 1905. According to his records in school, he seemed good in German, mathematics, Japanese and Chinese classics, but he was poor in apparatus gymnastics and swordplay. He seems to have been a man of brain rather than a man of brawn. He was remarkably inclined towards thoughts and ideas, and he read a lot of books about military history, philosophy, and sociology when he was a student of the Central Army Cadet School and the War Academy. Ishihara had studied in Germany for three years since 1922. He did research there on the great King Friedrich and the military history of Napoleon. He was under the tutelage of Delbruck and others at the University of Berlin. He thought about the reason for the German defeat in the First World War and discussed strategies of position warfare, especially in an economy blocked by the enemy during a continuing war.

When Ishihara came back from Germany, the USSR was in the midst of terrible confusion. In 1917, the Romanov dynasty was taken down and a communist regime was established. In 1924, Lenin, who had promoted the New Economic Policy (NEP), died of a disease, while Stalin and Trotskii were in the middle of a power struggle. Finally Stalin won; Trotskii was dismissed from the communist party in 1927 and was deported in 1929. Thanks to the NEP, the average person's life in the Soviet Union had attained the level it held just before the First World War, which reminds us of the drastic change in China brought about by Deng Xiaoping in the latter half of the 20th century. However, when Stalin seized power, the so-called NEP-men and Kulaks produced by NEP became targets of purges. Because of fierce collectivism in agriculture, millions of Kolkhoz farmers starved to death from 1932 to 1933. At that time, a passport system was promoted inside the Soviet Union. Ishihara regarded this situation in the Soviet Union as a chance for Japan to establish a solid footing in Manchuria and Inner Mongolia. He boasted that he needed only several divisions in order to defeat Russia completely. He was truly a man of great confidence.

Now that Russia was no longer a fearful enemy, it was only the United States that Japan should prepare itself to fight against. Ishihara's idea on this was quite remarkable. Theodore Roosevelt, who was the President of the U.S. at the beginning of the twentieth century, asked the military to propose international tactics to deal with possible opponent countries. This was called a "war simulation". Different colors were assigned to all possible opponent countries; by way of example, black for Germany and red for Great Britain. The war plan against Japan was called the Orange Plan. According to this plan, on the final stage of a possible war, by using the Navy against the Japanese Army and blocking the Japanese Army from both sea and sky, the military power of the Japanese Army, isolated in the continent, would be destroyed completely. Considering the scenario that really occurred thirty years later, I cannot help thinking that the plan was highly suggestive and even prophetic. Ishihara, following the historical facts of Napoleon's wars, insisted that Japan should fight the US by using its Army against the US Navy. He insisted further that the Japanese Army in the continent should not get any financial support from mainland Japan, but it should foster war by waging war, in the sense that soldiers should earn their living by collecting taxes, commodities and arms in the occupied territories. He finally concluded that, by doing so all over China, the Japanese Army could continue fighting, as long as it wanted. To my great surprise, his military strategy against China was quite simple and naïve. He claimed that if the Japanese Army restored peace by mopping up armed factions and clearing up native marauding bands, our superior, honest and clean-handed army would accomplish its lofty goal by winning the confidence of the Chinese aborigines. In 1911, the Chinese Revolution took place, and the despotic monarchy of the Qing Dynasty ended. After that, China entered the age of rival warlords, being far from a unified country, similar to that seen in contemporary Afghanistan. In the middle of the 1920s, Chiang Kai-shek started his Northern Expedition which was intended for the unification of China. In 1927, the Nationalist Chinese Government was established with Nanjing as its capital, and in the following year, major countries approved of it. The Northern Expedition was completed by Peter Hsueh Liang Chang's swearing of allegiance to the Nationalist Chinese Government. He was a son of Zhang Zuolin, murdered by the Japanese Army, and belonged to the Mukden armed faction. There is no doubt that the movement toward the establishment of a unified country was being fostered. I do not understand why Ishihara, who loved analysis, presupposed his prejudice against China in place of analyzing China as it was. Whatever led Ishihara to this flaw in his military plan, the flaw was undoubtedly crucial, if we take into consideration what really happened during the Second Sino-Japanese War (1937–1945) and the Pacific War (1941–1945). Such a kind of underrating of China was not restricted to Ishihara. In the middle of the 1920s, the Comintern supported the Kuomintang Party, but in 1932, in his letter to Molotov, Stalin concluded that the Nanjing Government consisted exclusively of frauds. Considering the anti-Japanese sentiment in China, it had only deteriorated since Japan leveled the notorious demands of 21 articles against China in 1915, in the midst of the First World War, which was a good model of imperialism. The anti-Japanese sentiment became stronger and stronger, and the anti-Japanese movements against the Japanese and their products were fierce. Especially, China fixed the ninth of May as a day of national shame. It was that day in 1915 when Japan forced China brutally to accept the demands of the 21 articles. This national day has been regarded as the origin of the anti-Japanese movement since then. In 1925, the anti-imperialism movement, which was to be called the 5.30 Incident, took place in Shanghai. Even if the imperial army consisted of the best and brightest, as Ishihara insisted, it seems to have been really difficult for the Chinese to welcome Japan unconditionally. In fact, after the establishment of Manchukuo, Japan was forced to fight a war of attrition against China before it began a war against the US.

Let alone the estimate of Ishihara as a theoretician, he was undoubtedly a genius in military affairs. The Kanto Army had about ten thousand soldiers at most until the time of the Manchurian Incident. On the other hand, after the death of his father Zhang Zuolin, the Mukden Army, which was led by Peter Hsueh Liang Chang, had more than 250 thousand soldiers and had excellent firearms and even aircraft bought from Europe or from the U.S. Firstly, Ishihara made soldiers hide 24-centimeter cannons beforehand, and fire one after another at the Northern Barracks, the hub of the Mukden Army. By doing so, he created a terrible scene of veritable pandemonium. At the same time, he made another group of soldiers occupy crucial stations of the railway. Deprived of means of transportation, the army with 250 thousand soldiers became little more than nothing. The Manchurian Incident, which started in September 1931, was ended in February of the following year by acquiring all of Manchuria. It was as fast as a streak of lightning. Ishihara, acquiring a domain as large again as his homeland in such a short period, was undoubtedly far better than many military men. Nagata Tetsuzan, who was the military section chief and sent out Ishihara as a staff officer of strategy in the Kanto Army to Manchuria, made a good choice, saying that Ishihara would fight only winning wars. Ishihara had begun action at the very time of the Manchurian Incident, before the USSR finished the first stage of the Five Year Plan, before the United States recovered from the Great Depression, and before the autonomy of Manchukuo was taken over completely by the Nationalist Chinese Government. Later, Peter Hsueh Liang Chang was ungraciously named the unresisting shogun and was treated with contempt both domestically and internationally. Probably, he had not even imagined that only about ten thousand soldiers could gain ascendancy over the whole of Manchuria, and he wanted merely to avoid unneeded troubles.

In March 1932, Puyi, who was the last emperor of the Qing Dynasty, was invited to become a chief executive of the newly established Manchukuo. Preceding the report by the Lytton inquiry party, which was sent off by the League of Nations in September of the same year, the Japan-Manchukuo Act was sewed up, and Japan admitted Manchukuo. In January 1933, the Kanto Army claimed that not only Liaoning, Jilin and Heilongjiang, but also Rehe was a part of Manchukuo, and invaded Rehe, which was located in the East of the Nei Mongol. This was what was called the Rehe Tactics. Japan intended to secure the backland of Manchukuo against China and the USSR, and to be able to watch Beijing and Tianjin very closely. This action was somewhat drastic and presumably too heroic, because not a year had passed since the establishment of Manchukuo. If the action had failed, Japan would have become mired in a prolonged war against China. It was nothing but a kind of gambling. Even the Showa Emperor reluctantly admitted this action on condition that the army would never invade Kannai (the southern part of the Great Wall), and it would never bomb Kannai. In 31 May, 1933, the Kanto Army and the Nationalist Chinese Government Forces signed the Tanggu Armistice. Having in mind "struggling against difficulties for the sake of vengeance" with its confidence being betrayed by the Japanese Army, the Nationalist Chinese Government strengthened their power over Northern China. In passing, we note that in January 1932, the Shanghai Incident took place, which was caused by the land engagement party of the Japanese Navy. This is what was called the first Shanghai Incident. This was a typical example of rivalry and jealousy between the army and the navy in Japan. Because the army succeeded in the Manchurian Incident, the navy tried simply to succeed in another incident.

In the eighteenth century, Great Britain started selling opium to China with great profit. Opium is taken from the fruits of opium poppies by being scratched. The discharged emulsion is made into an ointment-like chemical. By parching it and inhaling it, one can attain euphoria, but if continued, one can become a physical wreck. Opium is very dangerous. Great Britain had imported plenty of tea, china, silk and so on from the Qing Dynasty, but it had nothing to export in return. Because Great Britain needed to compensate for the cost of the American Independence War, and store up capital for the coming Industrial Revolution, it aimed to balance out excessive imports from China by secretly exporting opium to the leaders of the Qing Dynasty. This completed the famous triangular trade among Great Britain, India and China. In the middle of the nineteenth century, Great Britain fought the Opium War against the Qing Dynasty, which protested the secret exportation of opium into China by Great Britain. As a result of its victory, Great Britain acquired Hong

Kong. Great Britain should have had no compunction at that time. Interestingly enough, some of the history textbooks in Britain do not have any description of this war. More interestingly, China, which still complains loudly and inflexibly of Imperial Japan's diabolical actions, has never demanded any official apologies from Britain. I do not know where China's political intention lies, but China can not help being blamed for its double standard. It has become a common sense that the management of colonies does not pay, and Imperial Japan's management of Manchuria was no exception. Imitating Great Britain, Japan sold opium secretly to the Chinese in order to pay great expenses for the management of Manchuria. Rehe Province, which the Kanto Army invaded coercively in January 1933, was a well-known place for planting opium poppies. When Manchukuo was established, there was no main industry except for agriculture so that they had to rely on opium. It is said that even in 1939, one sixth of the national budget of Manchukuo came from opium. Imperial Japan was an exemplary student of Great Britain in terms of imperialism, and it concluded the Japan-Britain Alliance before the Russo-Japanese war. Japan made the most of its role as an exemplary student of Great Britain, even in illegal opium sales. This story, however, has not become public history. It was only in the International Military Tribunal for the Far East, which was held after Japan was defeated in the Second World War, that the story could have been brought to light, but it was not realized in consideration of Great Britain, which has the same dark past. This tribunal in the Far East corresponded to the Nuremberg Trials in Germany. China, which cries out about the matter of wartime comfort women and the Nanjing Massacre with facts and myths mixed, has never referred to this. What a strange story we have!

The first Five Year Plan by the USSR had been conducted from 1928 to 1932, and the second Five Year Plan, from 1933 to 1937. These plans concentrated upon raising heavy manufacturing industries. They proceeded steadily without being affected by the World Depression, which improved the USSR's armaments. When the USSR started the second Five Year Plan, it also started to make a fortification zone along the Manchukuo-USSR boarder in the summer of 1933. In terms of the number of aircrafts, the USSR began to gain a great advantage over Japan. To secure its domain in a possible war against the USSR, Japan established a puppet government called the anti-communist[1] East Hebei Autonomous Council[2] in October 1935. It governed the northern five provinces of China such as Hebei and Shangdong. By this action, which is usually called the northern China separating operation, the economic block of northern China was separated from the economic block of southern China, in which you could find Shanghai, and it was connected to

---

1 http://en.wikipedia.org/wiki/Anti-Comintern_Pact

2 http://en.wikipedia.org/wiki/East_Hebei_Autonomous_Council

the Japanese-Manchurian economic area. The Kanto Army intended that the northern five provinces in China should get rid of any influence from Chiang Kai-shek's Nanjing government, and these five provinces would be a buffer between Manchuria and Nationalist China. Moreover, these buffer provinces would function as a remover of a threat by the USSR in possible wars against the Soviet Union. It was possibly threatening wars against the Soviet Union that drove the Kanto Army, especially in the establishment of Manchukuo and the separation of the northern China from its southern part.

The imperial system in Japan has continued for more than ten centuries; we do not know any other dynasty of such a long continuation period. The reason for this long-continuing dynasty was that, with some possible exceptions, the imperial system in Japan has been subject to the principle of reigning without governing, and the emperors have been contented with this dual structure of reigning and governing. Since emperors never touched real politics, their political responsibility was never questioned. In the Heian Period from 794 to 1192, Japan enjoyed aristocracy. From the latter part of the twelfth century to the middle of the nineteenth century, samurais governed Japan. For the last 100 years in the Heian Period, former emperors enjoyed their cloister governments standing side by side with the formal governments of current emperors. They did not however act in the manner of true emperors. The position of an emperor, surprisingly, has quite a few restrictions. Hirobumi Ito, regarded generally as one of the few best and brightest politicians in the Meiji period, visited Germany and Austria to investigate constitutions in Europe in 1882 and returned to Japan the next year. In such a short stay in Europe, he realized that democracy in Europe was based on Christianity, and that democracy would not function without such a spiritual backbone. Ito adopted Japan's traditional imperialism in place of Europe's christianity. This shows how deeply Ito understood democracy. According to Ito, in Japan at that time, some people adhered so strictly and so literally to the works by famous European or American radicals as to overthrow Japan itself, while the right-wing movement towards absolute monarchy was also strong. Ito and his colleagues wrote a constitution which gathered the nation around the emperor, and which took parliamentary democracy well into consideration at least up to the standard at that time. The constitution was promulgated in 11 February, 1889. Prior to this constitution, the Midhat constitution was promulgated in Ottoman Turkey in 23 December, 1876, but this was suspended indefinitely on the pretext of the Russo-Turkish War in the following year. Accordingly, Ito's constitution was the first constitution in Asia that was really put into operation. This was the beginning of the first Asian constitutional monarchy.

In the Showa Period (1926–1989), until the Second World War ended, the term "Showa Restoration" was often used by the right-wingers. As is well known, Meiji Restoration put an end to the time of the samurais, and Japan ushered in the modern era with the emperor as its central figure. However, the right-wingers insisted that, because society had slacked off, they had to reform

Japan by establishing a kind of absolute monarchy with the Showa emperor at its center. The thought that the emperor should be absolute can be traced back to the battlecry "Revere the Emperor and oust foreign barbarians" in the end of the Edo Period, but it was in the May 15 Incident in 1932 that this extreme thought began to hover over all of Japan. The May 15 Incident was a coup in which young commissioned officers of the Navy attacked the office of the Prime Minister and the head office of the Seiyukai (one of the parties in power at that time), and assasinated the Prime Minister Tsuyoshi Inukai, regarded as a principal bearer of the pro-Constitution. Strictly speaking, this was not a coup but was to be called a terrorism act of assassins, because the weapon they used did not belong to the national military, and the commissioned officers did not call up their soldiers. After the May 15 Incident, once the trial started, written petitions for clemency arrived from all over Japan one after another. The number of received petitions was said to be one million. That such a large number of petitions had arrived in such a time without television was really remarkable. Some petitions were written with blood or sealed with blood. A petition from Niigata Prefecture was most atrocious: Nine fifth fingers cut out at the base were attached to the petition. If things like this should be sent by mail in the contemporary world, they would be considered a bizarre or eccentric incident. However, at that time, not a few people were moved to tears at this radical act. They seriously discussed how to preserve the cut fingers forever, for they thought that such behavior was a mirror of patriotism. We know that what was to be called "the May 15 incident fever" took place and should have had a great impact on the trial. But nobody was sentenced to death in the May 15 Incident. A fifteen-year imprisonment was the most severe punishment. The incident itself was hogwash. The only remarkable result was the murder of Inukai, and a less remarkable result was that one of the police officers on security died and another was injured seriously. The reason why the incident should be significant in Japanese history is that, as a result of the fever caused by the May 15 incident, the Makoto Saito Cabinet, which was supported by the whole nation, came into existence, and the army came to have a strong power in the arena of politics.

The emperor-as-an-organ theory, which was based on the theory of a nation as a juristic person by the German jurisconsult Georg Jellinek, claimed that sovereignty belonged to the nation, and the emperor is nothing but the supreme organ among many national institutions, including the assembly and the cabinet. This theory was proposed by Professor of the Tokyo Imperial University Kitokuro Ichiki. His follower Tatsukichi Minobe developed this theory further and claimed that the diet which was the representative of the people could circumscribe the intent of the emperor through the Cabinet. He gave theoretical supports to the Taisho democracy, in which democratic ideas flourished and Taisho was the name of the emperor (reign: 1912–1926). The emperor-as-an-organ theory was the theory of constitution admitted by the Japanese government, and it skillfully reconciled the imperial system of Japan and the modern political system. The idea by Hirobumi Ito, who took a central role in designing the Meiji Constitution, was very close to this theory. On the other hand, the doctorine that sovereignty rested with the emperor had collected some supporters. In this theory, the emperor was given a position of transcendence, to which sovereignty should belong. One of the adherents of this theory was the constitutionalist Hozumi Yatsuka. He protested against enforcement of the civil law written by an invited French jurist Boissonade, who came to Japan as a part-timer of the Department of Justice. Hozumi insisted that the new civil law would kill Japan's traditional loyalty and filial piety. His follower Shinkichi Uesugi furiously disputed Minobe on the Constitution. Uesugi was destined to live in the time of the Taisho democracy, but in 1929 he died of disease at age 50 without seeing his time come. Interestingly enough, the emperor-as-an-organ theory incident, which took place in February 1935, truly showed how drastically the atmosphere in Japan had changed after the occurrence of the May 15 incident. From the present point of view, this incident was no more than a silly talk. A right-wing assemblyman Takeo Kikuchi opposed himself to Minobe's emperor-as-an-organ theory in the House of Lords, insisting that the theory was a quiet rebellion against the sacred national system of Japan. This became an issue, and, as a result of this, Minobe, who was at that time a member of the House of Lords, was sued for lese majesty, and he resigned. The Okada Cabinet at that time announced twice, with the most scrupulous care, a clarification of the national system, denied Minobe's theory, and declared that the sovereignty belonged to the emperor. On the other hand, the Showa emperor was favorable to Minobe's theory. He claimed that Minobe's theory was completely correct and he found no problem in the emperor-as-an-organ theory. The emperor correctly understood the constitutional monarchy. Right-wingers, who insisted on reverence for the emperor, did not understand the emperor's true intention at all.

After the May 15 Incident, there was a severe confrontation between the control faction and the imperial way faction inside the army. The control faction was a group espousing modernization in the army, and its members aimed at a nation with a high-level national defense in the age of all-out wars. Hideki Tojo and Tetsuzan Nagata were two central figures in the faction. The imperial way faction, on the other hand, aimed at the Showa restoration with the emperor as an absolute monarch, and was ready to stage a coup d'etat, if necessary. The faction was strongly right-wing. The central figures in the faction were Sadao Araki and Jinzaburo Masaki. Curiously enough, the imperial way faction was liable to give radical or even fanatical assertions on domestic issues, but the faction was highly moderate on foreign matters, while the control faction was so radical on foreign affairs that it insisted on the inevitability of a fatal blow against China, regarding Britain and the U.S. as principal foes. Anyway, after the coup d'etat by the imperial way faction in February 26, 1936, which was to be called the February 26 incident, suppressed, military officers of the imperial way faction were transferred to the first reserve, and the control faction revitalized the once abolished system of military ministers being always active military officers, so as to secure its victory against the imperial way faction. In this way, the victory of the control faction against the imperial way faction prepared the coming Second Sino-Japanese War in the following year. History is really interesting! By the way, Tetsuzan Nagata, who was the central figure in the control faction, was killed by Commander Aizawa from the imperial way faction in August 12, 1935, that is, before the February 26 incident. This is known as the Commander Aizawa incident. Because of this, Hideki Tojo became the leading figure of the control faction, and led Japan to the Second Sino-Japanese War, then to the Pacific War and finally to downfall. Teiichi Suzuki, who worked as President of Ministry of Strategy in the Second Konoe Cabinet, claimed that if Nagata had been alive, the Pacific War would have never taken place. He was quite right in saying so. Kanji Ishiwara, who played a principal role in the Manchurian Incident, was backed by Tetsuzan Nagata, but Ishiwara was stalled by his death, only to lose the struggle with Tojo. Even without the above remark by Teiichi Suzuki, it is clear that Japan would have walked on a quite different road if the duo of Nagata and Ishiwara had continued to hold power for a decade further. We sense the touch of destiny to find that the last task by Ishihara in the military was to deal with the aftermath of the February 26 incident. Ishiwara was relegated to the position of vice chief of staff in the Kanto Army at staff headquarters. In Manchuria he was often opposed to Hideki Tojo, and he was finally dismissed from the post. In March 1941, he retired from military service in action, and was transferred to the first reserve. In April of the same year, he was invited as a lecturer to the study of national defense at the Ritsumeikan University in Kyoto, which reminds us that Ishihara was more academic than military. Because of Tojo's persistent harassments, he was again dismissed even from his academic position several years later. However, because he stayed away from the army, he was not sued in the International Military Tribunal for the Far East after the Second World War, though he played a central role in the Manchurian Incident. He died a natural death on 15 August, 1949 at the age of 60.

The Second Sino-Japanese War (1937–1945) stemmed from an accident on the Marco Polo Bridge in July 1937. This accident induced the battle of Shanghai (August 13–November 9, 1937), where the largest number of people had been injured or killed since the fierce battle in Verdun in the First World War. This battle was to be called the Second Shanghai Incident. The Forces of the Nationalist Chinese government, which was defeated and left Shanghai, concentrated on the defense of the capital Nanjing. Nanjing, however, fell on December 13; the genocide which had continued for six weeks by the Japanese Army was conducted, and the name Nanjing Massacre has gone down in history. The Second Sino-Japanese War was a war without any formal declaration. The reason why neither Japan nor China declared war was that both of them feared application of the neutrality law of the U.S. This law allowed the U.S. to be independent from foreign skirmishes, and this had the same effects as economic sanctions to those countries which wanted to have wars. Due to this law, China feared that the import of materials including weapons from the U.S would diminish, while Japan worried about possible financial sanctions by the U.S. As a result, the declaration of war was postponed until the beginning of the Pacific War in 1941. A war without declaration had to be crooked in the sense that any military occupation or any compensation claim could not be done lawfully, because it was not an ordinary war. Interestingly enough, it was Germany that supported China militarily at that time. China had prepared for a possible war against Japan under the direction of German advisers since 1934, and more than a half of weapons exported from Germany flowed into China in 1936. In contrast, Japan was so aware of a possible war against the Soviet Union that it did not put elite troops into the battles of Shanghai and Nanjing. The Japanese troops in the battles consisted mainly of reservists and secondary reservists, and even "the first class B uneducated supplementary soldiers" were collected. These soldiers were soldiers who had passed the first class B test of conscription, but they had not yet experienced any practical call-up for one term of three months. Since the Japanese troops in the two battles consisted of such soldiers, the proficiency of the troops was undoubtedly low and their morale should be minimal. These troops tended to commit war crimes such as robbery, sexual assault, and bloodletting, let alone the problem whether the number of victims in the Nanjing Massacre which the Chinese government insisted was really correct or not. Shinichi Tanaka, the chief of the military section of the Ministry of the Army at that time, criticized the Japanese government, claiming that Japan blundered in the early stages of the battle of Shanghai, which drove major foreign countries against Japan, transmogrifying the possibly short battle into a long drawn-out war. As Tanaka continued, it was the Japanese government and its nonaggravation policy that invited a military disaster. By the way, even in 1934, Chiang Kai-shek, the president of the Nationalist Chinese Government, assumed a two-stage strategy for a possible war against Japan. The strategy went as follows: In the first stage, China would have a war against Japan along the watershed area of Chang Jiang such as Shanghai, Nanjing, and Wuhan. If defeated there, in its second stage, China would go into a long drawn-out battle based on inland Yunnan and Sichuan. Chiang Kai-shek, who had a secret meeting with the USSR in the same year, claimed that the Japanese policy against China inevitably would lead to interference by powerful countries and consequently to a World War: He was a pretty good politician. In 1938, the Japanese Army flattened Xuzhou and in August of the same year, occupied Canton and Wuhan. As a result, the Nationalist Chinese government pulled back to the midland, Chungking, and the Second Sino-Japanese War became a long drawn-out war, which Japan had to avoid at any cost. This reminds me of Sun Tsu's classical words "A war should be finished as early as possible, possibly even badly. I have never heard that a long war was operated successfully."

In the Nomonhan Incident in 1939, Japan and the Soviet Union fought for the borderline between Manchukuo and the Mongolian People's Republic, behind which Japan and the Soviet Union lay as their patrons. Since there were grasslands and deserts adequate only for the purpose of nomadic herding along the borderline, and nomads around there freely came and went across it, this incident had a strong political flavor. Japan was definitely more aggressive than the Soviet Union. The Outline of Handling of the Struggle against the Borderline between Manchuria and the USSR, written by Masanobu Tsuji, who was a staff officer of strategy, stated that in the area, where there was no clear borderline, the defense commander should voluntarily certify it and that if there occurred a conflict, victory must be secured regardless of the number of available soldiers, and regardless of the actual place of the borderline. This outline was ordered by Kenkichi Ueda, who was the Command of the Kanto Army. The Nomonhan Incident can be divided into two incidents: In the first Nomonhan Incident in May, thousands of soldiers fought, and in the second Nomonhan Incident in July and August, tens of thousands of soldiers fought. In some parts of the battles, Japan did a good job but in the end lost completely. In the armistice agreement concluded on September 16, the borderline was decided completely on the lines that the USSR insisted. Since this defeat, the Kanto Army, which boasted that it consisted of the best and the brightest, had become military comrades in silence. In this battle, the supply line of the USSR was about 700 km away, but huge transportation of soldiers and weapons by cars was conducted. In contrast, Japan's supply source was only about 200 km away. This was much shorter than that of the USSR, but Japanese soldiers were required to walk until reacing the battlefield. This showed clearly the backwardness of the Imperial Army. Indeed, Japanese soldiers had walked a prodigous number of miles by the end of the Second World War. The USSR had put more than 500 tanks into this battle, but Japan had put only 38 middle-sized tanks and 35 small-sized tanks. Moreover, since Japan regarded its tanks as a mere support for walking soldiers, so that they were not designed as an essential tool for the battle against the tanks of enemies, Japan had been defeated completely in such a motorized battle: When its precious 30 tanks broke down, Japan got its remaining tanks back in a hurry. In the aerial battle, Japan had a totally successful battle in the first Nomonhan Incident, but in the second, the USSR sent up airplanes that were much better in quality than the Japanese ones, and the USSR adopted the strategy of shooting at Japanese airplanes and leaving immediately. Since a Japanese battle plane of type 97 was very good at circling, the USSR tried to avoid dueling in the air. These strategies of the USSR were so successful that Japan was fighting a desperate battle. Japan's behavior after the defeat in the war was terribly weak: Many of the surviving captains were questioned so severely about their responsibility for blunders that they resorted to suicide.

In contrast, such staff officers as Masanobu Tsuji and Takushiro Hattori, who insisted on their self-assertiveness in crucial decisions only to lead Japan to its defeat, were disciplined very lightly, and they played a decisive role again in devising a strategy for the army in the Pacific War, which took place only two years later. This was a terrifying system without a due sense of responsibility. Sad to say, this pattern was repeated again and again. In addition, despite the conclusion of the Anti-Comintern Pact between Japan, Germany and Italy in November 1937, the German-Soviet Nonaggression Pact was also concluded on 23 August, 1939. Because of this, the Kiichiro Hiranuma Cabinet resigned en bloc, saying notoriously that the political situation in Europe was messed up beyond all comprehension. In concert with this, on August 25 of the same year, Britain, France and Poland concluded the Treaty of Mutual Assistance. In September of the same year, the USSR and Germany started invading Poland from East and West respectively. Whether Kiichiro Hiranuma could understood or not, the political and military meaning of the German-Soviet Nonaggression Pact was very clear: This pact had a secret protocol, in which it was decided how to divide Eastern Europe between Germany and the USSR. On September 3, Britain and France, which were allied nations of Poland, declared war against Germany. Thus was the Second World War started.

In April 1940, Germany attacked Denmark and Norway. In May, Germany invaded the Netherlands and Belgium. On 14 June, Germany made Paris fall. The Japanese Forces, which were irritated at the stalemate in battle lines of the Second Sino-Japanese War, was fascinated by this German breakthrough. On 23 September 1940, Japan occupied the Northern part of French Indochina in a hurry to catch up with aggressive Germany. This was intended to shut off the French Indochina assistance route to China and to have a toehold to advance to the south. Stiffened by this, the U.S. banned the exportation of steel and iron scraps to Japan, and gradually introduced a license system for the exportation of products besides steel. In September 1940, the Japan-Germany-Italy Tripartite Military Alliance was concluded, and this made decisive the hostility between Japan on the one hand and Britain and the U.S. on the other. From August 1940, Hitler executed several military operations one after another, in order to prepare itself to land in mainland Britain. Failing to get good results, Hitler transmogrified his operations against Britain into ones against the USSR. He started to invade the USSR, which was called the Barbarossa operation. Although Japan concluded the Japan-USSR neutrality pact in April of the previous year, the Kanto Army responded to this operation and it called up a big force to occupy the eastern USSR. The army had a very myopic and selfish idea that the USSR Forces must be short of hands because of the beginning of the Germany-USSR War. However, the USSR Forces were not so short of hands as the Kanto Army expected. Because of opposition by the Japanese government and the Japanese Navy, the operation was cancelled right before the actual execution. This is what we call the Kanto Army Special

Exercise. At the same time, on 23 July of the same year, the army occupied southern French Indochina. The U.S. responded sensitively to this movement, and it imposed a freeze on all Japanese assets in the mainland of the U.S. on 23 July of the same year, and on the exportation of oil to Japan on 1 August. The Japanese government and the military, which were shocked at this strong attitude of the U.S., were inclined to destroy the enemy before the supply of oil would be cut off and the power of the nation would dwindle away into nothing. On 8 December 1941, the army started to invade Kota Bharu in the British-ruled Malay Peninsula and about an hour later, the Japanese navy attacked Pearl Harbor in Hawaii. In this way, the Japanese Forces started a war against Britain and the U.S. The U.S., whose public opinion tended toward isolationism, regarded the attack by Japan as a good opportunity, and succeeded in unifying all Americans under the catchword of "Remember Pearl Harbor." The Second World War, which had been carried out mainly in Europe until then, became literally the World War due to the participation of the U.S. It was the beginning of what was called the Pacific War. In the beginning of the war, Japanese Forces occupied various places with progressive increase of force: Japan occupied Manila in January 1942, Singapore in February, and Rangoon in March. However, because the Japanese Forces were not prepared for a military logistics for such a large area, the military finally came to a halt. It is generally said that 2.3 million military men and civilians died in the Second Sino-Japanese War and the Pacific War. Among them, an estimated sixty percent (1.4 million) were starved to death. The more the Japanese Forces won, the more land it controlled and the longer battle line it had. In this situation, it was very probable for the Japanese Forces to collapse completely, once it got off the track by chance. The Japanese Forces, which had continued to take the offensive, was forced to stand on the defensive since the Battle of Midway Island on 5 June 1942. In this battle, the Japanese Forces lost four main aircraft carriers and many skilled pilots. Once the situation declined like this, it continued to decline faster and faster. Especially, because the U.S. knew military logistics well, the U.S. conducted an operation to be called "island hopping" or "leapfrogging". In this operation, instead of attacking such fortified bases of the Japanese Forces as could be seen in Rabaul in the southern Pacific Ocean, the U.S. concentrated its attack on islands that were regarded as crucial in approaching to the mainland of Japan but relatively poor in armament. By occupying such islands and by sending Japanese transport ships to the bottom, the islands that were used as bases by the Japanese Forces became isolated. The reason why the Japanese Forces registered a huge number of starved soldiers was partly attributed to this successful operation by the U.S.

The Kanto Army had only about ten thousand soldiers when the Manchurian Incident happened in 1931, but came to have 350 thousand soldiers by the end of 1940. In the Kanto Army Special Exercise, which I previously referred to, the Japanese Forces sent 350 thousand more soldiers to Manchuria. As a result, the Kanto Army had 700 thousand soldiers in total. When Japan declared war against the U.S., Britain, and the Netherlands on 8 December 1941, the Kanto Army was strictly ordered by the Imperial Headquarters not to fight even under the greatest provocation of the USSR Forces and not to provoke the USSR Forces so as not to impede progress of the southern operation in the Pacific. Some soldiers were extracted from the Kanto Army for the southern operation. The extraction accelerated more and more as defeat in the war became more and more realistic. By the end of March 1945, the thirteen divisions, which stayed in Manchuria when the Pacific War broke up, disappeared completely. Instead of thirteen divisions, the Kanto Army formed fourteen new divisions by consolidating soldiers who were still in Manchuria and regrouping them.. Furthermore, four divisions were incorporated from China, and one division was incorporated from the Korean Peninsula. The Kanto Army, which did not feel confident even by this, summoned 250 thousand people, excepting 100 thousand people who were indispensable for the government or industry. The Kanto Army produced an additional eight divisions in this way. This is what was called notoriously the root and branch mobilization. Nominally, at the end of the July of 1945, the Kanto Army came to have 24 divisions containing 780 thousand soldiers and 230 airplanes. However, the number of guns decreased to one second or one third of that at the Kanto Army Special Exercise, and the airplanes were a medley of old types and training planes. Some divisions did not have a gun at all. With all these disadvantages, the Kanto Army waited for the invasion of the USSR with fear and trembling. On the other hand, in December 1944, the USSR Forces began in secret to transfer its army from Europe to the Far East. Especially in May 1945, when Germany surrendered, this transfer was accelerated. At the end of August of the same year, the USSR Forces in the Far East consisted of 80 land divisions, 32 plane divisions, and 1.74 million soldiers and 300 thousand guns. There was no hope for the Kanto Army to compete with such a huge military force. On 9 August of the same year, the USSR Forces invaded Manchuria.

As I have claimed before, the time of all-out wars had come with the First World War. To win an all-out war, leaders are required to define transparent national purposes and devise scrupulous strategies of how to begin the war and how to end the war. On 2 November 1941, when the Showa Emperor asked Hideki Tojo, the Prime Minister at that time, what was the principal purpose of the coming war (the Pacific war). Tojo answered insouciantly that it was now under study and a result would come someday, though it was held infallibly at that time that Japan would begin to fight the US on 8 December 1941. In fact, after the Pacific War had started, the stated purpose of the war vacillated between self-defense and establishment of the Greater East Asia Co-Prosperity Sphere so as to relieve Asia from the colonialism of Europe and the U.S. The fact that the purpose of the war was plucked from thin air after the actual war had already started was not exceptional in Japanese history.

It was seen not only in the Pacific War but also in the Second Sino-Japanese War. In November 1938, when it became clear that the Second Sino-Japanese War would not end in the foreseeable future, the Konoe Cabinet abruptly announced the New East Asia Regularity Statement, claiming that Asian races should revolt against the governance of Europe and the U.S. and be united to make a new East Asian community. After the First World War, imperialism or colonialism was no longer a dogma to be claimed in public, and a new logic or slogan for war was to be looked for. This was a desperate slogan taken under the pressure of necessity.

To the Japanese Forces, the principal enemy had been Russia or the USSR until the end of the Second World War. It was in September 1943, in the middle of the Pacific War, that the educational inspector general, who was the superintendent of the education of the army, finally announced that the army would shift its education and research from "against the USSR" to "against the U.S.". We are really flabbergasted to see such an extreme example of traditional thinking. It was well known that the army and the navy were to each other as cat to dog, but when it came to the war against the U.S., the navy had to lead the war, because battles would be waged in the Pacific Ocean. In this sense, the military men of the army worried whether or not the navy was as strong as it traditionally boasted. The navy, for its part, had bragged again and again, as a reason for increases in budgets, that the defense of the sea around Japan was perfect, or that the navy would gladly take responsibility for the offence and defense of the West Pacific Ocean, so that it could not say frankly, at such a crucial moment, that it did not have enough confidence. The army could not quite believe the statement of the navy, but it had to decide to fight a war against the US by simply believing the navy's exaggerated statements. The true story was really fearful.

In the pre-war Meiji Constitution, the power of a Prime Minister was very weak, aside from the independence of the supreme command. This is because it was decreed in the Constitution that every minister should support the emperor directly, i.e., not via the Prime Minister. The Prime Minister was only one of the cabinet of ministers and did not stand over the others. In this case, oneness or unity of the cabinet could not be expected. The reason why cabinets had tended to be short-lived until the Second World War ended lay here. Once the Second Sino-Japanese War started and Japan found itself in the middle of the war, the disunity of the cabinets was seen as a crucial problem. To strengthen the function of the cabinet, the Planning Board was established in October 1937 for investigating and planning a controlled economy in wartime, and the Information Board was established to direct and control public speeches and publications. Ironically, however, the establishment of these boards contributed not to the strengthening of the cabinet but to the advancement of men of the fighting services to officialdom. These new boards provided plenty of employment in official circles to military men. It was highly natural that by amending the Meiji Constitution the authority of the Prime

Minister should have been strengthened, which would have strengthened the function of the cabinet. Since the emperor granted the Meiji Constitution, and since the emperor was regarded as absolute, the constitution could not be changed at all, because it was a code of laws that would be in effect forever, just as the Bible is to innocent Christians.

By opening the Second Sino-Japanese War in 1937, Japan converted to a wartime economy. In September 1937, the provisional military special finances were established in the same way as in the First Sino-Japanese War (1894–1895), the Russo-Japanese War (1904–1905), and the First World War (1914–1918). These finances were intended to help wage the war, and one fiscal year ranged from the very beginning of the war to the very end of the war. They would be closed at the end of the war. The compilation of the budget went beyond any control of the cabinet or the assembly on the pretext that it was a military secret. As a result, both the army and the navy made every endeavor to expand their equipment and personnel as if they had been crazed. Thanks to this, Japan had an advantage over the US when the Pacific War started. However, Japan expanded only military production at a time of relatively low level in the heavy chemical industry, so that this production not only oppressed private demands but also impeded the production of industrial goods that were indispensable for the production of other military goods, which finally weakened the economy itself too much for Japan to continue the war. Moreover, uncontrolled military expenditures undoubtedly gave rise to inflation. The Japanese Forces published military currency called "scrips" in occupied East Asia and China, and exported inflation overseas. In Germany, because the government put much importance on a sufficient supply of daily necessaries, private living expenditures in 1944, even at the end of the war, were higher than those in 1932, which were the lowest then because of the mighty depression in 1932–1933. In case of Japan, the army took the position that the public could be squeezed indefinitely, so that the living standard of Japanese people had become lower and lower since the beginning of the Second Sino-Japanese War. According to the word of an investigation officer of the Planning Board, the private demands could be quashed indefinitely just as we could wring a wet hand towel again and again. This reminds us of the words in the Edo period, saying that farmers and oil of sesame seeds could be exploited indefinitely. Because of this policy, private living expenditure in 1940, the previous year of the beginning of the Pacific War, was lower than that in 1930, the year under the Showa depression. In 1942, private living expenditure was only 80 percent of that at the beginning of the Second Sino-Japanese War in 1937. The production of coal, which was the largest energy resource at that time, reached a peak in 1940 and declined. Production of textiles began to turn down at its peak in 1937. The production of food and chemical goods had its peak in 1939. The production of steel remained on the same level till 1943, but suddenly began to decrease. The lack of goods led inevitably to a system of allocation. In the first Konoe Cabinet in 1938, a national mobilization law was enacted that mandated the government to manage labor forces, materials, machines, and factories so as to wage the war. Based on this law, the government announced a flood of orders designed to officially control the economy. In 1939, to contain inflation, the government announced a law regulating wages so that it could decide the wages of laborers. As a result, the real level of wages became lower, as the general principles of economics tell us they will. In the same year, the government promulgated a national draft law and thus became able to force people to work in the military industry. The call-up paper for the military was called the red paper because of its color, but the call-up paper based on the national draft law was called the white paper. A foolish price control law was issued that fixed prices at the level on 18 September of the same year, but as was expected, black-market prices appeared. The government, being in trouble with this strategy, promulgated a daily necessities control act, and based on this, a rationing system and a ration ticket system were gradually introduced. In July of the same year, a regulation law for production and distribution of luxury goods, which was called the 7 July interdict, was issued to prohibit the production of luxury goods and noble metals. The slogan was "Luxury is our principal enemy". In June 1940, the rationing of sugar and matches began in six major cities like Tokyo and Osaka. In April 1941, the rationing of Japan's principal food, rice, began in the same six major cities. The amount of rice rationed had not changed very much till the end of the war, but wheat or potatoes came to be blended into rice, and in October 1944, the ratio of rice decreased to 66%. This was what was called the worsening of commodities allocated. The list of commodities rationed expanded: From November 1941, fish came to be rationed, in February 1942, clothes, miso and soy sauce, and in November of the same year, fruits and vegetables. Peoples' lives became worse. Naturally black markets became rampant, which gave rise to people who profited illegally. A song popular for singing in the dark said "our society is divided into two kinds of people, one class consisting of stars, anchors, and friends of black marketeers and the other class consisting of honest fools who stand in lines for allocated commodities". "Stars" here meant the army, which used stars as the indication of classes in the army, and "anchors" meant, of course, the navy.

In July 1944, Saipan fell and the Mariana Islands went down to the U.S completely. In the previous year, Japan fixed the absolute national defense line, which was along the Kurilian Islands, Bonin Islands, Saipan, Guam, Palau Islands, West New Guinea, and Burma. This perimeter was set to shrink the battle line. The fall of Saipan meant that a pivot of the defense line was broken. Therefore Japan could do nothing but take a defensive position. Since April 1942, the U.S. Forces had raided the mainland of Japan from the air but, by taking the Mariana Islands, they acquired a base from which to continue to do so more efficiently. From 14 November 1944, the U.S. Air Force raided Tokyo 104 times. Especially, the aerial bombing that started on 10 March

1945 was very fierce, and the number of airdropped bombs was about 380 thousand, which amounted to 1700 tons. At that time, due to the passing of an atmospheric depression, a strong northwest wind blew, which expanded fires and made the disaster more serious. Because of this, more than 100 thousand people were killed and more than one third of Tokyo was burned out. The US's air bombardments upon the mainland of Japan made many urban dwellers of the middle class lose their properties and forced them to live in shacks. In my opinion, the Japanese government could do nothing else but surrender after Saipan fell. Because the Japanese government delayed this inevitable decision for more than a year, many meaningless deaths followed. The war continued to be waged without any clear purpose or any unifying decision-making organization that could have tried to end it at the right time and at the right place. Although there are no exact statistics, at least more than fifty percent of the war dead during the fifteen-years war occurred after the fall of Saipan. Kamikaze, the synonym for Japanese suicide squads, was given birth after the fall of Saipan. The notorious atomic bombing in Hiroshima and Nagasaki by the US occurred more than one year after the fall of Saipan. The main purpose of the battle in Okinawa in January 1945 was not to defend Okinawa but to make the U.S. Forces take a heavy toll of lives and wounded as long as possible, so that the number of civilians killed in the Okinawa battle was almost the same as that of soldiers killed in the battle. Both numbers were more than 90 thousand. This was indeed a regrettable and pitiful story.

It could be said consistently during the Fifteen-Years War that there was no diplomacy on the side of Japan at all. China presented the case of the Manchurian invasion by the Japanese Forces to the League of Nations, which sent an investigation committee to Japan, China, and Manchuria from March 1932 through June of the same year. The committee consisted of members from the U.S., France, Germany and Italy, and its captain was a British Lord Lytton, who once served as a Viceroy of India. On 1 October, Lytton handed in the report written both in English and in Japanese to the Board of the League of Nations. The English report consisted of 148 pages, while the Japanese report contained 180 thousand letters This report, which had ten chapters, did not admit Manchukuo but asserted Japan's vested rights and interests in Manchuria, which is to be seen as the British way of realism. Japan had only to discard the shadow for the substance. The Major General of the U.S. Forces, McCoy, said that Japan would be satisfied with this report, but Japan, discontent with it, dropped out of the League of Nations in February 1933. In this manner, Japan became stuck in the mud of isolationism. In the time of the Second Sino-Japanese War, the first Konoe Cabinet declared in January 1938 that it would not negotiate with the Nationalist Chinese Government. Because there was no more possibility to talk with each other, the war had to be prolonged.

To the Japanese living in the mainland of Japan, the end of the war meant only accepting the Potsdam Declaration, discarding weapons, and coming under control of the landing U.S. Forces. However, to the Japanese living in China or Manchuria, the situation was not so simple. It was because they had shared a common foe, namely, imperial Japan, against which the Nationalist Chinese Government military and the Chinese Communist military could cooperate with each other. When Japan was defeated, they lost this common foe, so that they resumed fighting against each other in order to get the right to disarm the Japanese Army in China, which would surely affect the subsequent development in China. Finally the Japanese Army chose to be disarmed by the Nationalist Chinese Government. Because of this, in the northern part of China, there were fierce battles between the Chinese communist army desiring to disarm the Japanese Army on the one hand and the Japanese Army resisting it on the other, even after the Emperor announced the end of the war in 15 August 1945. The Korean Peninsula was handed to the Allied Forces on Japan's defeat in the war. By dividing the peninsula on the thirty-eighth parallel, the USSR was given jurisdiction on the northern part, and the U.S. on the southern part. The present north-south division of the Korean Peninsula began at that time. It seems that, though the USSR did not want the northern part of the Korean Peninsula, the U.S. conceded too much to the USSR, because Manchukuo collapsed more easily and earlier than the U.S. expected, so that they were muddled in a variety of bugbears. In the southern part of the Korean Peninsula, there was a movement by Yuh Woon-Hyung to build the Korean People's Republic, but this was opposed by the U.S. Forces. In Taiwan, the Nanjing Nationalist Chinese Government military led by Chiang Kai-shek disarmed the Japanese Forces. During this disarmament, people from the mainland of China, who were called "outer ministry people," occupied all important positions of the government. This action provoked a rebellion by people who were originally in Taiwan and called "inner ministry people." This was what is called the February 28 Incident, which occurred in 28 February 1947. In 1949, Chiang Kai-shek, who lost the Kuomintang-Communist civil war, moved to Taiwan with the Nanjing Nationalist Chinese Government. After this immigration, the Nationalist Chinese Government directly governed Taiwan until 1996.

In Southeast Asia, it was not the local powers having struggled to liberalize their countries from European countries but the British Forces that disarmed the Japanese Forces. This disarmament enabled France and the Netherlands to come back as suzerains. The principal purpose of the Greater East Asia Co-Prosperity Sphere was to liberalize Asia from domination by European countries and the U.S., but ironically the contrary was invited at the end of the war. Vietnam had to wage the First Indo-China War to attain independence from France. Sihanouk declared the independence of Cambodia on 12 March 1945, but because of Japan's defeat in the war, Cambodia was placed under the protection of France again. Complete independence was achieved in 1953 at last. Laos declared its independence on 8 April 1945, but because of Japan's defeat in the war, it was also placed under the protection of

France again. Its complete independence was achieved in 1953. The young Aung San, who is memorialized as the founding father of Burma, and who is the father of Nobel-prize-laureate Aung San Suu Kyi (who has been under house arrest for years) carried out an independence from Britain movement by getting support from the special institution of the Japanese Forces in the war. In March 1943, he was invited to Japan and honored with an Order of the Rising Sun when he was only 28. After a while, however, he split with the Japanese Forces and fought against fascism by getting support from the previous suzerain, Britain. After the Second World War, he clashed with Britain because Britain was not willing to admit the independence of Burma by breaking its word. His life had a really zigzag course. He was assassinated in 1947 without seeing the independence of Burma that was finally achieved in January 1948. In passing, we note that the "warship march," which is the official marching song of the former Imperial Japanese Navy and contemporary Japan Maritime Self Defense Force, is played in parades of the Burmese national army even now. Malaya, which Japan conquered by January 1942 so as to get rubber, tin, and petroleum, was placed under the control of Britain after Japan's defeat in the war. Hostilities among some ethnic groups within Malaya delayed its complete independence to as late as 1958. The Japanese Forces established a military administration in Indonesia in February 1942. The military asked such nationalists as Sukarno, who later became the founding president of Indonesia, to cooperate with Japan in order to secure natural resources. After the Second World War, Indonesia attained its independence by fighting against the military occupation of the Netherlands. In front of the building of the Indian Diet, there is a picture of Chandra Bose. He assisted Japan positively to attain the independence of India during the Pacific War. He attended the Great East Asia Conference in Tokyo as an observer in November 1943. The Japanese Forces recklessly tried to invade Imphal, which is located in the northeastern part of India, in March 1944, when the possibility of Japan's defeat in the war had become quite probable. The result was of course a complete blunder. Chandra Bose participated in the invasion of Imphal as the Supreme Command of the Indian National Army.

In the Second World War, the USSR got the eastern part of Europe. The USSR wanted to also to reap some fruits in the Far East, and took blatant actions at the end of the war, and after the war. To say nothing of the invasion into Manchukuo just a week before the end of the war, the USSR invaded Sakhalin on 11 August. The battle continued till 25 August. On 10 August, the Japanese government announced its preliminary acceptance of the Potsdam Declaration. On 14 August, the government announced final acceptance of the declaration. The invasion of Shumushu Island by USSR Forces in the Kurile Islands started in 18 August. The armistice was accepted on 21 August. Later, the USSR bloodlessly occupied Iturup, Ostrov Urup, and the Shikotan Islands. Kunashir Island was occupied on 2 September, when the signing of capitulation was done. The Habomai Islands were occupied on 3 September.

Since the time of Imperialism, Russia had used Siberia as a place of exile. After the establishment of the USSR, it sent many counter-revolutionists into camps in Siberia and some other places. The USSR sent about 600 thousand Japanese hostages to camps in Siberia and others, and it forced them into labor. More than two million hostages of German soldiers were also sent to these camps. A lot of hostages from the Axis countries such as Bulgaria and Romania were also sent to such camps and forced into labor. For one or two years since the hostages were captured by the USSR, the food situation in the USSR was extremely bad, because Ukraine, which was the granary of the USSR, took a massive hit from the poor harvest caused by the war. Confiscation of allocated food by officials often happened in camps, and this worsened the food situation for detainees. Moreover, because of the lack of doctors and nurses, good treatment was not expected when people became sick. But two or three years later, the situation in the USSR improved considerably. Forced labor came to be assigned according to the physical capability of each hostage. Taking this rapidly improving situation in the USSR into consideration, we can say that it was the fork in the road to their mother countries whether hostages could survive the first one or two years in the USSR or not.

Prior to the invasion of Manchuria by the USSR Forces, every useful man was called out, root and branch, as a soldier by the Kanto Army, so that only graybeards, children and women were to be seen in the settlement of Manchuria. The Kanto Army knew of the planned invasion of Manchuria by the USSR Forces beforehand, but the army did not announce this important fact to settlers there. In advance, the Kanto Army made their own plan in preparation for such an invasion. According to this plan, Japan would give up three fourths of Manchuria and would concentrate on the defense of such remaining places as southern Manchuria and the area bordering on the Korean Peninsula, in which one could find the nucleus of Manchukuo in politics and economics. In short, Japan had decided to turn its back to the Japanese Pilgrim Fathers in Manchuria. To the settlers there, the invasion by the USSR Forces appeared like a bolt out of the blue. The evacuation order was announced in the evening of the day when the USSR Forces were preparing to invade at midnight. In some places, people did not hear of the order until more than a week later, after the end of the war. Naturally, it was not easy for Japanese settlers in Manchuria to return to mainland Japan with their food and clothes on horse-drawn carriages. Because people living in the same village in Manchuria moved together as a group, their destiny varied from one village to another in totality, not from one individual to another. One group of settlers mistook another group of settlers as tanks of the USSR Forces and committed suicide en masse. The refugees had to fear not only the USSR Forces but also the indigenous Chinese, who used to be looked down on by the Japanese, and who had been deprived of their lands. Their fury was spent on the refugees who had nothing to protect themselves. In this situation, the difficulties for Japanese children and ladies left behind in China were many.

The left-behind children were infants who were adopted by kind local Chinese as their children, because infants were nothing but a heavy burden in escaping. The left-behind ladies were those who chose to live by becoming wives of Chinese men. In the same way, in the battle of Okinawa, the army did not put any importance on the protection of people's lives and properties. In 1972, Japan and China normalized their diplomatic relations. Nine years later at last, the left-behind children and ladies found ways to go back to Japan, though no way was by any means smooth.

Mathematics around Takeo Nakasawa

1. Linear Algebra

Nowadays every freshman or sophomore specialized in science, engineering and economics is required to study linear algebra as well as advanced calculus. Nevertheless we should not forget that linear algebra is a relatively new field in the very long history of mathematics. Linear algebra was once known as the theory of matrices and determinants. It took a long period for mathematicians to accept non-numerical entities as objects of their own research. Therefore it is not surprising to note that determinants were introduced long before matrices were exactly formulated, because determinants are numerical entities, though obtained through relatively complicated processes, but matrices are not. It was in the latter half of the 17th century that two great mathematicians discovered determinants independently, one in Japan and the other in Hanover, which lies in the Western part of Germany now. The former was called Takakazu Seki, and the other was called Gottfried Wilhelm Leibniz. Unfortunately their discoveries did not have a great influence on the main development of mathematics. What is now called the formula of Cramer was announced by Cramer in 1750. It was Cauchy that has established a systematic theory of determinants in 1812. In 1858 Cayley published a paper containing a result which is now called the Cayley-Hamilton theorem. What is now called the Jordan standard form of a square matrix was discovered by Jordan in 1871, when he was keenly aware of its applications in the theory of linear differential equations. It was Frobenius that has introduced the notion of linear independence in solutions for linear equations in 1879. The abstract notion of a vector space was introduced by Peano in 1888, though it attracted little attention at that time, and it was reintroduced by later mathematicians such as Steiniz in the context of algebraic extensions of fields in 1910. It is very interesting to note that it was only in the beginning of the 20th century that a textbook on linear algebra became available, so that linear algebra was presumably an exciting subject to Takeo Nakasawa in the 1930s. Stephan Banach introduced the notion of a Banach space in his doctoral thesis in 1920, and the term of a vector space has found itself in the vocabulary of every mathematician when Banach's pioneering book on the theory of linear operators was published a decade later. Strange to say, the burgeoning theory of infinite-dimensional vector spaces made the notion of a vector space accepted by the general mathematical community. This is presumably because finite-dimensional linear algebra can be formulated as the theory of matrices and determinants without any reference to vector spaces at all, but the theory of linear operators on infinite-dimensional vector spaces is forced to begin with the very definition of an infinite-dimensional vector space.

2. Axiomatic Methods

Axiomatic methods are now commonplace in mathematics, but they were considerably fresh at the beginning of the 20th century. Of course, axiomatic methods can be traced back to Euclid's famous Elements in ancient Greece, but modern axiomatics is quite different from that of ancient Greeks. Generally speaking, renaissance is always more than revival of old ideas. In modern axiomatics axioms are not evident facts but something like rules of a game. No modern mathematician would like to define such fundamental terms as a point or a line, but he or she is ready to accept such terms as indefinable entities. Nevertheless, Euclid's axiomatics can survive modern times with minor modifications, as Hilbert has shown at his Foundations of Geometry in 1899. Modern mathematicians are interested in the independence of axioms of an axiomatic system as well as their consistency and completeness. Takeo Nakasawa was concerned with the independence of the axioms of an axiomatic system in his second paper on matroid theory.

3. Projective Geometry

Projective geometry was introduced in the 17th century by Desargues, who was an engineer and an architect. To him, every conic section is projectively equivalent to a circle, so that any property of a circle preserved by every projective transformation holds for any conic section. Desargues' synthetic arguments were not welcome at that time, when Descartes' analytical arguments on figures by using coordinates were beginning to be appreciated. Inspired by Monge's descriptive geometry in the final days of the 18th century, such mathematicians of the 19th century as Poncelet, Steiner and von Staudt developed projective geometry., and it found its axiomatic formulation in Veblen and Young's monumental book published in 1909. Their book is undoubtedly in the tradition of Hilbert's Foundations of Geometry.

4. Lattice Theory

The first stage of lattice theory began in the middle of the 19th century with Boole's publication of Mathematical Analysis of Logic, which formalized classical propositional logic in the guise of Boolean algebras. Although some mathematicians such as Peirce and Schröder followed Boole's traditions at the end of the 19th century and they have found it useful to introduce the concept of a lattice, it was in the middle of the 1930s that Garrett Birkhoff's brilliant series of papers succeeded in selling the concept to the general mathematical community. In particular, influenced by Whitney's paper, Birkhoff introduced the notion of geometric lattice. In [3] Birkhoff called a relatively complemented semi-modular lattice a matroid lattice.

5. Predecessors and Contemporaries of Takeo Nakasawa

Whitney is now famous for his works in differential topology, but he began his academic career in such a combinatorial field as graph theory. He has shown in the 1930s that the famous four-color problem, i.e., the problem whether four colors suffice or does not suffice to color every possible map so that any two adjacent countries are given distinct colors, can be formulated as a problem of graph theory. It is well known that the problem was finally settled affirmatively by Kenneth Appel and Wolfgang Haken in 1976, which caused a still-reverberating dispute whether proofs verifiable only by computers are really mathematical proofs or not. Several mathematicians such as Birkhoff, MacLane, van der Waerden and, of course, Nakasawa gave axiomatic treatments of linear dependence as well as algebraic dependence in the 1930s, but it was Whitney that noticed first similarity between the notion of linear dependence in linear algebra and the dependence of edges of a graph in graph theory. The notion of dual graph available in graph theory enabled Whitney to go in a direction which other pioneers of such axiomatic treatments of linear dependence could not imagine. After writing some papers in graph theory in the first half of the 1930s, Whitney [13] introduced the notion of a matroid, independently from Nakasawa, in 1935 for treating linear independence of column vectors in a matrix and independence of edges of a graph simultaneously. As far as historical sources in matroid theory is concerned, no better book is to be desired than Kung's [7].

Bibliography

[1] Birkhoff, Garrett (1935): Abstract linear dependence and lattices, American Journal of Mathematics, 57, 800–804

[2] Birkhoff, Garrett (1935): Combinatorial relations in projective geometries, Annals of Mathematics, 36, 743–748

[3] Birkhoff, Garrett (1967) Lattice Theory (3rd edition), American Mathematical Society

[4] Derbyshire, John (2006): Unknown Qauntity, a Real and Imaginary History of Algebra, Joseph Henry Press

[5] Grätzer, George (1978): General Lattice Theory, Birkhäuser

[6] Katz, Victor (1998): History of Mathematics (2nd edition), Addison-Wesley

[7] Kung, Joseph P. S. (1986): A Source Book in Matroid Theory, Birkhäuser

[8] MacLane, Saunders (1937): A combinatorial condition for planar graphs, Fundamenta Mathematicae, 28, 22–32

[9] MacLane, Saunders (1938): A lattice formulation for transcendence degrees and p-bases, Duke Mathematical Journal, 4, 455–468

[10] Oxley, James G. (1992): Matroid Theory, Oxford University Press

[11] Van der Waerden (1937): Moderne Algebra, (2nd edition), Springer-Verlag

[12] Whitney, Hassler (1932): Non-separable and planar graphs, Transactions of the American Mathematical Society, 34, 339–362

[13] Whitney, Hassler (1935): On the abstract properties of linear dependence, American Journal of Mathematics, 57, 509–533

Chronological Tables

Chronological Table on Takeo Nakasawa

1913 Takeo Sogabe was born.
1924 He was adopted by Nakasawa family.
1935 Takeo Nakasawa graduated from the Tokyo University of Arts and Sciences.
He became an assistant of the Tokyo University of Arts and Sciences.
His first paper on matroid theory was published.
1936 His second and third papers on matroid theory were published.
1938 His paper on topology was published.
He was dismissed from the University on 22 Aug.
1939 He got married on 26 July.
1940 His first daughter was born.
1941 His first son was born.
1944 His second son was born.
1946 He died in Khabarovsk on 20 June.

Chronological Table on Japanese History

1868 Meiji Restoration (from the samurai world to the modern world).
The reign of Meiji emperor (reign: 1968–1912) began.
1873 The call-up law was proclaimed
1889 The Meiji constitution was proclaimed
1890 The first imperial congress was held

1894 The first Sino-Japanese war (1894–1895) began
1902 The first Anglo-Japanese Alliance was concluded
1904 The Russo-Japanese war (1904–1905) began
1905 The second Anglo-Japanese Alliance was concluded
1907 The first Russo-Japanese agreement was concluded.
1910 The second Russo-Japanese agreement was concluded.
Korea was formally annexed to Japan.
1911 The third Anglo-Japanese Alliance was concluded
1912 The third Russo-Japanese agreement was concluded.
Meiji emperor died, and Taisho emperor (reign:1912–1926) succeeded him.
1915 Japan offered the Twenty-One Demands to the Republic of China
1916 The fourth Russo-Japanese agreement was concluded.
1921 The Anglo-Japanese Alliance was abolished
1926 Taisho emperor died, and Showa emperor (reign:1926–1989) succeeded him.
1931 The Manchurian incident occurred. The fifteen years war (1931–1945) began
1932 Manchukuo (the puppet state of imperial Japan in Manchuria) was established.
The 5.15 incident occurred.
1933 Japan seceded from the League of Nations.
1935 Japan virtually dominated the northern part of China.
1936 The 2.26 incident occurred.
The Anti-Comintern Pact was concluded.
1937 The second Sino-Japanese war (1937–1945) began.
1939 The Nomonhan incident (the battle of Khalkhin Gol) occurred.
1940 Japan invaded the northern Indochina.
The Tripartite Pact was concluded.
1941 Japan invaded southern Indochina.
The government of Tojo was established.
The Pacific war began
1945 Japan surrendered in the Pacific war.

Chronological Table on World Affairs

1861 The Kingdom of Italy (1861–1946) began.
1871 The German Empire (1871–1918) was established.
1877 The Russo-Turkish war (1877–1878) began.
1905 Bloody Sunday (Russia)
1908 Young Turk Revolution
1912 The Republic of China was established.
1914 The First World War (1914–1918) began.

1917 The Russian revolution
1920 The League of Nations was established.
1922 The Soviet Union was established.
1924 Lenin died.
1925 Stalin grasped the power of the Soviet Union.
1928 The First Five-Year Plan (1928–1932) began (Soviet Union).
1930 London Naval Treaty
1932 The Second Five-Year Plan (1932–1937) began (Soviet Union).
1939 German-Soviet Nonaggression Pact
The Second World War (1939–1945) began.
1941 The United States of America participated in the Second World War.
1943 Italy surrendered.
Cairo Declaration
Tehran Conference
1944 Germany surrendered.
1945 Yalta Conference
Potsdam Conference

Works of Takeo Nakasawa

First published in: *Science Report of the Tokyo Bunrika Daigaku,* Section A.

Zur Axiomatik der linearen Abhängigkeit. I.

Zur Axiomatik der linearen Abhängigkeit. I.

Von Takeo NAKASAWA

(Eingegangen am 30. Juni, 1935)

Einleitung.

In der vorliegenden Untersuchung soll ein Axiomensystem für eine neue Formulierung der linearen Abhängigkeit des n-dimensionalen projektiven Raumes angegeben werden, indem wir hauptsächlich den *Zyklenkalkül*, den Herr G. Thomsen bei seiner Grundlegung der elementaren Geometrie hergestellt hat(1), hier in einem noch abstrakteren Sinne verwenden.

Zuerst wollen wir die Geometrie des ersten Verknüpfungsraumes aufbauen, dessen Definition später angegeben wird.

Bezeichnungen.

In dieser Schrift werden wir häufig folgende Bezeichnungen brauchen.

1. $A \rightarrow B$ bedeutet, dass aus A B folgt.
2. $A \leftrightarrow B$ bedeutet, dass $A \rightarrow B$ und $B \rightarrow A$.
3. $A \rightarrow W$. bedeutet, dass A zum Widerspruch gerät.
4. A, B bedeutet, dass A und B.
5. A oder B bedeutet, dass mindestens eins von A und B.
6. $A -(S)\rightarrow B$ bedeutet, dass auf Grund des Aussages S aus A B folgt.
7. $\left.\begin{matrix} A_1, \\ A_2, \\ \vdots, \\ A_k \end{matrix}\right\} \rightarrow B$ bedeutet, dass aus dem gleichzeitigen Bestehen der k Aussagen $A_1, A_2, \cdots, A_k$ B folgt.
8. $A \rightarrow \left\{\begin{matrix} B_1, \\ B_2, \\ \vdots, \\ B_k \end{matrix}\right.$ bedeutet, dass aus A gleichzeitig die k Aussagen $B_1, B_2, \cdots, B_k$ folgt.

(1) G. THOMSEN: Grundlagen der Elementargeometrie, (Leipzig 1933), (S. 67–S. 70).

9. A_i, $(i=1,\cdots,m)$ bedeutet, dass A_1, A_2, $\cdots$, und A_m.
10. Wir brauchen auch noch die folgenden Bezeichnungen von Mengenlehre, wie sie gewöhnlich bedeuten; $\geqq$, $\not\geqq$, $\supset$, $\not\supset$, $=$, $\neq$, $\ni$, $\not\ni$, $+$, u. s. w.

ERSTES KAPITEL

Der $\mathfrak{V}_1$-Raum.

§ 1. Axiome.

Grundannahme: Wir denken uns eine gewisse Menge der Elementen; $\mathfrak{V}_1 \ni a_1, a_2, \cdots, a_s, \cdots$. Für gewisse Reihen der Elementen, die wir *Zyklen* nennen wollen, denken wir dazu die Relationen "*gelten*" oder "*gültig sein*", in Zeichen $a_1 \cdots a_s = 0$ [(2)], bzw. "*nicht gelten*" oder "*nicht gültig sein*", in Zeichen $a_1 \cdots a_s \neq 0$. Diese Relationen sollen nun folgenden Axiomen genügen;

Axiom 1. (Reflexivität) : aa.

Axiom 2. (Folgerung) : $a_1 \cdots a_s \rightarrow a_1 \cdots a_s x$, $(s = 1, 2, \cdots)$.

Axiom 3. (Vertauschung) : $a_1 \cdots a_i \cdots a_s \rightarrow a_i \cdots a_1 \cdots a_s$, $(s = 2, 3, \cdots ; i = 2, \cdots, s)$.

Axiom 4. (Transitivität) : $a_1 \cdots a_s \neq 0$, $x a_1 \cdots a_s$, $a_1 \cdots a_s y \rightarrow x a_1 \cdots a_{s-1} y$, $(s = 1, 2, \cdots)$.

Definition I. Eine solche Menge $\mathfrak{V}_1$ heisst der erste Verknüpfungsraum, in kurzen Worten, **$\mathfrak{V}_1$-Raum.**

V_1. 1. Nehmen wir uns die Menge von allen Punkten des klassischen n-dimensionalen projektiven Raumes als $\mathfrak{V}_1$-Raum, und nennen wir von gewissen darin liegenden linearen abhängigen, bzw. unabhängigen Reihen von Punkten als gelten, bzw. nicht gelten, so sind die Grundannahme, sowie die Axiome 1, 2, 3, 4 erfüllt, wie leicht einzusehen[(3)].

V_1. 2. Da das obige Axiomensystem keine Existenzaussage enthält, so ist eine beliebige Teilmenge des $\mathfrak{V}_1$-Raumes auch ein $\mathfrak{V}_1$-Raum.

---

(2) Von jetzt an gewöhnlich, statt "$a_1 \cdots a_s = 0$", werden wir auch schreiben kurz "$a_1 \cdots a_s$", wie Thomsen in seiner Zyklenkalkül braucht (Vgl. a. a. O. [(1)]).

(3) Also von jetzt an, werden wir oft die Wörter "Punkt" statt "Element", oder "linear abhängig", bzw. "linear unabängig" statt "gelten", bzw. "nicht gelten" brauchen.

Ferner aus Axiomen 1, 2, 3, folgt leicht:

V_1. 3. Ein Zyklus, welcher mehr als zwei dasselbe Element enthält, ist stets gültig; folglich sind die Elemente in einem nicht gültigen Zyklus einander verschieden.

§ 2. Lineare Räume.

Definition II. Die Menge der allen Elementen x aus $\mathfrak{B}_1$ derart, dass $a_1 \cdots a_n \neq 0$ aber $a_1 \cdots a_n x$, nennen wir den von $a_1, \cdots, a_n$ erzeugten *linearen Raum in* $\mathfrak{B}_1$ *vom Range* n; in Zeichen: $\mathfrak{R}^n(a_1 \cdots a_n)$, und den Zyklus $a_1 \cdots a_n$ nennen wir die *Basis* des $\mathfrak{R}^n(a_1 \cdots a_n)$.

Definition II'. Die Menge der allen Elementen x aus $\mathfrak{B}_1$ derart, dass $x = 0$, nennen wir die *Nullstelle* von $\mathfrak{B}_1$; in Zeichen: $\mathfrak{N}$[4].

Dann, weil $x=0$ —(Axiom 2, 3)→ $a_1 \cdots a_n x$, so kann man folgendermassen behaupten:

V_1. 4. $\mathfrak{R} \geqq \mathfrak{N}$;

in Worten: Jeder lineare Raum enthält stets die Nullstelle in sich.

Und, weil (Axiom 1)→ $a_i a_i$ —(Axiom 2, 3)→ $a_1 \cdots a_n a_i$, $(i=1, \cdots, n)$, so folgt,

V_1. 5. $\mathfrak{R}^n(a_1 \cdots a_n) \ni a_1, \cdots, a_n$;

in Worten: Der lineare Raum enthält alle ihre Basispunkten in sich.

Ferner, weil $a_1 \cdots a_m x$, $m \leqq n$ —(Axiom 2, 3)→ $a_1 \cdots a_m \cdots a_n x$, so folgt,

V_1. 6. $n \geqq m \rightarrow \mathfrak{R}^n(a_1 \cdots a_n) \geqq \mathfrak{R}^m(a_1 \cdots a_m)$.

Satz 1. Ist $n \geqq 1$, so besteht die folgende Formel;

$$\left.\begin{array}{l} a_1 \cdots a_n \neq 0, \\ a_1 \cdots a_n x_1, \\ a_1 \cdots a_n x_2 \end{array}\right\} \rightarrow a_2 \cdots a_n x_1 x_2 .$$

Beweis:

$$\left.\begin{array}{l} a_1 \cdots a_n \neq 0, \\ a_1 \cdots a_n x_1, \\ a_1 \cdots a_n x_2 \end{array}\right\} \text{—(Axiom 3)}\rightarrow \left\{\begin{array}{l} a_2 \cdots a_n a_1 \neq 0, \\ x_1 a_2 \cdots a_n a_1, \\ a_2 \cdots a_n a_1 x_2 \end{array}\right\} \text{—(Axiom 4)}\rightarrow x_1 a_2 \cdots a_n x_2$$

$$\text{—(Axiom 3)}\rightarrow a_2 \cdots a_n x_1 x_2 .$$

(4) Dieselbe Menge ist anderseits auch als linearer Raum vom Range 0, d.h. $\mathfrak{N}^0$, zu betrachten, solange wir den leeren Zyklus als nicht gültig voraussetzen.

Daher

$$\left.\begin{array}{l} a_1 \cdots a_n \neq 0, \\ a_1 \cdots a_n x_1 , \\ a_1 \cdots a_n x_2 \end{array}\right\} \rightarrow a_2 \cdots a_n x_1 x_2 , \qquad \text{w. z. b. w.}$$

Satz 2. Ist $n \geqq 2$, so besteht die folgende Formel;

$$\left.\begin{array}{l} a_1 \cdots a_n \neq 0, \\ a_1 \cdots a_n x_1 , \\ a_1 \cdots a_n x_2 , \\ a_1 \cdots a_n x_3 \end{array}\right\} \rightarrow a_3 \cdots a_n x_1 x_2 x_3 .$$

Beweis: (i) Falls $a_2 \cdots a_n x_1 \neq 0$, so folgt,

$$\left.\begin{array}{l} a_1 \cdots a_n \neq 0, \\ a_1 \cdots a_n x_1 , \\ a_1 \cdots a_n x_2 , \\ a_1 \cdots a_n x_3 \end{array}\right\} \text{—(Satz 1)}\rightarrow \left\{\begin{array}{l} a_2 \cdots a_n x_1 \neq 0, \\ a_2 \cdots a_n x_1 x_2 , \\ a_2 \cdots a_n x_1 x_3 \end{array}\right\} \text{—(Satz 1)}\rightarrow a_3 \cdots a_n x_1 x_2 x_3 .$$

(ii) Falls $a_2 \cdots a_n x_2 \neq 0$, so folgt,

$$\left.\begin{array}{l} a_1 \cdots a_n \neq 0, \\ a_1 \cdots a_n x_1 , \\ a_1 \cdots a_n x_2 , \\ a_1 \cdots a_n x_3 \end{array}\right\} \text{—(Satz 1)}\rightarrow \left\{\begin{array}{l} a_2 \cdots a_n x_2 \neq 0, \\ a_2 \cdots a_n x_2 x_1 , \\ a_2 \cdots a_n x_2 x_3 \end{array}\right\} \text{—(Satz 1)}\rightarrow a_3 \cdots a_n x_2 x_1 x_3$$

$$\text{—(Axiom 3)}\rightarrow a_3 \cdots a_n x_1 x_2 x_3 .$$

(iii) Falls $a_2 \cdots a_n x_1$, $a_2 \cdots a_n x_2$, so folgt,

$$a_1 \cdots a_n \neq 0 \text{ —(Axiom 2, 3)}\rightarrow \left.\begin{array}{l} a_2 \cdots a_n \neq 0, \\ a_2 \cdots a_n x_1 , \\ a_2 \cdots a_n x_2 \end{array}\right\} \text{—(Satz 1)}\rightarrow a_3 \cdots a_n x_1 x_2$$

$$\text{—(Axiom 2)}\rightarrow a_3 \cdots a_n x_1 x_2 x_3 .$$

Daher .

$$\left.\begin{array}{l} a_1 \cdots a_n \neq 0, \\ a_1 \cdots a_n x_1 , \\ a_1 \cdots a_n x_2 , \\ a_1 \cdots a_n x_3 \end{array}\right\} \rightarrow a_3 \cdots a_n x_1 x_2 x_3 , \qquad \text{w. z. b. w.}$$

Satz 3. Ist $n \geqq m-1$, so besteht die folgende Formel;

$$\left.\begin{array}{l} a_1 \cdots a_n \neq 0, \\ a_1 \cdots a_n x_1, \\ a_1 \cdots a_n x_2, \\ \cdots\cdots\cdots\cdots, \\ a_1 \cdots a_n x_m \end{array}\right\} \rightarrow a_m \cdots a_n\, x_1 \cdots x_m .$$

Beweis: Wir beweisen dies durch die vollständige Induktion in Bezug auf m. Falls m gleich 1 ist, ist es trivial, und falls m gleich 2, bzw. 3 ist, ist es bereits im Satz 1, bzw. Satz 2 bewiesen worden. Also genügt es zu zeigen, dass aus dem Fall $[m-1]$ den Fall $[m]$ folgt.

(i) Falls unter m Elementen $x_1, \cdots, x_m$, mindestens ein solches Element z.B. x_1 existiert, dass $a_2 \cdots a_n x_1 \neq 0$, dann ergibt sich

$$\left.\begin{array}{l} a_1 \cdots a_n \neq 0, \\ a_1 \cdots a_n x_1, \\ a_1 \cdots a_n x_2, \\ \cdots\cdots\cdots\cdots, \\ a_1 \cdots a_n x_m \end{array}\right\} -(\text{Satz } 1)\rightarrow \left.\begin{array}{l} a_2 \cdots a_n x_1 \neq 0, \\ \left\{\begin{array}{l} a_2 \cdots a_n x_1 x_2, \\ a_2 \cdots a_n x_1 x_3, \\ \cdots\cdots\cdots\cdots\cdots, \\ a_2 \cdots a_n x_1 x_m \end{array}\right. \end{array}\right\} -[m-1]\rightarrow a_m \cdots a_n x_1 \cdots x_m .$$

(ii) Falls für alle Elementen $x_1, \cdots, x_m$ die Relationen $a_2 \cdots a_n x_i$, $(i = 1, \cdots, m)$, bestehen, dann ergibt sich

$$a_1 \cdots a_n \neq 0 -(\text{Axiom } 2, 3)\rightarrow \left.\begin{array}{l} a_2 \cdots a_n \neq 0, \\ a_2 \cdots a_n x_1, \\ a_2 \cdots a_n x_2, \\ \cdots\cdots\cdots\cdots, \\ a_2 \cdots a_n x_{m-1} \end{array}\right\} -[m-1]\rightarrow a_m \cdots a_n x_1 \cdots x_{m-1}$$

$$-(\text{Axiom } 2)\rightarrow a_m \cdots a_n\, x_1 \cdots x_m .$$

Daher

$$\left.\begin{array}{l} a_1 \cdots a_n \neq 0, \\ a_1 \cdots a_n x_1, \\ a_1 \cdots a_n x_2, \\ \cdots\cdots\cdots\cdots, \\ a_1 \cdots a_n x_m \end{array}\right\} \rightarrow a_m \cdots a_n\, x_1 \cdots x_m , \qquad \text{w. z. b. w.}^{(5)}$$

(5) Es sei hier bemerkt, dass wir in diesem Beweis das Axiom 1 gar nicht benutzen.

Setzt man, im Satz 3, besonders $n = m-1$, so folgt die folgende Behauptung;

$$\left.\begin{array}{l} a_1 \cdots a_n \neq 0, \\ a_1 \cdots a_n x_1, \\ a_1 \cdots a_n x_2, \\ \cdots\cdots\cdots\cdots, \\ a_1 \cdots a_n x_{n+1} \end{array}\right\} \to x_1 x_2 \cdots x_{n+1}.$$

Diesen Satz stellen wir durch die folgende Formel dar:

Satz 4. $\mathfrak{R}^n(a_1 \cdots a_n) \ni x_1, x_2, \cdots, x_{n+1} \to x_1 x_2 \cdots x_{n+1}$;

in Worten: Je $n+1$ im linearen Raum vom Range n enthaltenen Punkte sind linear abhängig.

Zusatz. $\mathfrak{R}^n(a_1 \cdots a_n) \ni x_1, \cdots, x_m$ und $n < m \to x_1 \cdots x_m$;

in Worten: Je mehr als $n+1$ im linearen Raum vom Range n enthaltenen Punkte sind linear abhängig.

Beweis: Nach Satz 4 folgt $x_1 \cdots x_{n+1}$.

Dann ist $n < m$, so folgt

$$x_1 \cdots x_{n+1} \text{ —(Axiom 2)}\to x_1 \cdots x_m.$$

Satz 5. $\mathfrak{R}_1^n \geqq \mathfrak{R}_2^n \to \mathfrak{R}_1^n \leqq \mathfrak{R}_2^n$;

in Worten: Wenn ein linearer Raum n^{ten} Ranges $\mathfrak{R}_1^n$ einen anderen linearen Raum n^{ten} Ranges $\mathfrak{R}_2^n$ enthält, so stimmen die zwei linearen Räume mit einander ein(6).

Beweis: Seien die Basen von $\mathfrak{R}_1$, bzw. $\mathfrak{R}_2$ $a_1 \cdots a_n$, bzw. $b_1 \cdots b_n$, so ist,

$$\mathfrak{R}_1 \geqq \mathfrak{R}_2 \ni b_1, \cdots, b_n \to \mathfrak{R}_1 \ni b_1, \cdots, b_n.$$

Also

$$\left.\begin{array}{l} \left\{\begin{array}{l} a_1 \cdots a_n \neq 0, \\ a_1 \cdots a_n b_1, \\ \cdots\cdots\cdots\cdots, \\ a_1 \cdots a_n b_n, \end{array}\right. \\ \text{Anderseits } \mathfrak{R}_1 \ni x \to a_1 \cdots a_n x \end{array}\right\} \text{(Satz 4)}\to b_1 \cdots b_n x \to \mathfrak{R}_2 \ni x.$$

(6) Falls zwei lineare Räume $\mathfrak{R}_1$, $\mathfrak{R}_2$ mit einander einstimmen, schreiben wir in Zeichen $\mathfrak{R}_1 = \mathfrak{R}_2$.

Daher $\mathfrak{R}_1 \ni x \rightarrow \mathfrak{R}_2 \ni x$.

Daher $\mathfrak{R}_1 \leqq \mathfrak{R}_2$, w. z. b. w.

Zusatz. $\mathfrak{R}_1^n \not\supset \mathfrak{R}_2^n$

in Worten: Der lineare Raum enthält keinen linearen Raum desselben Ranges als seine echte Teilmenge.

Satz 6. $n < m \rightarrow \mathfrak{R}_1^n \not\geqq \mathfrak{R}_2^m$;

in Worten: Jeder lineare Raum enthält keinen anderen linearen Raum höheren Ranges.

Beweis: Seien die Basen von $\mathfrak{R}_1$, bzw. $\mathfrak{R}_2$ je $a_1 \cdots a_n$, bzw. $b_1 \cdots b_m$, und wäre vorläufig $\mathfrak{R}_1 \geqq \mathfrak{R}_2$, so wäre

$$\mathfrak{R}_1^n(a_1 \cdots a_n) \geqq \mathfrak{R}_2^m(b_1 \cdots b_m) \ni b_1, \cdots, b_m \rightarrow \mathfrak{R}_1^n(a_1 \cdots a_n) \ni b_1, \cdots, b_m .$$

Daher $\left.\begin{array}{l}\mathfrak{R}_1^n(a_1 \cdots a_n) \ni b_1, \cdots, b_m \\ \text{und} \quad n < m\end{array}\right\}$ (Zusatz zum Satz 4) $\rightarrow b_1 \cdots b_m \rightarrow W$.

Daher $\mathfrak{R}_1 \not\geqq \mathfrak{R}_2$, w. z. b. w.

Zusatz. $n \neq m \rightarrow \mathfrak{R}_1^n \neq \mathfrak{R}_2^m$;

in Worten: Die zwei linearen Räume verschiedenen Ranges können nicht mit einander einstimmen.

Satz 7. $\mathfrak{R}_1^n(a_1 \cdots a_n) = \mathfrak{R}_2^m(b_1 \cdots b_m) \longleftrightarrow n = m,\ a_1 \cdots a_n b_i,\ (i=1, \cdots m)$;

in Worten: Zwei lineare Räume $\mathfrak{R}_1$ und $\mathfrak{R}_2$ stimmen dann und nur dann ein, wenn die Ränge derselben Räume gleich sind, und einer von ihnen z.B. $\mathfrak{R}_1$, die Basispunkten des anderen $\mathfrak{R}_2$ enthält.

Beweis: (i) Die Bedingung ist notwendig:

Aus Zusatz zum Satz 6, ist es notwendig $n = m$.

Und

$$\mathfrak{R}_1 = \mathfrak{R}_2 \ni b_1, \cdots, b_m \rightarrow \mathfrak{R}_1 \ni b_1, \cdots, b_m \rightarrow a_1 \cdots a_n b_i,\ (i=1, \cdots, m).$$

(ii) Die Bedingung ist hinreichend:

$$\left.\begin{array}{l} a_1 \cdots a_n \neq 0, \\ a_1 \cdots a_n b_1, \\ \cdots\cdots\cdots\cdots, \\ a_1 \cdots a_n b_n, \\ \mathfrak{R}_1 \ni x \rightarrow a_1 \cdots a_n x \end{array}\right\} \text{(Satz 4)} \rightarrow b_1 \cdots b_n x \rightarrow \mathfrak{R}_2 \ni x.$$

Daher $\mathfrak{R}_1 \subsetneqq \mathfrak{R}_2$.

Daher $\mathfrak{R}_1^n \subsetneqq \mathfrak{R}_2^n$ —(Satz 5)→ $\mathfrak{R}_1 = \mathfrak{R}_2$.

Zusatz 1. $\mathfrak{R}(a_1 \cdots a_n) \ni b_1, \cdots, b_n$ und $b_1 \cdots b_n \neq 0$

$$\to \mathfrak{R}(a_1 \cdots a_n) = \mathfrak{R}(b_1 \cdots b_n);$$

in Worten: Die lineare unabhängige Reihe von n Punkte in einem linearen Raume n^{ten} Ranges lässt sich als Basis desselben Raumes genommen werden.

Zusatz 2. Falls $a_1 \cdots a_n \neq 0$, $b_1 \cdots b_n \neq 0$, so besteht umkehrbar

$$\left.\begin{matrix} a_1 \cdots a_n b_1, \\ \cdots\cdots\cdots, \\ a_1 \cdots a_n b_n \end{matrix}\right\} \longleftrightarrow \left\{\begin{matrix} b_1 \cdots b_n a_1, \\ \cdots\cdots\cdots, \\ b_1 \cdots b_n a_n \end{matrix}\right\}.$$

Satz 8. $\mathfrak{R}^n(a_1 \cdots a_n) \ni b_1, \cdots, b_m$ und $b_1 \cdots b_m \neq 0$, $n > m$

→ Die Existenz von $a_{i_{m+1}}, \cdots, a_{i_n}$ in der Reihe $a_1, \cdots, a_n$ derart, dass $b_1 \cdots b_m a_{i_{m+1}} \cdots a_{i_n} \neq 0$;

in Worten: Im linearen Raume $\mathfrak{R}^n(a_1 \cdots a_n)$ existiert stets eine Basis desselben Raumes, welche die gegebene darin liegende linear unabhängige Reihe von m Punkte $b_1, \cdots, b_m$, $(m < n)$ als ihre Teilreihe enthält.

Beweis: Induktionell genügt es zu zeigen, dass es ein Element $a_{i_{m+1}}$ in der Reihe $a_1, \cdots, a_n$ gibt, derart, dass $b_1 \cdots b_m a_{i_{m+1}} \neq 0$.

Nun wäre $b_1 \cdots b_m a_i = 0$, $(i = 1, \cdots, n)$, so würde sich folgern,

$$\left.\begin{matrix} b_1 \cdots b_m \neq 0, \\ b_1 \cdots b_m a_1, \\ \cdots\cdots\cdots, \\ b_1 \cdots b_m a_n, \\ m < n \end{matrix}\right\} \text{(Zusatz zum Satz 4)} \to a_1 \cdots a_n \to W.$$

So muss $b_1 \cdots b_m a_i \neq 0$ für mindestens ein a_i. Setzt man dasselbe a_i als $a_{i_{m+1}}$, so kommt unser Beweis zum Schluss.

Satz 9. $\mathfrak{R}_1(a_1 \cdots a_n) \ni b_1, \cdots, b_m$ und $b_1 \cdots b_m \neq 0$

$$\to \mathfrak{R}_1(a_1 \cdots a_n) \geqq \mathfrak{R}_2(b_1 \cdots b_m);$$

in Worten: Der lineare-Raum, der die Basis eines anderen linearen Raumes enthält, enthält stets alle seine Punkte.

Beweis: Nach Satz 4, folgt, dass $n \geqq m$. Also nach Satz 8 existieren die Elemente $a_{i_{m+1}}, \cdots, a_{i_n}$ in der Reihe $a_1, \cdots, a_n$ derart, dass $b_1 \cdots b_m a_{i_{m+1}} \cdots a_{i_n} \neq 0$.

Dann wegen Zusatz 1 zum Satz 7,

$$\mathfrak{R}_1(a_1 \cdots a_n) = \mathfrak{R}(b_1 \cdots b_m a_{i_{m+1}} \cdots a_{i_n}) .$$

Daher $\mathfrak{R}_2(b_1 \cdots b_m) \ni x \rightarrow b_1 \cdots b_m x \rightarrow b_1 \cdots b_m a_{i_{m+1}} \cdots a_{i_n} x$
$\rightarrow \mathfrak{R}(b_1 \cdots b_m a_{i_{m+1}} \cdots a_{i_n}) \ni x \rightarrow \mathfrak{R}_1(a_1 \cdots a_n) \ni x$.

Daher $\mathfrak{R}_2 \ni x \rightarrow \mathfrak{R}_1 \ni x$.

Daher $\mathfrak{R}_1(a_1 \cdots a_n) \geqq \mathfrak{R}_2(b_1 \cdots b_m)$, w. z. b. w.

Satz 10. $\mathfrak{R}_1^n(a_1 \cdots a_n) > \mathfrak{R}_2^m(b_1 \cdots b_m) \longleftrightarrow$
$n > m, \; a_1 \cdots a_n b_i, \; (i = 1, \cdots, m)$;

in Worten: Der lineare Raum $\mathfrak{R}_1$ enthält einen anderen linearen Raum $\mathfrak{R}_2$ dann und nur dann, wenn der Rang von $\mathfrak{R}_1$ grösser als der von $\mathfrak{R}_2$ ist, und $\mathfrak{R}_1$ die Basispunkten von $\mathfrak{R}_2$ enthält.

Beweis: (i) Die Bedingung ist notwendig:

$$\left.\begin{array}{ll}\text{Zusatz zum Satz 5 lehrt, dass} & n \neq m, \\ \text{Satz 6 lehrt, dass} & n \nless m\end{array}\right\} \rightarrow n > m .$$

Ferner

$$\mathfrak{R}_1 > \mathfrak{R}_2 \ni b_1, \cdots, b_m \rightarrow \mathfrak{R}_1 \ni b_1, \cdots, b_m \rightarrow a_1 \cdots a_n b_i, \; (i = 1, \cdots, m).$$

(ii) Die Bedingung ist hinreichend:

$a_1 \cdots a_n b_i, \; (i = 1, \cdots, m) \rightarrow \mathfrak{R}_1(a_1 \cdots a_n) \ni b_1, \cdots, b_m$ und $b_1 \cdots b_m \neq 0$
$-(\text{Satz 9}) \rightarrow \mathfrak{R}_1(a_1 \cdots a_n) \geqq \mathfrak{R}_2(b_1 \cdots b_m)$.

$$\left.\begin{array}{r}\mathfrak{R}_1(a_1 \cdots a_n) \geqq \mathfrak{R}_2(b_1 \cdots b_m), \\ n > m\end{array}\right\} (\text{Satz 7}) \rightarrow \mathfrak{R}_1 > \mathfrak{R}_2 .$$

Daher $\mathfrak{R}_1^n(a_1 \cdots a_n) > \mathfrak{R}_2^m(b_1 \cdots b_m)$.

§ 3. Der Rang.

Definition III. Von einer gewissen Menge M der Elementen, nennen wir denjenigen nicht gültigen Zyklus, dessen Zahl der Elementen am grössten unter den allen zu M gehörigen nicht gültigen Zyklen ist, die *Basis* der Menge M, und die Zahl der Elementen in der Basis nennen wir den *Rang* der Menge M; in Zeichen: Rang M(7, 8).

So sind die folgenden Behauptungen leicht zu erhalten:

V_1. 7. Wenn man den linearen Raum als linearer Raum oder als Menge betrachtet, so stimmen auch die Bedeutung des Ranges in beiden Fällen einander ein. Gleiches gilt auch für die Bedeutung der Basis.

V_1. 8. $M_1 \geqq M_2 \rightarrow \text{Rang}\, M_1 \geqq \text{Rang}\, M_2$.

Satz 11. Sei die Basis der M $a_1 \cdots a_n$, so besteht die folgende Formel;

$$\mathfrak{R}(a_1 \cdots a_n) \geqq M .$$

Beweis: Sei x ein beliebiges Element von M, so folgt nach der Definition, dass

$$a_1 \cdots a_n \neq 0 , \quad a_1 \cdots a_n x .$$

Daher $\mathfrak{R}(a_1 \cdots a_n) \ni x$.

Daher $\mathfrak{R}(a_1 \cdots a_n) \geqq M$, w. z. b. w.

Definition IV. Den Durchschnitt der k Mengen $M_1, \cdots, M_k$, schreiben wir in Zeichen als $\mathfrak{D}(M_1, \cdots M_k)$.

Dann folgt leicht;

V_1. 9. $\mathfrak{D}(M_1 M_2) = \mathfrak{D}(M_2 M_1)$,

$$\mathfrak{D}(\mathfrak{D}(M_1 \cdots M_{k-1}), M_k) = \mathfrak{D}(M_1 \cdots M_k) ,$$

$$\mathfrak{D}(M_1, \mathfrak{D}(M_2 \cdots M_k)) = \mathfrak{D}(M_1 \cdots M_k) .$$

Satz 12. Der $\mathfrak{D}(\mathfrak{R}_1 \mathfrak{R}_2)$ ist ein linearer Raum.

Beweis: Seien die Basen von $\mathfrak{R}_1$, $\mathfrak{R}_2$, und $\mathfrak{D}(\mathfrak{R}_1 \mathfrak{R}_2)$ je $a_1 \cdots a_n$, $b_1 \cdots b_m$, und $c_1 \cdots c_k$, so folgt nach Satz 11,

$$\mathfrak{R}(c_1 \cdots c_k) \geqq \mathfrak{D}(\mathfrak{R}_1 \mathfrak{R}_2) .$$

(7) Anlog wird man auch die Bedeutung begreifen, dass der Rang M gleich ∞ oder 0 ist.

(8) Falls $M \ni a_1, a_2, \cdots$, werden wir oft statt "Rang M" auch "Rang $(a_1 a_2 \cdots)$" schreiben.

Anderseits ist

$$\mathfrak{R}_1, \mathfrak{R}_2 \geqq \mathfrak{D}(\mathfrak{R}_1\mathfrak{R}_2) \ni c_1, \cdots, c_k;$$

also $\left.\begin{array}{l}\mathfrak{R}_1 \ni c_1, \cdots, c_k \text{ und } c_1 \cdots c_k \neq 0 - (\text{Satz 9}) \rightarrow \mathfrak{R}_1 \geqq \mathfrak{R}(c_1 \cdots c_k), \\ \mathfrak{R}_2 \ni c_1, \cdots, c_k \text{ und } c_1 \cdots c_k \neq 0 - (\text{Satz 9}) \rightarrow \mathfrak{R}_2 \geqq \mathfrak{R}(c_1 \cdots c_k)\end{array}\right\}$

$$\rightarrow \mathfrak{D}(\mathfrak{R}_1\mathfrak{R}_2) \geqq \mathfrak{R}(c_1 \cdots c_k).$$

Daher $\mathfrak{D}(\mathfrak{R}_1\mathfrak{R}_2) = \mathfrak{R}(c_1 \cdots c_k)$.

Also ist $\mathfrak{D}(\mathfrak{R}_1\mathfrak{R}_2)$ ein linearer Raum.

Aus Satz 12 und V_1. 9 folgt der

Zusatz. Der $\mathfrak{D}(\mathfrak{R}_1 \cdots \mathfrak{R}_k)$ ist ein linearer Raum.

Und ferner aus V_1. 4, folgt,

V_1. 10. $\mathfrak{D}(\mathfrak{R}_1 \cdots \mathfrak{R}_k) \geqq \mathfrak{R}$.

Definition V. Unter allen k Mengen $M_1, \cdots, M_k$ enthaltenden linearen Räumen nennen wir denjenigen, dessen Rang am kleinsten ist, den *Vereinigungsraum* von $M_1, \cdots, M_k$, und wir schreiben in Zeichen als $\mathfrak{V}(M_1 \cdots M_k)$(9).

Dann folgt leicht,

V_1. 11. $\mathfrak{V}(M_1 M_2) = \mathfrak{V}(M_2 M_1)$,

V_1. 12. $\mathfrak{V}(M_1 + \cdots + M_k) = \mathfrak{V}(M_1 \cdots M_k)$.

Satz 13. Sei die Basis von M $a_1 \cdots a_n$, so besteht die folgende Gleichung;

$$\mathfrak{V}(M) = \mathfrak{R}(a_1 \cdots a_n);$$

in Worten: Sei der Rang der Menge M endlich, so bestimmt sich der einzige Vereinigungsraum $\mathfrak{V}(M)$, und derselbe ist nichts anders als linearer Raum, welcher die Basis von M als seine Basis besitzt.

Beweis: Nach Satz 11, folgt

$$\mathfrak{R}(a_1 \cdots a_n) \geqq M.$$

Anderseits folgt nach Definition V,

$$\mathfrak{V}(M) \geqq M \ni a_1, \cdots, a_n \rightarrow \mathfrak{V}(M) \ni a_1, \cdots, a_n$$
$$-(\text{Satz 9}) \rightarrow \mathfrak{V}(M) \geqq \mathfrak{R}(a_1 \cdots a_n).$$

(9) Falls $M \ni a_1, a_2, \cdots$, werden wir statt "$\mathfrak{V}(M)$" oft auch "$\mathfrak{V}(a_1 a_2 \cdots)$" schreiben.

Daher $\mathfrak{B}(M) \geqq \mathfrak{R}(a_1 \cdots a_n) \geqq M$.

Also ist nach Definition V, sowie Zusatz zum Satz 5,

$$\mathfrak{B}(M) = \mathfrak{R}(a_1 \cdots a_n), \qquad \text{w. z. b. w.}$$

Zusatz. Rang $\mathfrak{B}(M) =$ Rang M.

Satz 14. $\mathfrak{R} \geqq M \rightarrow \mathfrak{R} \geqq \mathfrak{B}(M)$;

in Worten: Ein solcher linearer Raum, der eine Menge M in sich enthält, enthält auch den Vereinigungsraum $\mathfrak{B}(M)$ in sich.

Beweis: Sei die Basis der M $a_1 \cdots a_n$, so folgt

$$\mathfrak{R} \geqq M \ni a_1, \cdots, a_n \rightarrow \mathfrak{R} \ni a_1, \cdots, a_n - (\text{Satz } 9) \rightarrow \mathfrak{R} \geqq \mathfrak{R}(a_1 \cdots a_n).$$

Nun ist $\mathfrak{R}(a_1 \cdots a_n) = \mathfrak{B}(M)$.

Daher $\mathfrak{R} \geqq \mathfrak{B}(M)$, w. z. b. w.

Satz 15. $\mathfrak{R} \geqq M_1, \cdots, M_k \rightarrow \mathfrak{R} \geqq \mathfrak{B}(M_1 \cdots M_k)$;

in Worten: Ein solcher linearer Raum, der k Mengen $M_1, \cdots, M_k$ in sich enthält, enthält auch den Vereinigungsraum $\mathfrak{B}(M_1 \cdots M_k)$ in sich.

Beweis: $\mathfrak{R} \geqq M_1, \cdots, M_k \rightarrow \mathfrak{R} \geqq M_1 + \cdots + M_k$

$$-(\text{Satz } 14) \rightarrow \mathfrak{R} \geqq \mathfrak{B}(M_1 + \cdots + M_k) - (\text{V}_1.\ 12) \rightarrow \mathfrak{R} \geqq \mathfrak{B}(M_1 \cdots M_k),$$

w. z. b. w.

Aus Satz 15 kann man leicht folgenden Zusatz herleiten:

Zusatz. $\mathfrak{B}(\mathfrak{B}(M_1 \cdots M_{k-1}), M_k) = \mathfrak{B}(M_1 \cdots M_k)$,
$\mathfrak{B}(M_1, \mathfrak{B}(M_2 \cdots M_k)) = \mathfrak{B}(M_1 \cdots M_k)$.

Satz 16. $\mathfrak{B}_1(\mathfrak{R}_1(a_1 \cdots a_n), \mathfrak{R}_2(b_1 \cdots b_m)) = \mathfrak{B}_2(a_1 \cdots a_n b_1 \cdots b_m)$.

Beweis: $\left.\begin{array}{l} \mathfrak{B}_1 \geqq \mathfrak{R}_1 \ni a_1, \cdots, a_n \ \text{und} \\ \mathfrak{B}_1 \geqq \mathfrak{R}_2 \ni b_1, \cdots, b_m \end{array}\right\} \rightarrow \mathfrak{B}_1 \ni a_1, \cdots, a_n, b_1, \cdots, b_m$

$$-(\text{Satz } 14) \rightarrow \mathfrak{B}_1 \geqq \mathfrak{B}_2(a_1 \cdots a_n b_1 \cdots b_m).$$

Anderseits,

$$\left.\begin{array}{l} \mathfrak{B}_2 \ni a_1, \cdots, a_n \ \text{und} \ a_1 \cdots a_n \neq 0 - (\text{Satz } 9) \rightarrow \mathfrak{B}_2 \geqq \mathfrak{R}_1(a_1 \cdots a_n), \\ \mathfrak{B}_2 \ni b_1, \cdots, b_m \ \text{und} \ b_1 \cdots b_m \neq 0 - (\text{Satz } 9) \rightarrow \mathfrak{B}_2 \geqq \mathfrak{R}_2(b_1 \cdots b_m) \end{array}\right\}$$

$$-(\text{Satz } 15) \rightarrow \mathfrak{B}_2 \geqq \mathfrak{B}_1(\mathfrak{R}_1 \mathfrak{R}_2).$$

Daher $\mathfrak{B}_1(\mathfrak{R}_1(a_1\cdots a_n), \mathfrak{R}_2(b_1\cdots b_m)) = \mathfrak{B}_2(a_1\cdots a_n b_1\cdots b_m)$, w. z. b. w.

Man kann analog folgenden Satz beweisen;

Satz 17. $\mathfrak{B}_1(\mathfrak{R}_1(a_1\cdots a_n), \cdots, \mathfrak{R}_k(l_1\cdots l_m)) = \mathfrak{B}_2(a_1\cdots a_n \cdots l_1\cdots l_m)$.

Satz 18. $\text{Rang}\,\mathfrak{D}(\mathfrak{R}_1^n\,\mathfrak{R}_2^m) + \text{Rang}\,\mathfrak{B}(\mathfrak{R}_1^n\,\mathfrak{R}_2^m) \leqq n+m$;

in Worten: Die Summe der Ränge des Durchschnittes und des Vereinigungsraumes zweier gegebenen linearen Räume des n^{ten}, bzw. m^{ten} Ranges ist höchstens gleich $n+m$[10].

Beweis: Seien die Basen von $\mathfrak{R}_1$, $\mathfrak{R}_2$, und $\mathfrak{D}(\mathfrak{R}_1\mathfrak{R}_2)$, bzw. je $a_1\cdots a_n$, $b_1\cdots b_m$, und $c_1\cdots c_k$, so folgt nach Satz 8, und Zusatz 1 zum Satz 7, z. B.,

$$\mathfrak{R}_1(a_1\cdots a_n) = \mathfrak{R}_1(c_1\cdots c_k a_{k+1}\cdots a_n),$$

$$\mathfrak{R}_2(b_1\cdots b_m) = \mathfrak{R}_2(c_1\cdots c_k b_{k+1}\cdots b_m).$$

Also folgt nach Satz 16

$$\begin{aligned}\mathfrak{B}(\mathfrak{R}_1\mathfrak{R}_2) &= \mathfrak{B}(c_1\cdots c_k a_{k+1}\cdots a_n c_1\cdots c_k b_{k+1}\cdots b_m)\\ &= \mathfrak{B}(c_1\cdots c_k a_{k+1}\cdots a_n b_{k+1}\cdots b_m).\end{aligned}$$

Daher
$$\begin{aligned}\text{Rang}\,\mathfrak{B}(\mathfrak{R}_1\mathfrak{R}_2) &= \text{Rang}\,\mathfrak{B}(c_1\cdots c_k a_{k+1}\cdots a_n b_{k+1}\cdots b_m)\\ &= \text{Rang}\,(c_1\cdots c_k a_{k+1}\cdots a_n b_{k+1}\cdots b_m)\\ &\leqq n+m-k.\end{aligned}$$

Und wegen Satz 12 folgt leicht

$$\text{Rang}\,\mathfrak{D}(\mathfrak{R}_1\mathfrak{R}_2) = \text{Rang}\,\mathfrak{R}(c_1\cdots c_k) = k.$$

Daher $\text{Rang}\,\mathfrak{D}(\mathfrak{R}_1\mathfrak{R}_2) + \text{Rang}\,\mathfrak{B}(\mathfrak{R}_1\mathfrak{R}_2) \leqq n+m$, w. z. b. w.

Satz 19. $\text{Rang}\,\mathfrak{D}(M_1M_2) + \text{Rang}\,\mathfrak{B}(M_1M_2) \leqq \text{Rang}\,M_1 + \text{Rang}\,M_2$[11].

Beweis: Da der Vereinigungsraum gegebener Menge auch ein linearer Raum ist, so besteht nach Satz 18 die folgende Ungleichung;

---

(10) Es sei hier bemerkt, dass in dieser Formel das Gleichzeichen nicht immer besteht.

(11) Der Satz 18 und Satz 19 sind gleichwertig im Inhalt, weil man, wie bereits im V_1. 2 erwähnt, die Summenmenge M_1+M_2 von M_1 und M_2 des Satz 19 als unser $\mathfrak{B}_1$-Raum betrachten kann.

$$\text{Rang}\,\mathfrak{D}(\mathfrak{V}(M_1),\,\mathfrak{V}(M_2))+\text{Rang}\,\mathfrak{V}(\mathfrak{V}(M_1),\,\mathfrak{V}(M_2))$$
$$\leqq \text{Rang}\,\mathfrak{V}(M_1)+\text{Rang}\,\mathfrak{V}(M_2)\,.$$

Nun folgt wegen Zusatz zum Satz 13

$$\text{Rang}\,\mathfrak{V}(M_1)=\text{Rang}\,M_1\,,$$
$$\text{Rang}\,\mathfrak{V}(M_2)=\text{Rang}\,M_2\,.$$

Und wegen Zusatz zum Satz 15 folgt

$$\text{Rang}\,\mathfrak{V}(\mathfrak{V}(M_1),\,\mathfrak{V}(M_2))=\text{Rang}\,\mathfrak{V}(M_1M_2)\,.$$

Ferner ist,

$$\left.\begin{array}{l}\mathfrak{V}(M_1)\geqq M_1\,,\\ \mathfrak{V}(M_2)\geqq M_2\,,\end{array}\right\}\rightarrow \mathfrak{D}(\mathfrak{V}(M_1),\,\mathfrak{V}(M_2))\geqq \mathfrak{D}(M_1M_2)\,.$$

Daher $\quad \text{Rang}\,\mathfrak{D}(\mathfrak{V}(M_1),\,\mathfrak{V}(M_2))\geqq \text{Rang}\,\mathfrak{D}(M_1M_2)\,.$

Also folgt

$$\text{Rang}\,\mathfrak{D}(M_1M_2)+\text{Rang}\,\mathfrak{V}(M_1M_2)\leqq \text{Rang}\,M_1+\text{Rang}\,M_2\,,\qquad \text{w. z. b. w.}$$

Satz 20. Sind alle Ränge der k Mengen $M_1,\cdots,M_k$ endlich, so besteht die folgende Ungleichung ;

$$\text{Rang}\,(M_1+\cdots+M_k)\leqq \text{Rang}\,M_1+\cdots+\text{Rang}\,M_k\,.$$

Beweis : Nach Satz 19 folgt leicht,

$$\text{Rang}\,\mathfrak{V}(M_1,\,\mathfrak{V}(M_2\cdots M_k))\leqq \text{Rang}\,M_1+\text{Rang}\,\mathfrak{V}(M_2\cdots M_k)$$
$$\leqq \text{Rang}\,M_1+\text{Rang}\,M_2+\text{Rang}\,\mathfrak{V}(M_3\cdots M_k)$$
$$\leqq\cdots\leqq \text{Rang}\,M_1+\text{Rang}\,M_2+\cdots+\text{Rang}\,M_k\,.$$

Also, weil

$$\text{Rang}\,(M_1+\cdots+M_k)=\text{Rang}\,\mathfrak{V}(M_1\cdots M_k)$$
$$=\text{Rang}\,\mathfrak{V}(M_1,\,\mathfrak{V}(M_2\cdots M_k))\,,$$

folgt

$$\text{Rang}\,(M_1+\cdots+M_k)\leqq \text{Rang}\,M_1+\cdots+\text{Rang}\,M_k\,,\qquad \text{w. z. b. w.}$$

§ 4. Die Reduktionsmethode.

Bei unserem Zyklenkalkül, brauchen wir oft, als eine der höchst wirksamen Beweismethoden, eine, sog. *Reduktionsmethode*. In diesem Paragraphen will ich dieselbe erklären, und einige durch diese Methode beweisbare Sätze angeben.

Satz 21. $a_1 \cdots a_n x,\ xy \to a_1 \cdots a_n y$ oder x .

Beweis: Man setze zunächst $a_2 \cdots a_n x \neq 0$; dann folgt

$$\left.\begin{array}{r} a_2 \cdots a_n x \neq 0\ , \\ a_1 a_2 \cdots a_n x\ , \\ xy \to a_2 \cdots a_n xy \end{array}\right\} (\text{Axiom } 4) \to a_1 a_2 \cdots a_n y\ .$$

Also folgt

$$a_1 \cdots a_n x\ , \quad xy \to a_1 \cdots a_n y \quad \text{oder} \quad a_2 \cdots a_n x\ .$$

Daher $\quad a_1 \cdots a_n x\ , \quad xy \to a_1 \cdots a_n y \quad$ oder $\quad \{a_2 \cdots a_n x,\ xy\}$.

Wir stellen diese Bedeutung mit der folgenden Bezeichnung dar;

$$\begin{array}{ccc} \left.\begin{array}{r} a_1 \cdots a_n x\ , \\ xy \end{array}\right\} & \longrightarrow & \left.\begin{array}{r} a_2 \cdots a_n x\ , \\ xy \end{array}\right\} . \\ \downarrow & & \\ a_1 \cdots a_n y\ . & & \end{array}$$

Dann bekommen wir die folgende logische Bezeichnungsformel, indem wir die obige Bezeichnungsoperation nach einander wiederholen.

$$\begin{array}{ccccccccc} \left.\begin{array}{r} a_1 \cdots a_n x\ , \\ xy \end{array}\right\} & \longrightarrow & \left.\begin{array}{r} a_2 \cdots a_n x\ , \\ xy \end{array}\right\} & \longrightarrow & \left.\begin{array}{r} a_3 \cdots a_n x\ , \\ xy \end{array}\right\} & \longrightarrow \cdots \longrightarrow & \left.\begin{array}{r} a_n x\ , \\ xy \end{array}\right\} & \longrightarrow & x\ . \\ \downarrow & & \downarrow & & \downarrow & & \downarrow & & \\ a_1 \cdots a_n y\ . & & a_2 \cdots a_n y\ . & & a_3 \cdots a_n y\ . & & a_n y\ . & & \end{array}$$

Da anderseits

$$a_1 \cdots a_n y \leftarrow a_2 \cdots a_n y \leftarrow a_3 \cdots a_n y \leftarrow \cdots \leftarrow a_n y$$

besteht, so kann man auch die obige Formel folgendermassen noch einmal umschreiben;

$$\begin{array}{ccccccccc}
\left.\begin{array}{r} a_1\cdots a_n x\,,\\ xy\end{array}\right\} & \longrightarrow & \left.\begin{array}{r} a_2\cdots a_n x\,,\\ xy\end{array}\right\} & \longrightarrow & \left.\begin{array}{r}a_3\cdots a_n x\,,\\ xy\end{array}\right\} & \longrightarrow\cdots\longrightarrow & \left.\begin{array}{r}a_n x\,,\\ xy\end{array}\right\} & \longrightarrow & x\,.\\
\downarrow && \downarrow && \downarrow && \downarrow && \\
a_1\cdots a_n y\,. & \leftarrow & a_2\cdots a_n y & \leftarrow & a_3\cdots a_n y & \leftarrow\cdots\leftarrow & a_n y &&
\end{array}$$

Daher
$$\begin{array}{ccc}
\left.\begin{array}{r} a_1\cdots a_n x\,,\\ xy\end{array}\right\} & \longrightarrow & x\,,\\
\downarrow && \\
a_1\cdots a_n y\,, &&
\end{array}\qquad \text{w. z. b. w.}$$

Das letzte zweite Schema nennen wir die *Reduktionsformel,* und die die Reduktionsformel in sich enthaltenden Beweismethode nennen wir die Reduktionsmethode.

Als methodisches Beispiel der durch dieselbe Beweisführung beweisbaren Sätze sei der folgende genommen.

Satz 22. $a_1\cdots a_n x_1\cdots x_k\,,\ x_1\cdots x_k y \rightarrow a_1\cdots a_n x_1\cdots x_{k-1} y$ oder $x_1\cdots x_k\,.$

Beweis :

$$\left.\begin{array}{r} a_2\cdots a_n x_1\cdots x_k \neq 0\,,\\ a_1\cdots a_n x_1\cdots x_k\,,\\ x_1\cdots x_k y\end{array}\right\}\rightarrow \left.\begin{array}{l} a_2\cdots a_n x_1\cdots x_k \neq 0\,,\\ a_1 a_2\cdots a_n x_1\cdots x_k\,,\\ a_2\cdots a_n x_1\cdots x_k y\end{array}\right\}(\text{Axiom 4})\rightarrow a_1\cdots a_n x_1\cdots x_{k-1} y.$$

Daher
$$\begin{array}{ccc}
\left.\begin{array}{r} a_1\cdots a_n x_1\cdots x_k\,,\\ x_1\cdots x_k y\end{array}\right\} & \longrightarrow & a_2\cdots a_n x_1\cdots x_k\,.\\
\downarrow && \\
a_1\cdots a_n x_1\cdots x_{k-1} y\,. &&
\end{array}$$

Daher
$$\begin{array}{ccc}
\left.\begin{array}{r} a_1\cdots a_n x_1\cdots x_k\,,\\ x_1\cdots x_k y\end{array}\right\} & \longrightarrow & \left.\begin{array}{r} a_2\cdots a_n x_1\cdots x_k\,,\\ x_1\cdots x_k y\end{array}\right\}.\\
\downarrow && \\
a_1\cdots a_n x_1\cdots x_{k-1} y\,. &&
\end{array}$$

$$\begin{array}{ccccccc}
\left.\begin{array}{r} a_1\cdots a_n x_1\cdots x_k\,,\\ x_1\cdots x_k y\end{array}\right\} & \longrightarrow & \left.\begin{array}{r} a_2\cdots a_n x_1\cdots x_k\,,\\ x_1\cdots x_k y\end{array}\right\} & \longrightarrow\cdots\longrightarrow & \left.\begin{array}{r} a_n x_1\cdots x_k\,,\\ x_1\cdots x_k y\end{array}\right\} & \longrightarrow & x_1\cdots x_k\,.\\
\downarrow && \downarrow && \downarrow && \\
a_1\cdots a_n x_1\cdots x_{k-1} y\,. & \leftarrow & a_2\cdots a_n x_1\cdots x_{k-1} y & \leftarrow\cdots\leftarrow & a_n x_1\cdots x_{k-1} y &&
\end{array}$$

Daher

$$\begin{array}{ll} \left.\begin{array}{r} a_1\cdots a_n x_1\cdots x_k\,, \\ x_1\cdots x_k y \end{array}\right\} & \longrightarrow x_1\cdots x_k\,, \\ \qquad\qquad\downarrow & \\ a_1\cdots a_n x_1\cdots x_{k-1} y\,, & \end{array}$$

w. z. b. w.[12]

Satz 23. $a_1\cdots a_n x\,,\ b_1\cdots b_m x \rightarrow a_1\cdots a_n b_1\cdots b_m$ oder x .

Beweis:

$$\left.\begin{array}{l} a_1\cdots a_n x\,, \\ b_1\cdots b_m x \end{array}\right\} \longrightarrow \left.\begin{array}{r} a_1\cdots a_n b_2\cdots b_m x\,, \\ b_1 b_2\cdots b_m x \end{array}\right\} \text{(Satz 22)} \begin{array}{l} \longrightarrow b_2\cdots b_m x\,. \\ \downarrow \\ a_1\cdots a_n b_1\cdots b_m\,. \end{array}$$

Daher

$$\begin{array}{ll} \left.\begin{array}{l} a_1\cdots a_n x\,, \\ b_1\cdots b_m x \end{array}\right\} \longrightarrow & \left.\begin{array}{l} a_1\cdots a_n x\,, \\ b_2\cdots b_m x \end{array}\right\}. \\ \qquad\downarrow & \\ a_1\cdots a_n b_1\cdots b_m\,. & \end{array}$$

Daher

$$\begin{array}{ccccccc} \left.\begin{array}{l} a_1\cdots a_n x\,, \\ b_1\cdots b_m x \end{array}\right\} & \longrightarrow & \left.\begin{array}{l} a_1\cdots a_n x\,, \\ b_2\cdots b_m x \end{array}\right\} & \longrightarrow \cdots \longrightarrow & \left.\begin{array}{r} a_1\cdots a_n x, \\ b_m x \end{array}\right\} & \longrightarrow & x\,. \\ \downarrow & & \downarrow & & \downarrow & & \\ a_1\cdots a_n b_1\cdots b_m\,. & \leftarrow & a_1\cdots a_n b_2\cdots b_m & \leftarrow \cdots \leftarrow & a_1\cdots a_n b_m & & \end{array}$$

Daher

$$\begin{array}{ll} \left.\begin{array}{l} a_1\cdots a_n x\,, \\ b_1\cdots b_m x \end{array}\right\} & \longrightarrow x\,, \\ \qquad\downarrow & \\ a_1\cdots a_n b_1\cdots b_m\,, & \end{array}$$

w. z. b. w.

Satz 24. $a_1\cdots a_n x_1\cdots x_k\,,\ b_1\cdots b_m x_1\cdots x_k$
$\rightarrow a_1\cdots a_n b_1\cdots b_m x_1\cdots x_{k-1}$ oder $x_1\cdots x_k$.

Beweis:

$$\left.\begin{array}{l} a_1\cdots a_n x_1\cdots x_k\,, \\ b_1\cdots b_m x_1\cdots x_k \end{array}\right\} \longrightarrow \left.\begin{array}{r} a_1\cdots a_n b_2\cdots b_m x_1\cdots x_k\,, \\ b_1 b_2\cdots b_m x_1\cdots x_k \end{array}\right\} \text{(Satz 22)} \begin{array}{l} \longrightarrow b_2\cdots b_m x_1\cdots x_k\,. \\ \downarrow \\ a_1\cdots a_n b_1\cdots b_m x_1\cdots x_{k-1}\,. \end{array}$$

(12) Der Beweis des Satz 22 wäre ein typisches Beispiel der Reduktionsmethode.

Daher
$$\begin{array}{ccc} \left.\begin{array}{l} a_1\cdots a_n x_1\cdots x_k, \\ b_1\cdots b_m x_1\cdots x_k \end{array}\right\} & \longrightarrow & \left.\begin{array}{l} a_1\cdots a_n x_1\cdots x_k, \\ b_2\cdots b_m x_1\cdots x_k \end{array}\right\}. \\ \big\downarrow & & \\ a_1\cdots a_n b_1\cdots b_m x_1\cdots x_{k-1}. & & \end{array}$$

Daher
$$\begin{array}{ccccccc} \left.\begin{array}{l} a_1\cdots a_n x_1\cdots x_k, \\ b_1\cdots b_m x_1\cdots x_k \end{array}\right\} & \longrightarrow & \left.\begin{array}{l} a_1\cdots a_n x_1\cdots x_k, \\ b_2\cdots b_m x_1\cdots x_k \end{array}\right\} & \longrightarrow \cdots \longrightarrow & \left.\begin{array}{l} a_1\cdots a_n x_1\cdots x_k, \\ b_m x_1\cdots x_k \end{array}\right\} & \longrightarrow & x_1\cdots x_k. \\ \big\downarrow & & \big\downarrow & & \big\downarrow & & \\ a_1\cdots a_n b_1\cdots b_m x_1\cdots x_{k-1}, & \longleftarrow & a_1\cdots a_n b_2\cdots b_m x_1\cdots x_{k-1} & \longleftarrow \cdots \longleftarrow & a_1\cdots a_n b_m x_1\cdots x_{k-1} & & \end{array}$$

Daher
$$\begin{array}{ccc} \left.\begin{array}{l} a_1\cdots a_n x_1\cdots x_k, \\ b_1\cdots b_m x_1\cdots x_k \end{array}\right\} & \longrightarrow & x_1\cdots x_k, \\ \big\downarrow & & \\ a_1\cdots a_n b_1\cdots b_m x_1\cdots x_{k-1}, & & \end{array}$$
w. z. b. w.

Zusatz 1. $a_1 \cdots a_n x,\ bx,\ x \neq 0 \rightarrow a_1 \cdots a_n b$.

Zusatz 2. $a_1 \cdots a_n x,\ b_1 b_2 x,\ x \neq 0 \rightarrow a_1 \cdots a_n b_1 b_2$.

Zusatz 3. $a_1 \cdots a_n x,\ b_1 \cdots b_m x,\ x \neq 0 \rightarrow a_1 \cdots a_n b_1 \cdots b_m$.

Zusatz 4. $a_1\cdots a_n b_1\cdots b_m \neq 0 \rightarrow \mathfrak{D}(\mathfrak{R}(a_1\cdots a_n),\ \mathfrak{R}(b_1\cdots b_m)) = \mathfrak{N}$.

Zusatz 5. $a_1 \cdots a_n x_1 \cdots x_k,\ b_1 \cdots b_m x_1 \cdots x_k,\ x_1 \cdots x_k \neq 0$
$$\rightarrow a_1 \cdots a_n b_1 \cdots b_m x_1 \cdots x_{i-1} x_{i+1} \cdots x_k,\ (i = 1, \cdots, k).$$

Zusatz 6. $a_1 \cdots a_n b_1 \cdots b_m d_1 \cdots d_k \neq 0$
$$\mathfrak{D}(\mathfrak{R}(a_1\cdots a_n d_1\cdots d_k),\ \mathfrak{R}(b_1\cdots b_m d_1\cdots d_k)) = \mathfrak{R}(d_1\cdots d_k).$$

Bemerkung :

Wenn die Prämisse eines Satzes, wie oben, aus zwei Zyklen besteht, so gibt es in diesem Fall ausser obiger Reduktionsmethode noch eine stets brauchbare Beweismethode, welche sich auf Satz 18 zurückführen lässt. Um dieselbe Methode zu erklären wollen wir einen neuen Beweis des Satzes 23 angeben.

(Satz 23.) $a_1 \cdots a_n x,\ b_1\cdots b_m x \rightarrow a_1\cdots a_n b_1\cdots b_m$ oder x.

Beweis: (i) Sei $a_1\cdots a_n$, so folgt trivialerweise $a_1\cdots a_n b_1\cdots b_m$.

(ii) Sei $b_1\cdots b_m$, so folgt trivialerweise $a_1\cdots a_n b_1\cdots b_m$.

(iii) Nun sei $a_1\cdots a_n \neq 0$, $b_1\cdots b_m \neq 0$, so folgt wegen des Satz 18,
$$\text{Rang}\ \mathfrak{D}(\mathfrak{R}(a_1\cdots a_n),\ \mathfrak{R}(b_1\cdots b_m)) + \text{Rang}\ \mathfrak{V}(\mathfrak{R}(a_1\cdots a_n),\ \mathfrak{R}(b_1\cdots b_m)) \leqq n + m.$$

Dabei ist

$$a_1 \cdots a_n x \,,\; b_1 \cdots b_m x \rightarrow \mathfrak{D}(\mathfrak{R}(a_1 \cdots a_n),\, \mathfrak{R}(b_1 \cdots b_m)) \ni x \,.$$

Und ist weiter

$$\mathfrak{V}(\mathfrak{R}(a_1 \cdots a_n)\,,\; \mathfrak{R}(b_1 \cdots b_m)) \ni a_1\,,\; \cdots,\; a_n\,,\; b_1\,,\; \cdots,\; b_m\,.$$

Also folgt

$$\text{Rang}\; \mathfrak{D}(\mathfrak{R}(a_1 \cdots a_n)\,,\; \mathfrak{R}(b_1 \cdots b_m)) \geqq \text{Rang}\,(x)\,,$$

$$\text{Rang}\; \mathfrak{V}(\mathfrak{R}(a_1 \cdots a_n)\,,\; \mathfrak{R}(b_1 \cdots b_m)) \geqq \text{Rang}\,(a_1 \cdots a_n b_1 \cdots b_m)\,.$$

Daher $\quad \text{Rang}\,(x) + \text{Rang}\,(a_1 \cdots a_n b_1 \cdots b_m) \leqq n + m\,.$

Daher $\quad x \neq 0 \rightarrow a_1 \cdots a_n b_1 \cdots b_m\,,$

$\quad\quad a_1 \cdots a_n b_1 \cdots b_m \neq 0 \rightarrow x\,.$

Daher $\quad a_1 \cdots a_n b_1 \cdots b_m$ oder x.

Also ist

$$a_1 \cdots a_n x \,,\; b_1 \cdots b_m x \rightarrow a_1 \cdots a_n b_1 \cdots b_m \quad \text{oder} \quad x\,, \qquad \text{w. z. b. w.}$$

Satz 25. $a_1 \cdots a_n\,,\; a_i b_i\,,\; a_i \neq 0\,,\; (i = 1\,,\; \cdots,\; n) \rightarrow b_1 \cdots b_n\,.$

Beweis:

$$a_1 \cdots a_n\,,\; a_i b_i - (\text{Satz 21}) \rightarrow a_1 \cdots a_{i-1} b_i a_{i+1} \cdots a_n \quad \text{oder} \quad a_i\,.$$

Daher

$$\begin{array}{ccccccc} \left.\begin{array}{r} a_1 \cdots a_n\,, \\ a_1 b_1 \end{array}\right\} & \longrightarrow & \left.\begin{array}{r} b_1 a_2 \cdots a_n\,, \\ a_2 b_2 \end{array}\right\} & \longrightarrow & \left.\begin{array}{r} b_1 b_2 a_3 \cdots a_n\,, \\ a_3 b_3 \end{array}\right\} & \longrightarrow & \cdots \\ & \downarrow & & \downarrow & & \downarrow & \\ & a_1 \rightarrow W\,. & & a_2 \rightarrow W\,. & & a_3 \rightarrow W\,. & \end{array}$$

$$\begin{array}{ccc} \cdots \longrightarrow \left.\begin{array}{r} b_1 \cdots b_{n-1} a_n\,, \\ a_n b_n \end{array}\right\} & \longrightarrow & b_1 \cdots b_n\,. \\ & \downarrow & \\ & a_n \rightarrow W\,. & \end{array}$$

Daher $\quad a_1 \cdots a_n\,,\; a_i b_i\,,\; a_i \neq 0\,,\; (i = 1,\; \cdots,\; n) \rightarrow b_1 \cdots b_n\,, \quad$ w. z. b. w.

Zusatz. $a_1 \cdots a_n \neq 0\,,\; a_i b_i\,,\; b_i \neq 0\,,\; (i = 1\,,\; \cdots,\; n) \rightarrow b_1 \cdots b_n \neq 0\,.$

Beweis: Wäre vorläufig $b_1 \cdots b_n$, so wäre

$$b_1 \cdots b_n\,,\; a_i b_i\,,\; b_i \neq 0\,,\; (i = 1\,,\; \cdots,\; n) - (\text{Satz 25}) \rightarrow a_1 \cdots a_n \rightarrow W.$$

Daher $\quad a_1 \cdots a_n \neq 0\,,\; a_i b_i\,,\; b_i \neq 0\,,\; (i = 1\,,\; \cdots,\; n) \rightarrow b_1 \cdots b_n \neq 0\,,$

w. z. b. w.

Satz 26.

$$\left.\begin{array}{l} *a_2\cdots a_n x,\\ a_1*\cdots a_n x,\\ \cdots\cdots\cdots\cdots,\\ a_1a_2\cdots * x\end{array}\right\} \to a_1a_2\cdots a_n \quad \text{oder} \quad x .$$

Beweis :

$$\left.\begin{array}{l}*a_2\cdots a_n x,\\ a_1*\cdots a_n x\end{array}\right\} \text{(Satz 22)} \begin{array}{l}\longrightarrow **a_3\cdots a_n x .\\ \downarrow\\ a_1a_2\cdots a_n .\end{array}$$

Daher

$$\left.\begin{array}{l}*a_2\cdots a_n x,\\ a_1*\cdots a_n x\end{array}\right\} \begin{array}{c}\longrightarrow\\ \downarrow\\ a_1a_2\cdots a_n .\end{array} \left.\begin{array}{l}**a_3\cdots a_n x,\\ a_1a_2*\cdots a_n x\end{array}\right\} \begin{array}{c}\longrightarrow\\ \downarrow\\ a_1a_2\cdots a_n .\end{array} \left.\begin{array}{l}***a_4\cdots a_n x,\\ a_1a_2a_3*\cdots a_n x\end{array}\right\} \begin{array}{c}\longrightarrow \cdots\\ \downarrow\\ a_1a_2\cdots a_n .\end{array}$$

$$\cdots\longrightarrow \left.\begin{array}{l}*\cdots * a_n x,\\ a_1\cdots a_{n-1} * x\end{array}\right\} \begin{array}{l}\longrightarrow x .\\ \downarrow\\ a_1a_2\cdots a_n .\end{array}$$

Daher

$$\left.\begin{array}{l} *a_2\cdots a_n x,\\ a_1*\cdots a_n x,\\ \cdots\cdots\cdots,\\ a_1a_2\cdots * x\end{array}\right\} \begin{array}{l}\longrightarrow x ,\\ \downarrow\\ a_1a_2\cdots a_n ,\end{array}$$

w. z. b. w.

Zusatz. $a_1\cdots a_n \neq 0 \to \mathfrak{D}(\mathfrak{R}_1(*a_2\cdots a_n),\ \mathfrak{R}_2(a_1*\cdots a_n),\ \cdots,\ \mathfrak{R}_n(a_1\cdots a_{n-1}*)) = \mathfrak{N}$;

in Worten: Es gibt keinen Punkt, welcher zugleich in aller $(n-1)$-Seitensimplexen eines n-Simplexes enthalten ist.

Satz 27.

$$\left.\begin{array}{l} *a_2\cdots a_n x_1\cdots x_k,\\ a_1*\cdots a_n x_1\cdots x_k,\\ \cdots\cdots\cdots\cdots\cdots,\\ a_1a_2\cdots * x_1\cdots x_k\end{array}\right\} \to a_1a_2\cdots a_n x_1\cdots x_{k-1} \quad \text{oder} \quad x_1\cdots x_k .$$

Beweis :

$$\left.\begin{array}{l}*a_2\cdots a_i\cdots a_n x_1\cdots x_k,\\ a_1a_2\cdots *_i\cdots a_n x_1\cdots x_k\end{array}\right\} \text{(Satz 22)} \begin{array}{l}\longrightarrow *a_2\cdots *_i\cdots a_n x_1\cdots x_k .\\ \downarrow\\ a_1a_2\cdots a_n x_1\cdots x_{k-1} .\end{array}$$

Daher
$$\left.\begin{array}{l} * a_2\cdots a_n x_1\cdots x_k , \\ a_1 * \cdots a_n x_1\cdots x_k , \\ \cdots\cdots\cdots\cdots , \\ a_1 a_2\cdots * x_1\cdots x_k \end{array}\right\} \longrightarrow \left.\begin{array}{l} * a_3\cdots a_n x_1\cdots x_k , \\ a_2 * \cdots a_n x_1\cdots x_k , \\ \cdots\cdots\cdots\cdots , \\ a_2 a_3\cdots * x_1\cdots x_k \end{array}\right\} \longrightarrow \cdots$$
$$\downarrow \qquad\qquad\qquad \downarrow$$
$$a_1 a_2\cdots a_n x_1\cdots x_{k-1} . \longleftarrow a_2 a_3\cdots a_n x_1\cdots x_{k-1} \longleftarrow \cdots$$

$$\cdots \longrightarrow \left.\begin{array}{l} * a_n x_1\cdots x_k , \\ a_{n-1} * x_1\cdots x_k \end{array}\right\} \longrightarrow x_1\cdots x_k .$$
$$\downarrow$$
$$\cdots \longleftarrow a_{n-1} a_n x_1\cdots x_{k-1}$$

Daher
$$\left.\begin{array}{l} * a_2\cdots a_n x_1\cdots x_k , \\ a_1 * \cdots a_n x_1\cdots x_k , \\ \cdots\cdots\cdots\cdots , \\ a_1 a_2\cdots * x_1\cdots x_k \end{array}\right\} \longrightarrow x_1\cdots x_k ,$$
$$\downarrow$$
$$a_1 a_2\cdots a_n x_1\cdots x_{k-1} , \qquad \text{w. z. b. w.}$$

Zusatz. $a_1\cdots a_n \neq 0 , \quad m \leqq n$

$$\longrightarrow \mathfrak{D}(\mathfrak{R}_1(* a_2\cdots a_m a_{m+1}\cdots a_n) , \mathfrak{R}_2(a_1 * \cdots a_m a_{m+1}\cdots a_n) ,$$
$$\cdots , \mathfrak{R}_m(a_1 a_2\cdots * a_{m+1}\cdots a_n)) = \mathfrak{R}(a_{m+1}\cdots a_n) .$$

§ 5. Einige Bemerkungen.

Betreffs des § 4 sei nun besonders erwähnt, dass alle Sätze in dem § 4 lauter Folgerungen sind von den Axiomen ausser Axiom 1. Dies wollen wir besonders betonen, da wir die Absicht haben später Geometrien aufzubauen mit den Axiomen 2, 3, 4, ohne Benutzung des Axiom 1.

Betreffs ganzer dieser Arbeit sei bemerkt, dass die alle Sätze keinerlei Existenzforderung enthalten. Dies steht auch im tiefen Zusammenhang mit einer Fortsetzung dieser Arbeit, s.g. Theorie des zweiten Verknüpfungsraumes, welche demnächst erscheinen wird.

Zur Axiomatik der linearen Abhängigkeit. II.

Zur Axiomatik der linearen Abhängigkeit. II.

Von Takeo NAKASAWA

(Eingegangen am 28. November, 1935)

Einleitung.

In der vorliegenden Schrift, welche eine Fortsetzung meiner früheren Arbeit[13] ist, soll die Geometrie des zweiten Verknüpfungsraumes $\mathfrak{V}_2$ (Def. VI, §1) aufgebaut werden, und zwar, nach der Herstellung der Hilfsbegriffen, wie Primzyklen (Def. VII, § 4), linearer Primraum (Def. VIII, § 3), Trennung der linearen Räume (Def. IX, § 4), Trennung der Elemente (Def. XI, § 5), wollen wir als Hauptresultat dieser Arbeit einen Zerlegungssatz des linearen Raumes (Satz 62, § 4), sowie unseren $\mathfrak{V}_2$-Raumes (Satz 63, § 5), bzw. in die direkte Summe von linearen Primräummen, sowie von Primverknüpfungsräumen angeben.

Bezeichnungen.

Ausser den Bezeichnungen, die wir in meiner früheren Arbeit[13] benutzen, wollen wir einige neue, 1, 7, 8, 11, hinzufügen:

1.	$\therefore$	bedeutet "also".
2.	$A \rightarrow B$	bedeutet, dass aus A B folgt.
3.	$A \longleftrightarrow B$	bedeutet, dass $A \rightarrow B$ und $B \rightarrow A$.
4.	A, B	bedeutet, dass A und B.
5.	A_i, $(i = 1, \cdots, k)$	bedeutet, dass A_1, A_2, $\cdots$, und A_k.
6.	A oder B	bedeutet, dass mindestens eins von A und B.
7.	A_1 od. $\cdots$ od. A_k	bedeutet, dass mindestens eins von A_1, A_2, $\cdots$ und A_k.
8.	z. B. A_1	bedeutet, dass A_1 od. A_2 od. $\cdots$ od. A_k.
9.	$A \rightarrow W.$	bedeutet, dass A zum Widerspruch gerät.
10.	A —(S)→ B	bedeutet, dass auf Grund des Aussages S aus A B folgt.

(13) T. NAKASAWA, „*Zur Axiomatik der linearen Abhängigkeit. I*", Science Reports of the Tokyo Bunrika Daigaku, Section A, Volume 2, No. 43, (p. 235 - p. 255).

11. Ez, A_z bedeutet, dass ein Element z mit der Eigenschaft A_z existiert.

12. $\left.\begin{matrix} A_1, \\ \vdots, \\ A_k \end{matrix}\right\} \to B$ bedeutet, dass aus dem gleichzeitigen Bestehen der k Aussagen $A_1, A_2, \cdots, A_k$ B folgt.

13. $A \to \left\{\begin{matrix} B_1, \\ \vdots, \\ B_k \end{matrix}\right.$ bedeutet, dass aus A gleichzeitig die k Aussagen $B_1, B_2, \cdots, B_k$ folgt.

14. Die folgenden Bezeichnungen von Mengenlehre werden auch noch benutzt, wie sie gewöhnlich bedeuten; $\geqq$, $\not\geqq$, $>$, $\not>$, $=$, $\neq$, $\ni$, $\not\ni$, $+$, u. s. w.

ZWEITES KAPITEL

Der $\mathfrak{V}_2$-Raum.

Das vorliegende Axiomensystem unterscheidet sich von dem in meiner früheren Arbeit[13] angenommenen dadurch, dass erstens im Axiom 1* die Bedingung $a \neq 0$ erfüllt werden soll, und dass zweitens noch das letzte Axiom, d. h. die Existenz des Durchschnittselementes, zugefügt wird. Daher ist der $\mathfrak{V}_2$-Raum eine Verengung des früher betrachteten $\mathfrak{V}_1$-Raums.

§ 1. Axiome.

Grundannahme: Wir denken uns eine gewisse Menge von Elementen; $\mathfrak{V}_2 \ni a_1, a_2, \cdots, a_s, \cdots$. Für gewisse Reihen der Elemente, die wir *Zyklen* nennen wollen, denken wir dazu die Relationen "*gelten*", in Zeichen $a_1 \cdots a_s = 0$, gewöhnlich in kurzen Zeichen, $a_1 \cdots a_s$, bzw. "*nicht gelten*", in Zeichen $a_1 \cdots a_s \neq 0$. Diese Relationen sollen nun folgenden Axiomen genügen;

Axiom 1.* (Reflexivität) : $a \neq 0$, aa.

Axiom 2. (Folgerung) : $a_1 \cdots a_s \to a_1 \cdots a_s x$, $(s = 1, 2, \cdots)$.

Axiom 3. (Vertauschung): $a_1 \cdots a_i \cdots a_s \to a_i \cdots a_1 \cdots a_s$, $(s = 2, 3, \cdots; i = 2, \cdots, s)$.

Axiom 4. (Transitivität) : $a_1 \cdots a_s \neq 0$, $xa_1 \cdots a_s$, $a_1 \cdots a_s y \to xa_1 \cdots a_{s-1} y$, $(s = 1, 2, \cdots)$.

Axiom 5. (Durchschnitt): $a_1 \cdots a_s xy \rightarrow Ez,\ a_1 \cdots a_s z,\ xyz,$
$(s = 2, 3, \cdots).$

Definition VI. Eine solche Menge $\mathfrak{V}_2$ heisst der *zweite Verknüpfungsraum*, in kurzen Worten, **$\mathfrak{V}_2$-Raum.**

V_2. 1. Nehmen wir uns die Menge von allen Punkten des klassichen n-dimensionalen projektiven Raumes als $\mathfrak{V}_2$-Raum, und nennen wir von gewissen darin liegenden linearen abhängigen, bzw. unabhängigen Reihen von Punkten als gelten, bzw. nicht gelten, so sind ersichtlich die Grundannahme, sowie die Axiome 1^*, 2, 3, 4, 5 erfüllt. Daher zeigt sich der projektive Raum eine typische Darstellung des $\mathfrak{V}_2$-Raumes in Bezug auf der linearen Abhängigkeit der Punktreihen. Also von jetzt an, werden wir oft die Wörter "Punkt" statt "Element", oder "linear abhängig", bzw. "linear unabhängig" statt "gelten" bzw. "nicht gelten" Gebrauch machen.

V_2. 2. Wie oben gesagt ist ein beliebiger $\mathfrak{V}_2$-Raum auch ein $\mathfrak{V}_1$-Raum. Daher gelten alle im $\mathfrak{V}_1$-Raum bestehenden Sätze auch im $\mathfrak{V}_2$-Raum.

V_2. 3. Ferner aus Axiom 1^* folgt leicht, dass die Nullstelle(14) von $\mathfrak{V}_2$ eine leere Menge ist. Also von jetzt an, werden wir häufig die leere Menge mit dem Zeichen $\mathfrak{N}$ bezeichnen.

§ 2. Der Durchschnitt.

Satz 28. $a_1 \cdots a_n b_1 \cdots b_m \rightarrow Ez,\ a_1 \cdots a_n z,\ b_1 \cdots b_m z.$

Beweis: Wir beweisen dies durch die vollständige Induktion in Bezug auf m.

(i) Falls m gleich 1 ist, dann ergibt sich

$$a_1 \cdots a_n b_1 \text{ —(Axiom 1*)→ } Eb_1,\ a_1 \cdots a_n b_1,\ b_1 b_1.$$

(ii) Falls m gleich 2 ist, dann ergibt sich

$$a_1 \cdots a_n b_1 b_2 \text{ —(Axion 5)→ } Ez,\ a_1 \cdots a_n z,\ b_1 b_2 z.$$

(iii) Nun genügt es zu zeigen, dass aus dem Fall $[m-1]$ der Fall $[m]$ folgt.

$$a_1 \cdots a_n b_1 \cdots b_m \rightarrow a_1 \cdots a_n b_1 \cdots b_{m-2} b_{m-1} b_m$$
$$\text{—(Axiom 5)→ } Ey,\ a_1 \cdots a_n b_1 \cdots b_{m-2} y,\ b_{m-1} b_m y.$$

(14) Vgl. a. a. O. p. 237!

$$a_1 \cdots a_n b_1 \cdots b_{m-2} y \text{ —}[m-1]\text{→ } Ez\,,\; a_1 \cdots a_n z\,,\; b_1 \cdots b_{m-2} yz\,.$$

Dann besteht

$$b_1 \cdots b_{m-2} yz\,,\; b_{m-1} b_m y \text{ —(Zusatz 2 zum Satz 24)→ } b_1 \cdots b_{m-2} b_{m-1} b_m z.$$

Daher $a_1 \cdots a_n b_1 \cdots b_m \to Ez\,,\; a_1 \cdots a_n z\,,\; b_1 \cdots b_m z\,,$

w. z. b. w.

Satz 29. Rang $\mathfrak{D}(\mathfrak{R}_1\mathfrak{R}_2)$ + Rang $\mathfrak{B}(\mathfrak{R}_1\mathfrak{R}_2)$ = Rang $\mathfrak{R}_1$ + Rang $\mathfrak{R}_2$; in Worten : Die Summe der Ränge des Durchschnittes und des Vereinigungsraumes zwei gegebener linearer Räume ist gleich der Summe der Ränge der zwei linearen Räume.

Beweis : Sei die Basis von $\mathfrak{D}(\mathfrak{R}_1\mathfrak{R}_2)$ $c_1 \cdots c_k$, und seien die Basen von $\mathfrak{R}_1$, und $\mathfrak{R}_2$ bzw. je $c_1 \cdots c_k a_{k+1} \cdots a_n$, und $c_1 \cdots c_k b_{k+1} \cdots b_m$, so folgt nach Satz 18,

$$\mathfrak{B}(\mathfrak{R}_1\mathfrak{R}_2) = \mathfrak{B}\,(c_1 \cdots c_k a_{k+1} \cdots a_n b_{k+1} \cdots b_m)\,.$$

Nun wäre $c_1 \cdots c_k a_{k+1} \cdots a_n b_{k+1} \cdots b_m = 0$, so würde sich folgern,

$$c_1 \cdots c_k a_{k+1} \cdots a_n b_{k+1} \cdots b_m \text{ —(Satz 28)→ } Ez,\; c_1 \cdots c_k a_{k+1} \cdots a_n z,\; b_{k+1} \cdots b_m z.$$

$$\left.\begin{array}{ll} c_1 \cdots c_k a_{k+1} \cdots a_n z \to \mathfrak{R}_1\,(c_1 \cdots c_k a_{k+1} \cdots a_n) \ni z, \\ b_{k+1} \cdots b_m z \qquad\quad \to \mathfrak{R}_2\,(c_1 \cdots c_k b_{k+1} \cdots b_m) \ni z \end{array}\right\} \to \mathfrak{D}(\mathfrak{R}_1\mathfrak{R}_2) \ni z \to c_1 \cdots c_k z\,.$$

$$c_1 \cdots c_k z\,,\; b_{k+1} \cdots b_m z \text{ —(Zusatz 3 zum Satz 24)→ } c_1 \cdots c_k b_{k+1} \cdots b_m \to W.$$

So muss $c_1 \cdots c_k a_{k+1} \cdots a_n b_{k+1} \cdots b_m \neq 0$.

Daher Rang $\mathfrak{B}(c_1 \cdots c_k a_{k+1} \cdots a_n b_{k+1} \cdots b_m) = n+m-k$.

Daher Rang $\mathfrak{B}(\mathfrak{R}_1\mathfrak{R}_2) = n+m-k$.

Anderseits,

Rang $\mathfrak{R}_1(c_1 \cdots c_k a_{k+1} \cdots a_n) = n$,

Rang $\mathfrak{R}_2(c_1 \cdots c_k b_{k+1} \cdots b_m) = m$,

Rang $\mathfrak{D}\,(\mathfrak{R}_1\mathfrak{R}_2)$ = Rang $\mathfrak{R}_1\,(c_1 \cdots c_k) = k$.

Daher Rang $\mathfrak{D}(\mathfrak{R}_1\mathfrak{R}_2)$ + Rang $\mathfrak{B}(\mathfrak{R}_1\mathfrak{R}_2)$ = Rang $\mathfrak{R}_1$ + Rang $\mathfrak{R}_2$,

w. z. b. w.

Dies ist nichts anders als der Satz 18, beidem das Bestehen einer Ungleichung behauptet wurde. Infolge der Voraussetzung des Axioms 5 lässt sich hier das Bestehen der Gleichung beweisen.

Zusatz 1. Ist $a_1 \cdots a_n \neq 0$, $b_1 \cdots b_m \neq 0$, und Rang $(a_1 \cdots a_n b_1 \cdots b_m) = s$, so existiert der Durchschnitt zweier linearer Räume $\mathfrak{R}(a_1 \cdots a_n)$, $\mathfrak{R}(b_1 \cdots b_m)$, dessen Rang gleich $n+m-s$ ist.

Zusatz 2. $\mathfrak{D}(\mathfrak{R}(a_1 \cdots a_n d_1 \cdots d_k),\ \mathfrak{R}(b_1 \cdots b_m d_1 \cdots d_k)) = \mathfrak{R}(d_1 \cdots d_k)$
$\rightarrow a_1 \cdots a_n b_1 \cdots b_m d_1 \cdots d_k \neq 0$.

Zusatz 3. $\mathfrak{D}(\mathfrak{R}(a_1 \cdots a_n),\ \mathfrak{R}(b_1 \cdots b_m)) = \mathfrak{N}$
$\rightarrow a_1 \cdots a_n b_1 \cdots b_m \neq 0$.

§ 3. Primzyklen.

Definition VII. Derjenige Zyklus $a_1 \cdots a_{n+1}$ derart, dass $a_1 \cdots a_{n+1} = 0$, aber $a_1 \cdots *_i \cdots a_{n+1} \neq 0$[15], $(i = 1, \cdots, n+1)$ ist, heisst der *Primzyklus* vom Range n , und wird mit $\overline{a_1 \cdots a_{n+1}}$ bezeichnet.

Satz 30. $\left.\begin{array}{l} a_1 \cdots a_n b_1 \cdots b_{m-1} \neq 0, \\ \overline{b_1 \cdots b_m} \end{array}\right\} \rightarrow \left\{\begin{array}{l} a_1 \cdots a_n b_1 \cdots *_i \cdots b_m \neq 0 , \\ \quad (i = 1, \cdots, m) . \end{array}\right.$

Beweis: Wäre vorläufig $a_1 \cdots a_n b_1 \cdots *_i \cdots b_m = 0$, so wäre nach Satz 22,

$$\left.\begin{array}{l} a_1 \cdots a_n b_1 \cdots *_i \cdots b_m , \\ \quad b_1 \cdots b_m \end{array}\right\} \rightarrow a_1 \cdots a_n b_1 \cdots b_{m-1} \text{ oder } b_1 \cdots *_i \cdots b_m .$$

Die beiden Ergebnisse widersprechen aber unseren Voraussetzungen;

$$a_1 \cdots a_n b_1 \cdots b_{m-1} ,\ a_1 \cdots a_n b_1 \cdots b_{m-1} \neq 0 \rightarrow W .$$
$$b_1 \cdots *_i \cdots b_m ,\ \overline{b_1 \cdots b_m} \rightarrow W .$$

Daher $a_1 \cdots a_n b_1 \cdots b_{m-1}$ oder $b_1 \cdots *_i \cdots b_m \rightarrow W$.

Daher muss $a_1 \cdots a_n b_1 \cdots *_i \cdots b_m \neq 0$, $(i = 1, \cdots, m)$.

Satz 31.

$$\left.\begin{array}{l} \overline{d_1 \cdots d_s a_1 \cdots a_n} , \\ \overline{d_1 \cdots d_s b_1 \cdots b_m} , \\ d_1 \cdots d_s a_1 \cdots a_{n-1} b_1 \cdots b_{m-1} \neq 0 \end{array}\right\} \rightarrow \left\{\begin{array}{l} d_1 \cdots *_i \cdots d_s a_1 \cdots *_j \cdots a_n b_1 \cdots b_m \neq 0 , \\ d_1 \cdots d_s a_1 \cdots *_j \cdots a_n b_1 \cdots *_k \cdots b_m \neq 0 , \\ d_1 \cdots *_i \cdots d_s a_1 \cdots a_n b_1 \cdots *_k \cdots b_m \neq 0 , \\ \quad (i = 1, \cdots, s;\ j = 1, \cdots, n; \\ \quad k = 1, \cdots, m). \end{array}\right.$$

Beweis:

$$\left.\begin{array}{l} \overline{d_1 \cdots d_s a_1 \cdots a_n} , \\ d_1 \cdots d_s a_1 \cdots a_{n-1} b_1 \cdots b_{m-1} \neq 0 \end{array}\right\} -(\text{Satz } 30) \rightarrow d_1 \cdots d_s a_1 \cdots * \cdots a_n b_1 \cdots b_{m-1} \neq 0 .$$

$$\left.\begin{array}{l} \overline{d_1 \cdots d_s b_1 \cdots b_m} , \\ d_1 \cdots d_s a_1 \cdots * \cdots a_n b_1 \cdots b_{m-1} \neq 0 \end{array}\right\} -(\text{Satz } 30) \rightarrow \left\{\begin{array}{l} d_1 \cdots * \cdots d_s a_1 \cdots * \cdots a_n b_1 \cdots b_m \neq 0, \\ d_1 \cdots d_s a_1 \cdots * \cdots a_n b_1 \cdots * \cdots b_m \neq 0. \end{array}\right.$$

$$\left.\begin{array}{l} \overline{d_1 \cdots d_s a_1 \cdots a_n} , \\ d_1 \cdots d_s a_1 \cdots * \cdots a_n b_1 \cdots * \cdots b_m \neq 0 \end{array}\right\} -(\text{Satz } 30) \rightarrow d_1 \cdots * \cdots d_s a_1 \cdots a_n b_1 \cdots * \cdots b_m \neq 0.$$

---

(15) $a_1 \cdots *_i \cdots a_{n+1}$ bedeutet $a_1 \cdots a_{i-1} a_{i+1} \cdots a_{n+1}$.

$$\therefore \left.\begin{array}{l}\overline{d_1 \cdots d_s a_1 \cdots a_n}\,, \\ \overline{d_1 \cdots d_s b_1 \cdots b_m}\,, \\ d_1 \cdots d_s a_1 \cdots a_{n-1} b_1 \cdots b_{m-1} \neq 0\end{array}\right\} \rightarrow \left\{\begin{array}{l} d_1 \cdots * \cdots d_s a_1 \cdots * \cdots a_n b_1 \cdots b_m \neq 0\,, \\ d_1 \cdots d_s a_1 \cdots * \cdots a_n b_1 \cdots * \cdots b_m \neq 0\,, \\ d_1 \cdots * \cdots d_s a_1 \cdots a_n b_1 \cdots * \cdots b_m \neq 0\,,\end{array}\right.$$

w. z. b. w.

Satz 32. $\overline{a_1 \cdots a_n x} \rightarrow Ey\,,\ \overline{a_1 \cdots a_{n-1} y}\,.$

Beweis: $a_1 \cdots a_n x$ —(Axiom 5)→ $Ey\,,\ a_1 \cdots a_{n-1} y\,,\ a_n xy\,.$

Nun wäre $a_1 \cdots *_i \cdots a_{n-1} y = 0$, so würde sich folgern,

$a_1 \cdots *_i \cdots a_{n-1} y\,,\ a_n xy$—(Zusatz 2 zum Satz 24)→ $a_1 \cdots *_i \cdots a_n x\,.$

$a_1 \cdots *_i \cdots a_n x\,,\ \overline{a_1 \cdots a_n x} \rightarrow W\,.$

So muss $\quad a_1 \cdots *_i \cdots a_{n-1} y \neq 0\,, \quad (i = 1, \cdots, n-1)\,.$

Daher $\quad \overline{a_1 \cdots a_{n-1} y}\,.$

Daher $\quad \overline{a_1 \cdots a_n x} \rightarrow Ey\,,\ \overline{a_1 \cdots a_{n-1} y}\,,$ w. z. b. w.

Es ergibt sich auch der folgende Satz durch Wiederholung der Operationen des Satzes 32.

Satz 33. $n \geqq m\,,\ \overline{a_1 \cdots a_n x} \rightarrow Ez\,,\ \overline{a_1 \cdots a_m z}\,.$

Satz 34. $\overline{a_1 \cdots a_n b_1 \cdots b_m} \rightarrow Ez\,,\ \overline{a_1 \cdots a_n z}\,,\ \overline{b_1 \cdots b_m z}\,.$

Beweis: $a_1 \cdots a_n b_1 \cdots b_m$ —(Satz 28)→ $Ez\,,\ a_1 \cdots a_n z\,,\ b_1 \cdots b_m z\,.$

Nun wäre $\quad a_1 \cdots *_i \cdots a_n z = 0$, so würde sich folgern,

$a_1 \cdots *_i \cdots a_n z\,,\ b_1 \cdots b_m z$ —(Zusatz 3 zum Satz 24)→ $a_1 \cdots *_i \cdots a_n b_1 \cdots b_m.$

$a_1 \cdots *_i \cdots a_n b_1 \cdots b_m\,,\ \overline{a_1 \cdots a_n b_1 \cdots b_m} \rightarrow W\,.$

So muss $\quad a_1 \cdots *_i \cdots a_n z \neq 0\,, \quad (i = 1, \cdots, n)\,.$

Daher $\quad \overline{a_1 \cdots a_n z}\,.$

Analog erhält man $\quad \overline{b_1 \cdots b_m z}\,.$

Daher $\quad \overline{a_1 \cdots a_n b_1 \cdots b_m} \rightarrow Ez\,,\ \overline{a_1 \cdots a_n z}\,,\ \overline{b_1 \cdots b_m z}\,,$

w. z. b. w.

Satz 35. $\left.\begin{array}{l}\overline{a_1 \cdots a_n z}\,,\ \overline{b_1 \cdots b_m z}\,, \\ a_1 \cdots a_{n-1} b_1 \cdots b_{m-1} z \neq 0\end{array}\right\} \rightarrow \overline{a_1 \cdots a_n b_1 \cdots b_m}\,.$

Beweis:

Einerseits ist

$a_1 \cdots a_n z,\ b_1 \cdots b_m z$ —(Zusatz 3 zum Satz 24)→ $a_1 \cdots a_n b_1 \cdots b_m\,.$

Anderseits ist

$$\left.\begin{array}{l}\overline{za_1 \cdots a_n}\,,\ \overline{zb_1 \cdots b_m}\,, \\ za_1 \cdots a_{n-1} b_1 \cdots b_{m-1} \neq 0\end{array}\right\} \text{—(Satz 31)}\rightarrow \left\{\begin{array}{l} a_1 \cdots * \cdots a_n b_1 \cdots b_m \neq 0, \\ a_1 \cdots a_n b_1 \cdots * \cdots b_m \neq 0\,.\end{array}\right.$$

Daher $\overline{a_1 \cdots a_n b_1 \cdots b_m}$.

Satz 36. $\left.\begin{array}{l} \overline{da_1 \cdots a_n x}\,, \\ \overline{dby}\,, \\ da_1 \cdots a_n b \neq 0 \end{array}\right\} \rightarrow Ez\,,\ \overline{da_1 \cdots a_n bz}\,.$

Beweis: $\left.\begin{array}{l} \overline{da_1 \cdots a_n x}\,,\ \overline{dby}\,, \\ da_1 \cdots a_n b \neq 0 \end{array}\right\} -(\text{Satz } 35) \rightarrow \overline{a_1 \cdots a_n bxy}$

$-(\text{Satz } 33) \rightarrow Ez\,,\ \overline{a_1 \cdots a_n yz}\,,\ \text{ und } a_1 \cdots a_n by \neq 0\,.$

$\left.\begin{array}{l} \overline{a_1 \cdots a_n yz}\,,\ \overline{dby}\,, \\ a_1 \cdots a_n by \neq 0 \end{array}\right\} -(\text{Satz } 35) \rightarrow \overline{da_1 \cdots a_n bz}\,.$

$\therefore \left.\begin{array}{l} \overline{da_1 \cdots a_n x}\,,\ \overline{dby}\,, \\ da_1 \cdots a_n b \neq 0 \end{array}\right\} \rightarrow Ez\,,\ \overline{da_1 \cdots a_n bz}\,,$ w. z. b. w.

Satz 37. $\left.\begin{array}{l} \overline{d_1 \cdots d_s a_1 \cdots a_n x}\,, \\ \overline{d_1 \cdots d_s b_1 \cdots b_m y}\,, \\ d_1 \cdots d_s a_1 \cdots a_n b_1 \cdots b_m \neq 0 \end{array}\right\} \rightarrow Ez\,,\ \overline{d_1 \cdots d_s a_1 \cdots a_n b_1 \cdots b_m z}\,.$

Beweis:

$$\overline{d_1 \cdots d_s b_1 \cdots b_m y} \ -(\text{Satz } 33) \rightarrow \left\{\begin{array}{l} Ey_1\,,\ \overline{d_1 b_1 y_1}\,, \\ Ey_2\,,\ \overline{d_1 b_2 y_2}\,, \\ \cdots\cdots\cdots\cdots\,, \\ Ey_m\,,\ \overline{d_1 b_m y_m}\,. \end{array}\right.$$

$\left.\begin{array}{l} \overline{d_1 \cdots d_s a_1 \cdots a_n x}\,,\ \overline{d_1 b_1 y_1}\,, \\ d_1 \cdots d_s a_1 \cdots a_n b_1 \neq 0 \end{array}\right\} -(\text{Satz } 36) \rightarrow Ez_1\,,\ \overline{d_1 \cdots d_s a_1 \cdots a_n b_1 z_1}\,.$

$\left.\begin{array}{l} \overline{d_1 \cdots d_s a_1 \cdots a_n b_1 z_1}\,,\ \overline{d_1 b_2 y_2}\,, \\ d_1 \cdots d_s a_1 \cdots a_n b_1 b_2 \neq 0 \end{array}\right\} -(\text{Satz } 36) \rightarrow Ez_2\,,\ \overline{d_1 \cdots d_s a_1 \cdots a_n b_1 b_2 z_2}\,.$

....

....

$\left.\begin{array}{l} \overline{d_1 \cdots d_s a_1 \cdots a_n b_1 \cdots b_{m-1} z_{m-1}}\,,\ \overline{d_1 b_m y_m}\,, \\ d_1 \cdots d_s a_1 \cdots a_n b_1 \cdots b_m \neq 0 \end{array}\right\} -(\text{Satz } 36) \rightarrow Ez,\ \overline{d_1 \cdots d_s a_1 \cdots a_n b_1 \cdots b_m z}\,.$

$\therefore \left.\begin{array}{l} \overline{d_1 \cdots d_s a_1 \cdots a_n x}, \\ \overline{d_1 \cdots d_s b_1 \cdots b_m y}, \\ d_1 \cdots d_s a_1 \cdots a_n b_1 \cdots b_m \neq 0 \end{array}\right\} \rightarrow Ez\,,\ \overline{d_1 \cdots d_s a_1 \cdots a_n b_1 \cdots b_m z}\,,$
w. z. b. w.

Satz 38. $\left.\begin{array}{l} \overline{a_1 \cdots a_n b_1 x}\,, \\ \overline{a_1 \cdots a_n b_1 b_2}\,, \\ a_1 \cdots a_n b_2 \neq 0 \end{array}\right\} \rightarrow Ez\,,\ \overline{a_1 \cdots a_n b_2 z}\,.$

Beweis: Wir beweisen dies durch die vollständige Induktion in Bezug auf n.

(i) Falls n gleich 0 ist, dann ergibt sich,

$$\overline{b_1 x}\,,\quad b_1 b_2 \;\rightarrow\; Eb_1\,,\quad \overline{b_2 b_1}\,.$$

(ii) Falls n gleich 1 ist, dann ergibt sich,

$$\left.\begin{array}{l} \overline{a_1 b_1 x}\,, \\ a_1 b_1 b_2\,, \\ a_1 b_2 \neq 0 \end{array}\right\} \rightarrow \left\{\begin{array}{l} \text{Sei } b_1 b_2 \neq 0 \rightarrow Eb_1\,,\ \overline{a_1 b_2 b_1}\,. \\ \text{Sei } b_1 b_2 = 0 \rightarrow Ex\,,\ \overline{a_1 b_2 x}\,. \end{array}\right.$$

(iii) Also genügt es zu zeigen, dass aus dem Fall $[n-1]$ der Fall $[n]$ folgt.

Sei nun $a_1 \cdots *_i \cdots a_n b_1 b_2 \neq 0$, $(i = 1, \cdots, n)$, so folgt,

$$\left.\begin{array}{r} a_1 \cdots *_i \cdots a_n b_1 b_2 \neq 0\,,\ (i = 1, \cdots, n), \\ a_1 \cdots a_n b_2 \neq 0 \qquad\qquad\quad , \\ \overline{a_1 \cdots a_n b_1 x} \rightarrow a_1 \cdots a_n b_1 \neq 0 \qquad\qquad\quad \end{array}\right\} \rightarrow Eb_1\,,\ \overline{a_1 \cdots a_n b_2 b_1}\,.$$

Sei mindestens für ein Element z. B. a_n $a_1 \cdots a_{n-1} b_1 b_2 = 0$, so folgt,

$$\left.\begin{array}{r} \overline{a_1 \cdots a_n b_1 x} -(\text{Satz } 32) \rightarrow Ex_1,\ \overline{a_1 \cdots a_{n-1} b_1 x_1}\,, \\ a_1 \cdots a_{n-1} b_1 b_2\,, \\ a_1 \cdots a_n b_2 \neq 0 \rightarrow\ a_1 \cdots a_{n-1} b_2 \neq 0 \end{array}\right\} - [n-1] \rightarrow Ey,\ \overline{a_1 \cdots a_{n-1} b_2 y}.$$

$$\left.\begin{array}{r} \overline{a_1 \cdots a_n b_1 x} -(\text{Satz } 32) \rightarrow Ex_2,\ \overline{a_1 \cdots a_{n-1} a_n x_2}, \\ \overline{a_1 \cdots a_{n-1} b_2 y}\,, \\ a_1 \cdots a_n b_2 \neq 0 \end{array}\right\} - (\text{Satz } 37) \rightarrow Ez,\ \overline{a_1 \cdots a_n b_2 z}\,.$$

$$\therefore\quad \left.\begin{array}{l} \overline{a_1 \cdots a_n b_1 x}\,, \\ a_1 \cdots a_n b_1 b_2\,, \\ a_1 \cdots a_n b_2 \neq 0 \end{array}\right\} \rightarrow Ez\,,\ \overline{a_1 \cdots a_n b_2 z}\,, \qquad \text{w. z. b. w.}$$

Definition VIII. Der lineare Raum vom Range n, der mindestens einen Primzyklus vom Range n in sich enthält, heisst der *lineare Primraum* vom Range n.

Satz 39. $\Re(a_1 \cdots a_n) \ni b_1, \cdots, b_{n+1}$, und $\overline{b_1 \cdots b_{n+1}}$

$$\rightarrow Ea_{n+1}\,,\ \Re(a_1 \cdots a_n) \ni a_{n+1}\,,\ \overline{a_1 \cdots a_{n+1}}\,.$$

Beweis:

Unter n Elementen $a_1, \cdots, a_n$, muss die Relation $b_1 \cdots b_{n-1} a_1 \neq 0$ mindestens für ein Element z. B. a_1 bestehen.

$$\therefore \left.\begin{array}{l}\overline{b_1 \cdots b_{n-1} b_n b_{n+1}},\\ b_1 \cdots b_{n-1} b_n a_1,\\ b_1 \cdots b_{n-1} a_1 \neq 0\end{array}\right\} -(\text{Satz } 38) \rightarrow Ez_1,\ \overline{b_1 \cdots b_{n-1} a_1 z_1}.$$

Unter $n-1$ Elementen $a_2, \cdots, a_n$, muss die Relation $b_1 \cdots b_{n-2} a_1 a_2 \neq 0$ mindestens für ein Element z. B. a_2 bestehen.

$$\therefore \left.\begin{array}{l}\overline{b_1 \cdots b_{n-2} a_1 b_{n-1} z_1},\\ b_1 \cdots b_{n-2} a_1 b_{n-1} a_2,\\ b_1 \cdots b_{n-2} a_1 a_2 \neq 0\end{array}\right\} -(\text{Satz } 38) \rightarrow Ez_2,\ \overline{b_1 \cdots b_{n-2} a_1 a_2 z_2}.$$

$$\cdots \quad \cdots \quad \cdots \quad \cdots \quad \cdots \quad \cdots$$

$$\cdots \quad \cdots \quad \cdots \quad \cdots \quad \cdots \quad \cdots$$

Zwischen 2 Elementen a_{n-1}, a_n, muss die Relation $b_1 a_1 \cdots a_{n-2} a_{n-1} \neq 0$ mindestens für ein Element z. B. a_{n-1} bestehen.

$$\therefore \left.\begin{array}{l}\overline{b_1 a_1 \cdots a_{n-2} b_2 z_{n-2}},\\ b_1 a_1 \cdots a_{n-2} b_2 a_{n-1},\\ b_1 a_1 \cdots a_{n-2} a_{n-1} \neq 0\end{array}\right\} -(\text{Satz } 38) \rightarrow Ez_{n-1},\ \overline{b_1 a_1 \cdots a_{n-1} z_{n-1}}.$$

Für 1 Element a_n, muss die Relation $a_1 \cdots a_{n-1} a_n \neq 0$ mindestens für ein Element z. B. a_n bestehen.

$$\therefore \left.\begin{array}{l}\overline{a_1 \cdots a_{n-1} b_1 z_{n-1}},\\ a_1 \cdots a_{n-1} b_1 a_n,\\ a_1 \cdots a_{n-1} a_n \neq 0\end{array}\right\} -(\text{Satz } 38) \rightarrow Ea_{n+1},\ \overline{a_1 \cdots a_n a_{n+1}}.$$

Daher $\Re(a_1 \cdots a_n) \ni b_1, \cdots, b_{n+1}$, und $\overline{b_1 \cdots b_{n+1}}$

$$\rightarrow Ea_{n+1},\ \overline{a_1 \cdots a_{n+1}},$$ w. z. b. w.

Satz 40. Jeder in einem linearen Primraume enthaltenen lineare Raum ist stets ein linearer Primraum.

Beweis: Seien die Basen vom enthaltenen linearen Raume bzw. enthaltenden linearen Raume je $a_1 \cdots a_m$, bzw. $a_1 \cdots a_m \cdots a_n$, so ist $\Re(a_1 \cdots a_m \cdots a_n)$ wegen unserer Voraussetzung ein linearer Primraum.

Also folgt nach Satz 39,

$$Ex,\ \Re(a_1 \cdots a_m \cdots a_n) \ni x,\ \overline{a_1 \cdots a_m \cdots a_n x}.$$

$$\overline{a_1 \cdots a_m \cdots a_n x} -(\text{Satz } 33) \rightarrow Ey,\ \overline{a_1 \cdots a_m y}.$$

$$\therefore\ Ey,\ \Re(a_1 \cdots a_m) \ni y,\ \overline{a_1 \cdots a_m y}.$$

Daher ist $\Re(a_1 \cdots a_m)$ auch ein linearer Primraum.

Satz 41. Der lineare Raum derart, dass alle darin liegenden linearen Räume vom Range 2 stets lineare Primräume sind, ist auch ein linearer Primraum.

Beweis : Sei die Basis des linearen Raumes $a_1 \cdots a_n$, so folgt nach unserer Voraussetzung,

$$Ex_1,\ \overline{a_1a_2x_1}\,;\ Ex_2,\ \overline{a_2a_3x_2}\,;\ \cdots\,;\ Ex_{n-1},\ \overline{a_{n-1}a_nx_{n-1}}\,.$$

Dann folgt,

$$\overline{a_1a_2x_1},\ \overline{a_2a_3x_2},\ a_1a_2a_3 \neq 0\ -(\text{Satz }37)\rightarrow\ Ey_3,\ \overline{a_1a_2a_3y_3}\,.$$

$$\overline{a_1a_2a_3y_3},\ \overline{a_3a_4x_3},\ a_1a_2a_3a_4 \neq 0\ -(\text{Satz }37)\rightarrow Ey_4,\ \overline{a_1a_2a_3a_4y_4}\,.\ .$$

$$\cdots \quad \cdots \quad \cdots \quad \cdots \quad \cdots \quad \cdots$$

$$\cdots \quad \cdots \quad \cdots \quad \cdots \quad \cdots \quad \cdots$$

$$\overline{a_1 \cdots a_{n-1}y_{n-1}},\ \overline{a_{n-1}a_nx_{n-1}},\ a_1 \cdots a_n \neq 0\ -(\text{Satz }37)\rightarrow Ey_n,\ \overline{a_1 \cdots a_ny_n}.$$

$$\therefore\ Ey_n,\ \mathfrak{R}(a_1 \cdots a_n) \ni y_n,\ \overline{a_1 \cdots a_ny_n}\,.$$

Daher ist $\mathfrak{R}(a_1 \cdots a_n)$ ein linearer Primraum.

§ 4. Die Zerlegung des linearen Raumes.

Definition IX. Wir sagen, dass der lineare Raum $\mathfrak{R}$ sich in die *direkte Summe* von k linearen Räumen $\mathfrak{R}_1, \cdots, \mathfrak{R}_k$ *zerlegt*, und schreiben in Zeichen

$$\mathfrak{R} = \mathfrak{R}_1 \overset{*}{+} \cdots \overset{*}{+} \mathfrak{R}_k\,,$$

wenn folgende zwei Bedingungen erfüllt sind :

$$\mathfrak{R} = \mathfrak{R}_1 + \cdots + \mathfrak{R}_k\,,$$

$$\text{Rang}\,\mathfrak{R} = \text{Rang}\,\mathfrak{R}_1 + \cdots + \text{Rang}\,\mathfrak{R}_k\,.$$

Ferner in diesem Falle, heissen die k linearen Räume $\mathfrak{R}_1, \cdots, \mathfrak{R}_k$ sich *mit einander trennen*, und wir bezeichnen dies in Zeichen

$$\mathfrak{R}_1 * \cdots * \mathfrak{R}_k\,.$$

Dann folgt leicht ;

V_2. 4. (i) $\mathfrak{R}_1 * \cdots * \mathfrak{R}_i * \cdots * \mathfrak{R}_k \rightarrow \mathfrak{R}_i * \cdots * \mathfrak{R}_1 * \cdots * \mathfrak{R}_k\,.$

(ii) $\mathfrak{R}_1 \overset{*}{+} \cdots \overset{*}{+} \mathfrak{R}_i \overset{*}{+} \cdots \overset{*}{+} \mathfrak{R}_k = \mathfrak{R}_i \overset{*}{+} \cdots \overset{*}{+} \mathfrak{R}_1 \overset{*}{+} \cdots \overset{*}{+} \mathfrak{R}_k\,.$

Satz 42. $\mathfrak{R} = \mathfrak{R}_1(a_1 \cdots a_n) \overset{*}{+} \cdots \overset{*}{+} \mathfrak{R}_k(l_1 \cdots l_m)$
$\rightarrow \mathfrak{R} = \mathfrak{R}(a_1 \cdots a_n \cdots l_1 \cdots l_m)$.

Beweis: $\mathfrak{R} = \mathfrak{R}_1 \overset{*}{+} \cdots \overset{*}{+} \mathfrak{R}_k \rightarrow \mathfrak{R} = \mathfrak{B}(\mathfrak{R}_1, \cdots, \mathfrak{R}_k)$
—(Satz 17)→ $\mathfrak{R} = \mathfrak{B}(a_1 \cdots a_n \cdots l_1 \cdots l_m)$.

Rang $\mathfrak{R}$ = Rang $\mathfrak{R}_1 + \cdots +$ Rang $\mathfrak{R}_k = n + \cdots + m$.

$\therefore$ Rang $\mathfrak{B}(a_1 \cdots a_n \cdots l_1 \cdots l_m) = n + \cdots + m$.

$\therefore$ $a_1 \cdots a_n \cdots l_1 \cdots l_m \neq 0$.

$\therefore$ $\mathfrak{R} = \mathfrak{R}(a_1 \cdots a_n \cdots l_1 \cdots l_m)$, w. z. b. w.

Zusatz. $\mathfrak{R}(a_1 \cdots a_n) * \cdots * \mathfrak{R}(l_1 \cdots l_m) \rightarrow a_1 \cdots a_n \cdots l_1 \cdots l_m \neq 0$.

Satz 43. $\left.\begin{array}{l} \mathfrak{R}_1(a_1 \cdots a_n) * \cdots * \mathfrak{R}_k(l_1 \cdots l_m), \\ a_1 \cdots a_n x \neq 0, \cdots, l_1 \cdots l_m x \neq 0 \end{array}\right\} \rightarrow a_1 \cdots a_n \cdots l_1 \cdots l_m x \neq 0.$

Beweis: Nach Satz 42 folgt,

$$\mathfrak{R}(a_1 \cdots a_n \cdots l_1 \cdots l_m) = \mathfrak{R}_1(a_1 \cdots a_n) \overset{*}{+} \cdots \overset{*}{+} \mathfrak{R}_k(l_1 \cdots l_m).$$

$$\left.\begin{array}{l} \mathfrak{R}(a_1 \cdots a_n \cdots l_1 \cdots l_m) = \mathfrak{R}_1 + \cdots + \mathfrak{R}_k, \\ a_1 \cdots a_n x \neq 0 \rightarrow \mathfrak{R}_1(a_1 \cdots a_n) \not\ni x, \\ \cdots \quad \cdots \quad \cdots, \\ l_1 \cdots l_m x \neq 0 \rightarrow \mathfrak{R}_k(l_1 \cdots l_m) \not\ni x \end{array}\right\} \begin{array}{l} \rightarrow \mathfrak{R}(a_1 \cdots a_n \cdots l_1 \cdots l_m) \not\ni x \\ \rightarrow a_1 \cdots a_n \cdots l_1 \cdots l_m x \neq 0. \end{array}$$

Zusatz. $\left.\begin{array}{l} \mathfrak{R}(a_1 \cdots a_n) * \cdots * \mathfrak{R}(l_1 \cdots l_m), \\ a_1 \cdots a_n \cdots l_1 \cdots l_m x \end{array}\right\} \rightarrow$ Es gilt genau eine der Aussagen $\{a_1 \cdots a_n x, \cdots, \text{ und } l_1 \cdots l_m x\}$.

Satz 44. $\mathfrak{R}_1 * \cdots * \mathfrak{R}_i * \cdots * \mathfrak{R}_k \rightarrow \mathfrak{B}(\mathfrak{R}_1 \cdots \mathfrak{R}_i) = \mathfrak{R}_1 + \cdots + \mathfrak{R}_i$.

Beweis: Seien die Basen von $\mathfrak{R}_1, \cdots, \mathfrak{R}_i \cdots, \mathfrak{R}_k$ bzw. je $a_1 \cdots a_n$, $\cdots$, $b_1 \cdots b_m$, $\cdots$, $l_1 \cdots l_s$, so folgt,

$\mathfrak{R}_1(a_1 \cdots a_n) * \cdots * \mathfrak{R}_i(b_1 \cdots b_m) * \cdots * \mathfrak{R}_k(l_1 \cdots l_s)$ — (Zusatz zum Satz 42)—
$\rightarrow a_1 \cdots a_n \cdots b_1 \cdots b_m \cdots l_1 \cdots l_s \neq 0 \rightarrow a_1 \cdots a_n \cdots b_1 \cdots b_m \neq 0$
$\rightarrow \mathfrak{B}(\mathfrak{R}_1 \cdots \mathfrak{R}_i) = \mathfrak{R}(a_1 \cdots a_n \cdots b_1 \cdots b_m)$.

Daher folgt,

$\mathfrak{B}(\mathfrak{R}_1 \cdots \mathfrak{R}_i) \ni x \rightarrow \mathfrak{R}(a_1 \cdots a_n \cdots b_1 \cdots b_m) \ni x \rightarrow a_1 \cdots a_n \cdots b_1 \cdots b_m x$
—(Zusatz zum Satz 43)→ $a_1 \cdots a_n x$ od. ... od. $b_1 \cdots b_m x$
$\rightarrow \mathfrak{R}_1 + \cdots + \mathfrak{R}_i \ni x$.

Daher $\mathfrak{B}(\mathfrak{N}_1 \cdots \mathfrak{N}_i) \leqq \mathfrak{N}_1 + \cdots + \mathfrak{N}_i$.

Anderseits, $\mathfrak{B}(\mathfrak{N}_1 \cdots \mathfrak{N}_i) \geqq \mathfrak{N}_1 + \cdots + \mathfrak{N}_i$.

Daher $\mathfrak{B}(\mathfrak{N}_1 \cdots \mathfrak{N}_i) = \mathfrak{N}_1 + \cdots + \mathfrak{N}_i$, w. z. b. w.

Zusatz. $\mathfrak{N}_1(a_1 \cdots a_n) * \cdots * \mathfrak{N}_i(b_1 \cdots b_m) * \cdots * \mathfrak{N}_k(l_1 \cdots l_s)$

$$\rightarrow \mathfrak{B}\left\{\mathfrak{N}_1(a_1 \cdots a_n), \cdots, \mathfrak{N}_i(b_1 \cdots b_m)\right\} = \mathfrak{N}(a_1 \cdots a_n \cdots b_1 \cdots b_m).$$

Satz 45. $\mathfrak{N}_1 * \cdots * \mathfrak{N}_i * \mathfrak{N}_{i+1} * \cdots * \mathfrak{N}_k$

$$\rightarrow \mathfrak{D}\left\{\mathfrak{B}(\mathfrak{N}_1 \cdots \mathfrak{N}_i), \mathfrak{B}(\mathfrak{N}_{i+1} \cdots \mathfrak{N}_k)\right\} = \mathfrak{N}.$$

Beweis: Seien die Basen von $\mathfrak{N}_1, \cdots, \mathfrak{N}_i, \mathfrak{N}_{i+1}, \cdots, \mathfrak{N}_k$ bzw. je $a_1 \cdots a_n, \ldots, b_1 \cdots b_m, c_1 \cdots c_r, \ldots, d_1 \cdots d_s$, so folgt nach Zusatz zum Satz 44,

$$\mathfrak{B}(\mathfrak{N}_1 \cdots \mathfrak{N}_i) = \mathfrak{N}(a_1 \cdots a_n \cdots b_1 \cdots b_m),$$

$$\mathfrak{B}(\mathfrak{N}_{i+1} \cdots \mathfrak{N}_k) = \mathfrak{N}(c_1 \cdots c_r \cdots d_1 \cdots d_s).$$

Daher

$$\mathfrak{D}\left\{\begin{matrix}\mathfrak{B}(\mathfrak{N}_1 \cdots \mathfrak{N}_i), \\ \mathfrak{B}(\mathfrak{N}_{i+1} \cdots \mathfrak{N}_k)\end{matrix}\right\} \ni x \rightarrow \mathfrak{D}\left\{\begin{matrix}\mathfrak{N}(a_1 \cdots a_n \cdots b_1 \cdots b_m), \\ \mathfrak{N}(c_1 \cdots c_r \cdots d_1 \cdots d_s)\end{matrix}\right\} \ni x$$

$$\rightarrow \left.\begin{matrix}a_1 \cdots a_n \cdots b_1 \cdots b_m x, \\ c_1 \cdots c_r \cdots d_1 \cdots d_s x\end{matrix}\right\} \rightarrow a_1 \cdots a_n \cdots b_1 \cdots b_m c_1 \cdots c_r \cdots d_1 \cdots d_s$$

$$-(\text{Zusatz zum Satz 42}) \rightarrow W.$$

Also ist $\mathfrak{D}\left\{\mathfrak{B}(\mathfrak{N}_1 \cdots \mathfrak{N}_i), \mathfrak{B}(\mathfrak{N}_{i+1} \cdots \mathfrak{N}_k)\right\} = \mathfrak{N}$.

Satz 46. $\mathfrak{N}_1 * \cdots * \mathfrak{N}_i * \cdots * \mathfrak{N}_k \rightarrow \mathfrak{N}_1 * \cdots * \mathfrak{N}_i$.

Beweis:

$$\mathfrak{N}_1 * \cdots * \mathfrak{N}_i * \cdots * \mathfrak{N}_k \;-(\text{Satz 44}) \rightarrow \mathfrak{B}(\mathfrak{N}_1 \cdots \mathfrak{N}_i) = \mathfrak{N}_1 + \cdots + \mathfrak{N}_i.$$

$$\begin{aligned}\text{Rang}\,\mathfrak{N}_1 + \cdots + \text{Rang}\,\mathfrak{N}_k &= \text{Rang}\,(\mathfrak{N}_1 + \cdots + \mathfrak{N}_k) \\ &\leqq \text{Rang}\,(\mathfrak{N}_1 + \cdots + \mathfrak{N}_i) + \text{Rang}\,\mathfrak{N}_{i+1} + \cdots + \text{Rang}\,\mathfrak{N}_k \\ &= \text{Rang}\,\mathfrak{B}(\mathfrak{N}_1 \cdots \mathfrak{N}_i) + \text{Rang}\,\mathfrak{N}_{i+1} + \cdots + \text{Rang}\,\mathfrak{N}_k \\ &\leqq \text{Rang}\,\mathfrak{N}_1 + \cdots + \text{Rang}\,\mathfrak{N}_k.\end{aligned}$$

$\therefore$ Rang $\mathfrak{B}(\mathfrak{N}_1 \cdots \mathfrak{N}_i) =$ Rang $\mathfrak{N}_1 + \cdots +$ Rang $\mathfrak{N}_i$.

$\therefore$ $\mathfrak{N}_1 * \cdots * \mathfrak{N}_i$, w. z. b. w.

Satz 47. $\mathfrak{N} = \mathfrak{N}_1 \overset{*}{+} \cdots \overset{*}{+} \mathfrak{N}_i \overset{*}{+} \cdots \overset{*}{+} \mathfrak{N}_k,\ \mathfrak{N}_i = \mathfrak{N}_{i1} \overset{*}{+} \cdots \overset{*}{+} \mathfrak{N}_{is}$

$$\rightarrow \mathfrak{N} = \mathfrak{N}_1 \overset{*}{+} \cdots \overset{*}{+} \mathfrak{N}_{i1} \overset{*}{+} \cdots \overset{*}{+} \mathfrak{N}_{is} \overset{*}{+} \cdots \overset{*}{+} \mathfrak{N}_k.$$

Beweis:

$$\left.\begin{array}{l}\mathfrak{R} = \mathfrak{R}_1 + \cdots + \mathfrak{R}_i + \cdots + \mathfrak{R}_k,\\ \mathfrak{R}_i = \mathfrak{R}_{i1} + \cdots + \mathfrak{R}_{is}\end{array}\right\} \rightarrow \mathfrak{R} = \mathfrak{R}_1 + \cdots + \mathfrak{R}_{i1} + \cdots + \mathfrak{R}_{is} + \cdots + \mathfrak{R}_k.$$

$$\left.\begin{array}{l}\text{Rang}\,\mathfrak{R} = \text{Rang}\,\mathfrak{R}_1 + \cdots + \text{Rang}\,\mathfrak{R}_i + \cdots + \text{Rang}\,\mathfrak{R}_k,\\ \text{Rang}\,\mathfrak{R}_i = \text{Rang}\,\mathfrak{R}_{i1} + \cdots + \text{Rang}\,\mathfrak{R}_{is}\end{array}\right\} -$$

$$\rightarrow \text{Rang}\,\mathfrak{R} = \text{Rang}\,\mathfrak{R}_1 + \cdots + \text{Rang}\,\mathfrak{R}_{i1} + \cdots + \text{Rang}\,\mathfrak{R}_{is} + \cdots + \text{Rang}\,\mathfrak{R}_k.$$

$$\therefore\quad \mathfrak{R} = \mathfrak{R}_1 \overset{*}{+} \cdots \overset{*}{+} \mathfrak{R}_{i1} \overset{*}{+} \cdots \overset{*}{+} \mathfrak{R}_{is} \overset{*}{+} \cdots \overset{*}{+} \mathfrak{R}_k, \qquad \text{w. z. b. w.}$$

Satz 48. $\mathfrak{R} = \mathfrak{R}_1 \overset{*}{+} \cdots \overset{*}{+} \mathfrak{R}_{i1} \overset{*}{+} \cdots \overset{*}{+} \mathfrak{R}_{is} \overset{*}{+} \cdots \overset{*}{+} \mathfrak{R}_k$

$$\rightarrow \mathfrak{R} = \mathfrak{R}_1 \overset{*}{+} \cdots \overset{*}{+} (\mathfrak{R}_{i1} \overset{*}{+} \cdots \overset{*}{+} \mathfrak{R}_{is}) \overset{*}{+} \cdots \overset{*}{+} \mathfrak{R}_k.$$

Beweis:

$$\mathfrak{R} = \mathfrak{R}_1 + \cdots + \mathfrak{R}_{i1} + \cdots + \mathfrak{R}_{is} + \cdots + \mathfrak{R}_k$$

$$\rightarrow \mathfrak{R} = \mathfrak{R}_1 + \cdots + (\mathfrak{R}_{i1} + \cdots + \mathfrak{R}_{is}) + \cdots + \mathfrak{R}_k.$$

$$\mathfrak{R}_1 * \cdots * \mathfrak{R}_{i1} * \cdots * \mathfrak{R}_{is} * \cdots * \mathfrak{R}_k \text{ —(Satz 46)}\rightarrow \mathfrak{R}_{i1} * \cdots * \mathfrak{R}_{is}$$

$$\rightarrow \mathfrak{R}_{i1} + \cdots \overset{*}{+} \mathfrak{R}_{is} \overset{*}{=} \mathfrak{R}_{i1} + \cdots + \mathfrak{R}_{is} \rightarrow$$

$$\text{Rang}\,(\mathfrak{R}_{i1} + \cdots + \mathfrak{R}_{is}) = \text{Rang}\,(\mathfrak{R}_{i1} \overset{*}{+} \cdots \overset{*}{+} \mathfrak{R}_{is}) = \text{Rang}\,\mathfrak{R}_{i1} + \cdots + \text{Rang}\,\mathfrak{R}_{is}.$$

$$\begin{aligned}\therefore\quad \text{Rang}\,\mathfrak{R} &= \text{Rang}\,\mathfrak{R}_1 + \cdots + \text{Rang}\,\mathfrak{R}_{i1} + \cdots + \text{Rang}\,\mathfrak{R}_{is} + \cdots + \text{Rang}\,\mathfrak{R}_k\\ &= \text{Rang}\,\mathfrak{R}_1 + \cdots + \text{Rang}\,(\mathfrak{R}_{i1} + \cdots + \mathfrak{R}_{is}) + \cdots + \text{Rang}\,\mathfrak{R}_k.\end{aligned}$$

$$\begin{aligned}\therefore\quad \mathfrak{R} &= \mathfrak{R}_1 \overset{*}{+} \cdots \overset{*}{+} (\mathfrak{R}_{i1} + \cdots + \mathfrak{R}_{is}) \overset{*}{+} \cdots \overset{*}{+} \mathfrak{R}_k\\ &= \mathfrak{R}_1 \overset{*}{+} \cdots \overset{*}{+} (\mathfrak{R}_{i1} \overset{*}{+} \cdots \overset{*}{+} \mathfrak{R}_{is}) \overset{*}{+} \cdots \overset{*}{+} \mathfrak{R}_k.\end{aligned}$$

Satz 49. $\left.\begin{array}{l}\mathfrak{R} = \mathfrak{R}_1 + \cdots + \mathfrak{R}_k,\\ \mathfrak{D}\left\{\mathfrak{R}(\mathfrak{R}_1 \cdots \mathfrak{R}_{i-1}),\ \mathfrak{R}_i\right\} = \mathfrak{N},\ i = 2, \cdots, k\end{array}\right\} \rightarrow \mathfrak{R} = \mathfrak{R}_1 \overset{*}{+} \cdots \overset{*}{+} \mathfrak{R}_k.$

Beweis:

$$\mathfrak{D}\left\{\mathfrak{B}(\mathfrak{R}_1 \cdots \mathfrak{R}_{i-1}),\ \mathfrak{R}_i\right\} = \mathfrak{N} \text{ —(Satz 29)}\rightarrow$$

$$\text{Rang}\,\mathfrak{B}\left\{\mathfrak{B}(\mathfrak{R}_1 \cdots \mathfrak{R}_{i-1}),\ \mathfrak{R}_i\right\} = \text{Rang}\,\mathfrak{B}(\mathfrak{R}_1 \cdots \mathfrak{R}_{i-1}) + \text{Rang}\,\mathfrak{R}_i.$$

$$\therefore\quad \text{Rang}\,(\mathfrak{R}_1 + \cdots + \mathfrak{R}_i) = \text{Rang}\,(\mathfrak{R}_1 + \cdots + \mathfrak{R}_{i-1}) + \text{Rang}\,\mathfrak{R}_i.$$

$$\begin{aligned}\therefore\quad \text{Rang}\,(\mathfrak{R}_1 + \cdots + \mathfrak{R}_k) &= \text{Rang}\,(\mathfrak{R}_1 + \cdots + \mathfrak{R}_{k-1}) + \text{Rang}\,\mathfrak{R}_k,\\ \text{Rang}\,(\mathfrak{R}_1 + \cdots + \mathfrak{R}_{k-1}) &= \text{Rang}\,(\mathfrak{R}_1 + \cdots + \mathfrak{R}_{k-2}) + \text{Rang}\,\mathfrak{R}_{k-1},\\ \cdots\quad \cdots\quad &\cdots\quad \cdots\quad \cdots\\ \text{Rang}\,(\mathfrak{R}_1 + \mathfrak{R}_2) &= \text{Rang}\,\mathfrak{R}_1 + \text{Rang}\,\mathfrak{R}_2.\end{aligned}$$

$$\therefore\quad \text{Rang}\,(\mathfrak{R}_1 + \cdots + \mathfrak{R}_k) = \text{Rang}\,\mathfrak{R}_1 + \cdots + \text{Rang}\,\mathfrak{R}_k.$$

$$\therefore \qquad \operatorname{Rang} \mathfrak{R} = \operatorname{Rang} \mathfrak{R}_1 + \cdots + \operatorname{Rang} \mathfrak{R}_k .$$

$$\therefore \qquad \mathfrak{R} = \mathfrak{R}_1 \overset{*}{+} \cdots \overset{*}{+} \mathfrak{R}_k , \qquad \text{w. z. b. w.}$$

Definition X. Derjenige lineare Raum $\mathfrak{R}(a_1 \cdots a_n)$, der sich unmöglich in die direkte Summe von mehr als zwei linearen Räumen zerlegt, heisst der *unzerlegbare lineare Raum,* und wird bezeichnet mit: $\overline{\mathfrak{R}}(a_1 \cdots a_n)$.

Satz 50.

$$\left.\begin{array}{l} a_1 \cdots a_n b_1 \cdots b_m c_1 \cdots c_k z, \\ a_1 \cdots a_n b_1 \cdots b_m z \neq 0 , \\ b_1 \cdots b_m c_1 \cdots c_k z \neq 0 , \\ c_1 \cdots c_k a_1 \cdots a_n z \neq 0 \end{array}\right\} \rightarrow \left\{\begin{array}{l} E z_1,\ a_1 \cdots a_n b_1 \cdots b_m z_1,\ a_1 \cdots a_n z_1 \neq 0,\ b_1 \cdots b_m z_1 \neq 0, \\ E z_2,\ b_1 \cdots b_m c_1 \cdots c_k z_2,\ b_1 \cdots b_m z_2 \neq 0,\ c_1 \cdots c_k z_2 \neq 0, \\ E z_3,\ c_1 \cdots c_k a_1 \cdots a_n z_3,\ c_1 \cdots c_k z_3 \neq 0,\ a_1 \cdots a_n z_3 \neq 0. \end{array}\right.$$

Beweis: $a_1 \cdots a_n b_1 \cdots b_m c_1 \cdots c_k z \rightarrow E z_1,\ a_1 \cdots a_n b_1 \cdots b_m z_1,\ c_1 \cdots c_k z z_1 .$

$c_1 \cdots c_k a_1 \cdots a_n z \neq 0 ,\ c_1 \cdots c_k z z_1 \rightarrow a_1 \cdots a_n z_1 \neq 0 .$

$b_1 \cdots b_m c_1 \cdots c_k z \neq 0 ,\ c_1 \cdots c_k z z_1 \rightarrow b_1 \cdots b_m z_1 \neq 0 .$

Daher folgt $E z_1,\ a_1 \cdots a_n b_1 \cdots b_m z_1 ,\ a_1 \cdots a_n z_1 \neq 0 ,\ b_1 \cdots b_m z_1 \neq 0.$

Ebenso folgt $E z_2 ,\ b_1 \cdots b_m c_1 \cdots c_k z_2 ,\ b_1 \cdots b_m z_2 \neq 0 ,\ c_1 \cdots c_k z_2 \neq 0 ,$

$E z_3 ,\ c_1 \cdots c_k a_1 \cdots a_n z_3 ,\ c_1 \cdots c_k z_3 \neq 0 ,\ a_1 \cdots a_n z_3 \neq 0.$

Satz 51.

$$\left.\begin{array}{l} a b_1 \cdots b_m c_1 \cdots c_k \neq 0 , \\ a b_1 \cdots b_m z_1 ,\ a z_1 \neq 0 ,\ b_1 \cdots b_m z_1 \neq 0, \\ a c_1 \cdots c_k z_2 ,\ a z_2 \neq 0 ,\ c_1 \cdots c_k z_2 \neq 0 \end{array}\right\} \rightarrow \left\{\begin{array}{l} E z,\ a b_1 \cdots b_m c_1 \cdots c_k z , \\ a b_1 \cdots b_m z \neq 0,\ a c_1 \cdots c_k z \neq 0 . \\ b_1 \cdots b_m c_1 \cdots c_k z \neq 0 . \end{array}\right.$$

Beweis: $a b_1 \cdots b_m z_1 ,\ a c_1 \cdots c_k z_2 \rightarrow b_1 \cdots b_m z_1 c_1 \cdots c_k z_2$

$\rightarrow E z ,\ b_1 \cdots b_m z_2 z ,\ c_1 \cdots c_k z_1 z .$

$a b_1 \cdots b_m z_1 ,\ c_1 \cdots c_k z_1 z \rightarrow a b_1 \cdots b_m c_1 \cdots c_k z .$

$a b_1 \cdots b_m c_1 \cdots c_k \neq 0 ,\ b_1 \cdots b_m z_1 \neq 0 ,\ a b_1 \cdots b_m z_1 \rightarrow b_1 \cdots b_m c_1 \cdots c_k z_1 \neq 0 .$

$b_1 \cdots b_m c_1 \cdots c_k z_1 \neq 0 ,\ c_1 \cdots c_k z_1 z \rightarrow b_1 \cdots b_m z \neq 0 .$

$a b_1 \cdots b_m c_1 \cdots c_k \neq 0 ,\ a c_1 \cdots c_k z_2 ,\ a z_2 \neq 0 \rightarrow a b_1 \cdots b_m z_2 \neq 0 .$

$b_1 \cdots b_m z \neq 0 ,\ a b_1 \cdots b_m z_2 \neq 0 ,\ b_1 \cdots b_m z_2 z \rightarrow a b_1 \cdots b_m z \neq 0 .$

$a b_1 \cdots b_m c_1 \cdots c_k \neq 0,\ c_1 \cdots c_k z_2 \neq 0,\ a c_1 \cdots c_k z_2 \rightarrow b_1 \cdots b_m c_1 \cdots c_k z_2 \neq 0.$

$b_1 \cdots b_m c_1 \cdots c_k z_2 \neq 0 ,\ b_1 \cdots b_m z_2 z \rightarrow c_1 \cdots c_k z \neq 0 .$

$a b_1 \cdots b_m c_1 \cdots c_k \neq 0,\ a b_1 \cdots b_m z_1 ,\ a z_1 \neq 0 \rightarrow a c_1 \cdots c_k z_1 \neq 0 .$

$c_1 \cdots c_k z \neq 0 ,\ a c_1 \cdots c_k z_1 \neq 0 ,\ c_1 \cdots c_k z_1 z \rightarrow a c_1 \cdots c_k z \neq 0 .$

$b_1 \cdots b_m c_1 \cdots c_k z_2 \neq 0 ,\ b_1 \cdots b_m z \neq 0 ;\ b_1 \cdots b_m z_2 z \rightarrow b_1 \cdots b_m c_1 \cdots c_k z \neq 0 .$

$\therefore\quad Ez,\ ab_1\cdots b_m c_1\cdots c_k z,\ ab_1\cdots b_m z \neq 0,\ ac_1\cdots c_k z\neq 0,\ b_1\cdots b_m c_1\cdots c_k z\neq 0,$ w. z. b. w.

Satz 52. $\overline{\mathfrak{R}}^s > \mathfrak{R}^{s-1} \rightarrow \overline{\mathfrak{R}}^{s-1}$.

Beweis: Seien $\overline{\mathfrak{R}}^s = \overline{\mathfrak{R}}(ab_1\cdots b_m c_1\cdots c_k),\ \mathfrak{R}^{s-1} = \mathfrak{R}(b_1\cdots b_m c_1\cdots c_k)$, und wäre vorläufig $\mathfrak{R}(b_1\cdots b_m c_1\cdots c_k) = \mathfrak{R}(b_1\cdots b_m)\overset{*}{+}\mathfrak{R}(c_1\cdots c_k)$, so wäre,

$\overline{\mathfrak{R}}(ab_1\cdots b_m c_1\cdots c_k)\rightarrow$

$Ex,\ \mathfrak{R}(ab_1\cdots b_m c_1\cdots c_k)\ni x,\ ab_1\cdots b_m x\neq 0,\ c_1\cdots c_k x\neq 0,$

$Ey,\ \mathfrak{R}(ab_1\cdots b_m c_1\cdots c_k)\ni y,\ ac_1\cdots c_k y\neq 0,\ b_1\cdots b_m y\neq 0.$

$$\left.\begin{array}{l}\mathfrak{R}(b_1\cdots b_m) * \mathfrak{R}(c_1\cdots c_k),\\ b_1\cdots b_m x\neq 0,\ c_1\cdots c_k x\neq 0,\\ b_1\cdots b_m y\neq 0,\ c_1\cdots c_k y\neq 0\end{array}\right\}-(\text{Satz 43})\rightarrow\left\{\begin{array}{l}b_1\cdots b_m c_1\cdots c_k x\neq 0,\\ b_1\cdots b_m c_1\cdots c_k y\neq 0.\end{array}\right.$$

Nun wäre $ac_1\cdots c_k x\neq 0$, so wäre,

$$\left.\begin{array}{l}ab_1\cdots b_m c_1\cdots c_k x,\\ ab_1\cdots b_m x\neq 0,\\ b_1\cdots b_m c_1\cdots c_k x\neq 0,\\ c_1\cdots c_k a x\neq 0\end{array}\right\}-(\text{Satz 50})\rightarrow\left\{\begin{array}{l}Ez,\ b_1\cdots b_m c_1\cdots c_k z,\\ b_1\cdots b_m z\neq 0,\ c_1\cdots c_k z\neq 0.\end{array}\right.$$

$$\left.\begin{array}{l}\mathfrak{R}(b_1\cdots b_m) * \mathfrak{R}(c_1\cdots c_k),\\ b_1\cdots b_m z\neq 0,\ c_1\cdots c_k z\neq 0,\\ b_1\cdots b_m c_1\cdots c_k z\end{array}\right\}-(\text{Satz 43})\rightarrow W.$$

So müsste $ac_1\cdots c_k x = 0$.

Ebenso müsste $ab_1\cdots b_m y = 0$.

$$\left.\begin{array}{l}ab_1\cdots b_m c_1\cdots c_k\neq 0,\\ ab_1\cdots b_m y,\ ay\neq 0,\ b_1\cdots b_m y\neq 0,\\ ac_1\cdots c_k x,\ ax\neq 0,\ c_1\cdots c_k x\neq 0\end{array}\right\}-(\text{Satz 51})\rightarrow\left\{\begin{array}{l}Ez,\ ab_1\cdots b_m c_1\cdots c_k z,\\ ab_1\cdots b_m z\neq 0,\ ac_1\cdots c_k z\neq 0,\\ b_1\cdots b_m c_1\cdots c_k z\neq 0\end{array}\right.$$

$$-(\text{Satz 50})\rightarrow\left\{\begin{array}{l}Ew,\ b_1\cdots b_m c_1\cdots c_k w,\\ b_1\cdots b_m w\neq 0,\ c_1\cdots c_k w\neq 0.\end{array}\right.$$

$$\left.\begin{array}{l}\mathfrak{R}(b_1\cdots b_m) * \mathfrak{R}(c_1\cdots c_k),\\ b_1\cdots b_m w\neq 0,\ c_1\cdots c_k w\neq 0,\\ b_1\cdots b_m c_1\cdots c_k w\end{array}\right\}-(\text{Satz 43})\rightarrow W.$$

$\therefore\quad \overline{\mathfrak{R}}(ab_1\cdots b_m c_1\cdots c_k),\ \mathfrak{R}(b_1\cdots b_m) * \mathfrak{R}(c_1\cdots c_k)\rightarrow W.$

$\therefore\quad \overline{\mathfrak{R}}(ab_1\cdots b_m c_1\cdots c_k)\rightarrow\overline{\mathfrak{R}}(b_1\cdots b_m c_1\cdots c_k),$ w. z. b. w.

Satz 53. Jeder in einem unzerlegbaren linearen Raume enthaltene lineare Raum ist stets ein unzerlegbarer linearer Raum.

Beweis: Sei $\overline{\mathfrak{R}}(a_1 \cdots a_m \cdots a_n) \gtrapprox \mathfrak{R}(a_1 \cdots a_m)$, so folgt nach Satz 52,

$$\overline{\mathfrak{R}}(a_1 \cdots a_m \cdots a_n) \rightarrow \overline{\mathfrak{R}}(a_1 \cdots a_m \cdots a_{n-1}) \rightarrow \cdots \rightarrow \overline{\mathfrak{R}}(a_1 \cdots a_m).$$

$$\therefore \quad \overline{\mathfrak{R}}(a_1 \cdots a_m \cdots a_n) \rightarrow \overline{\mathfrak{R}}(a_1 \cdots a_m),$$ w. z. b. w.

Satz 54. Der unzerlegbare lineare Raum vom Range 2 ist linearer Primraum.

Beweis: Sei den unzerlegbaren linearen Raum vom Range 2 $\mathfrak{R}(a_1 a_2)$, so folgt,

$$\overline{\mathfrak{R}}(a_1 a_2) \rightarrow Ez,\ \mathfrak{R}(a_1 a_2) \ni z,\ a_1 z \neq 0,\ a_2 z \neq 0$$
$$\rightarrow Ez,\ \mathfrak{R}(a_1 a_2) \ni z,\ \overline{a_1 a_2 z}\,.$$

Daher ist $\mathfrak{R}(a_1 a_2)$ ein linearer Primraum.

Satz 55. Der unzerlegbare lineare Raum ist linearer Primraum.

Beweis: Falls der Rang gleich 1 ist, ist es trivial, also genügt es zu zeigen, im Fall der Rang mehr als 2 ist. Dann ergibt sich, nach Satz 53, Satz 54, dass jeder im unzerlegbaren linearen Raume enthaltene lineare Raum vom Range 2 stets ein linearer Primraum ist. Daher ist nach Satz 41 der unzerlegbare lineare Raum auch ein linearer Primraum, w. z. b. w.

Satz 56. Jeder lineare Primraum ist unzerlegbarer linearer Raum.

Beweis: Wäre $\mathfrak{R}(a_1 \cdots a_n b_1 \cdots b_m) = \mathfrak{R}_1(a_1 \cdots a_n) \overset{*}{+} \mathfrak{R}_2(b \cdots b_m)$, während doch $\mathfrak{R}(a_1 \cdots a_n b_1 \cdots b_m)$ ein linearer Primraum ist, so wäre nach der Definition VIII, Satz 39,

$$\underline{Ez,\ \mathfrak{R}(a_1 \cdots a_n b_1 \cdots b_m) \ni z,\ \overline{a_1 \cdots a_n b_1 \cdots b_m z}.}$$

$$\overline{a_1 \cdots a_n b_1 \cdots b_m z} \rightarrow a_1 \cdots a_n z \neq 0,\ b_1 \cdots b_m z \neq 0.$$

$$\left.\begin{array}{l}\mathfrak{R}(a_1 \cdots a_n) * \mathfrak{R}(b_1 \cdots b_m),\\ a_1 \cdots a_n z \neq 0,\ b_1 \cdots b_m z \neq 0\end{array}\right\} - (\text{Satz } 43) \rightarrow a_1 \cdots a_n b_1 \cdots b_m z \neq 0.$$

$$\overline{a_1 \cdots a_n b_1 \cdots b_m z},\ a_1 \cdots a_n b_1 \cdots b_m z \neq 0 \rightarrow W.$$

Daher ist $\mathfrak{R}(a_1 \cdots a_n b_1 \cdots b_m)$ ein unzerlegbarer linearer Raum, w. z. b. w.

Aus Satz 55 und Satz 56 kann man folgendermassen behaupten:

Satz 57. Die Definitionen des linearen Primraumes und des unzerlegbaren linearen Raumes sind gleichwertig im Inhalt.

Satz 58. Der lineare Raum derart, dass alle darin liegenden linearen Räume vom Range 2 stets unzerlegbare lineare Räume sind, ist auch ein unzerlegbarer linearer Raum.

Beweis: Nach Satz 41, Satz 57, ist es klar.

Satz 59. $\left.\begin{array}{l}\overline{\mathfrak{R}}(d_1 \cdots d_s a_1 \cdots a_n),\\ \overline{\mathfrak{R}}(d_1 \cdots d_s b_1 \cdots b_m),\\ d_1 \cdots d_s a_1 \cdots a_n b_1 \cdots b_m \neq 0\end{array}\right\} \to \overline{\mathfrak{R}}(d_1 \cdots d_s a_1 \cdots a_n b_1 \cdots b_m).$

Beweis:

$$\overline{\mathfrak{R}}(d_1 \cdots d_s a_1 \cdots a_n) - (\text{Satz } 57,\ \text{Satz } 39) \to Ex,\ \overline{d_1 \cdots d_s a_1 \cdots a_n x}.$$

$$\overline{\mathfrak{R}}(d_1 \cdots d_s b_1 \cdots b_m) - (\text{Satz } 57,\ \text{Satz } 39) \to Ey,\ \overline{d_1 \cdots d_s b_1 \cdots b_m y}.$$

$$\left.\begin{array}{l}\overline{d_1 \cdots d_s a_1 \cdots a_n x},\\ \overline{d_1 \cdots d_s b_1 \cdots b_m y},\\ d_1 \cdots d_s a_1 \cdots a_n b_1 \cdots b_m \neq 0\end{array}\right\} - (\text{Satz } 37) \to Ez,\ \overline{d_1 \cdots d_s a_1 \cdots a_n b_1 \cdots b_m z}.$$

$$\therefore\quad Ez,\ \mathfrak{R}(d_1 \cdots d_s a_1 \cdots a_n b_1 \cdots b_m) \ni z,\ \overline{d_1 \cdots d_s a_1 \cdots a_n b_1 \cdots b_m z}.$$

$$\therefore\quad (\text{Satz } 57) \to \overline{\mathfrak{R}}(d_1 \cdots d_s a_1 \cdots a_n b_1 \cdots b_m),$$

w. z. b. w.

Satz 60. $\mathfrak{R}_1 * \cdots * \mathfrak{R}_k,\ \mathfrak{R}_1 \geqq \mathfrak{R}'_1,\ \cdots,\ \mathfrak{R}_k \geqq \mathfrak{R}'_k \ \to\ \mathfrak{R}'_1 * \cdots * \mathfrak{R}'_k.$

Beweis: Seien die Basen von $\mathfrak{R}'_1$, $\mathfrak{R}_1$, $\cdots$, $\mathfrak{R}'_k$, $\mathfrak{R}_k$ bzw. je $a_1 \cdots a_m$, $a_1 \cdots a_m \cdots a_n$, $\cdots$, $l_1 \cdots l_r$, $l_1 \cdots l_r \cdots l_s$, so folgt nach Zusatz zum Satz 42,

$$\mathfrak{R}_1 * \cdots * \mathfrak{R}_k \to a_1 \cdots a_m \cdots a_n \cdots l_1 \cdots l_r \cdots l_s \neq 0 \to a_1 \cdots a_m \cdots l_1 \cdots l_r \neq 0.$$

$$\therefore\quad \text{Rang } \mathfrak{R}'_1(a_1 \cdots a_m) + \cdots + \text{Rang } \mathfrak{R}'_k(l_1 \cdots l_r) = \text{Rang } \mathfrak{R}(a_1 \cdots a_m \cdots l_1 \cdots l_r).$$

$$\mathfrak{R}(a_1 \cdots a_m \cdots l_1 \cdots l_r) \ni x \to a_1 \cdots a_m \cdots l_1 \cdots l_r x \to a_1 \cdots a_m \cdots a_n \cdots l_1 \cdots l_r \cdots l_s x.$$

$$\left.\begin{array}{l}a_1 \cdots a_m \cdots a_n \cdots l_1 \cdots l_r \cdots l_s x,\\ \mathfrak{R}_1(a_1 \cdots a_n) * \cdots * \mathfrak{R}_k(l_1 \cdots l_s)\end{array}\right\} - (\text{Zusatz zum Satz } 43) \to a_1 \cdots a_m \cdots a_n x \ \text{od.} \cdots \text{od.}\ l_1 \cdots l_r \cdots l_s x.$$

$$\left.\begin{array}{l}a_1 \cdots a_m \cdots a_n x \ \text{od.} \cdots \text{od.}\ l_1 \cdots l_r \cdots l_s x,\\ a_1 \cdots a_m \cdots l_1 \cdots l_r x,\ a_1 \cdots a_m \cdots a_n \cdots l_1 \cdots l_r \cdots l_s \neq 0\end{array}\right\} \to a_1 \cdots a_m x \ \text{od.} \cdots \text{od.}\ l_1 \cdots l_r x.$$

$$a_1 \cdots a_m x \ \text{od.} \cdots \text{od.}\ l_1 \cdots l_r x \to \mathfrak{R}'_1 + \cdots + \mathfrak{R}'_k \ni x.$$

$$\therefore\quad \mathfrak{R}(a_1 \cdots a_m \cdots l_1 \cdots l_r) \ni x \to \mathfrak{R}'_1 + \cdots + \mathfrak{R}'_k \ni x.$$

$$\therefore\quad \mathfrak{R}(a_1 \cdots a_m \cdots l_1 \cdots l_r) \leqq \mathfrak{R}'_1 + \cdots + \mathfrak{R}'_k.$$

Anderseits folgt trivialerweise,

$$\mathfrak{R}(a_1 \cdots a_m \cdots l_1 \cdots l_r) \geqq \mathfrak{R}'_1 + \cdots + \mathfrak{R}'_k.$$

$$\therefore\quad \mathfrak{R}(a_1 \cdots a_m \cdots l_1 \cdots l_r) = \mathfrak{R}'_1 + \cdots + \mathfrak{R}'_k.$$

$$\therefore\quad \mathfrak{R}(a_1 \cdots a_m \cdots l_1 \cdots l_r) = \mathfrak{R}'_1 \overset{*}{+} \cdots \overset{*}{+} \mathfrak{R}'_k.$$

$$\therefore\quad \mathfrak{R}'_1 * \cdots * \mathfrak{R}'_k.$$

$$\therefore\quad \mathfrak{R}_1 * \cdots * \mathfrak{R}_k,\ \mathfrak{R}_1 \geqq \mathfrak{R}'_1,\ \cdots,\ \mathfrak{R}_k \geqq \mathfrak{R}'_k \ \to\ \mathfrak{R}'_1 * \cdots \mathfrak{R}'_k,$$

w. z. b. w.

Satz 61. $\mathfrak{R} = \mathfrak{R}_1 \overset{*}{+} \cdots \overset{*}{+} \mathfrak{R}_k$, $\mathfrak{R} \geqq \overline{\mathfrak{R}}'$ $\rightarrow$ Es gilt genau eine der Aussagen $\{\mathfrak{R}_1 \geqq \overline{\mathfrak{R}}', \cdots,$ und $\mathfrak{R}_k \geqq \overline{\mathfrak{R}}'\}$.

Beweis :

Seien $\mathfrak{D}(\overline{\mathfrak{R}}', \mathfrak{R}_1) = \mathfrak{R}_1'$, $\cdots$, $\mathfrak{D}(\overline{\mathfrak{R}}', \mathfrak{R}_k) = \mathfrak{R}_k'$, so folgt,

$$\mathfrak{R}_1 + \cdots + \mathfrak{R}_k \geqq \overline{\mathfrak{R}}' \rightarrow \mathfrak{R}_1' + \cdots + \mathfrak{R}_k' = \overline{\mathfrak{R}}' .$$

$$\mathfrak{R}_1 * \cdots * \mathfrak{R}_k, \; \mathfrak{R}_1 \geqq \mathfrak{R}_1', \cdots, \mathfrak{R}_k \geqq \mathfrak{R}_k' \text{—(Satz 60)}\rightarrow \mathfrak{R}_1' * \cdots * \mathfrak{R}_k' .$$

$\therefore$ $\overline{\mathfrak{R}}' = \mathfrak{R}_1' \overset{*}{+} \cdots \overset{*}{+} \mathfrak{R}_k'$.

So müssen z. B. $\overline{\mathfrak{R}}' = \mathfrak{R}_1'$, $\mathfrak{R}_2' = \cdots = \mathfrak{R}_k' = \mathfrak{N}$.

$\therefore$ $\overline{\mathfrak{R}}' = \mathfrak{R}_1' \rightarrow \mathfrak{R}_1 \geqq \overline{\mathfrak{R}}'$.

Daher gilt genau eine der Aussagen

$$\{\mathfrak{R}_1 \geqq \overline{\mathfrak{R}}', \cdots, \text{ und } \mathfrak{R}_k \geqq \overline{\mathfrak{R}}'\} .$$

Zusatz. $\mathfrak{R}_1 \overset{*}{+} \cdots \overset{*}{+} \mathfrak{R}_k \ni a_1, \cdots, a_{n+1}$, $\overline{a_1 \cdots a_{n+1}}$ $\rightarrow$ Es gilt genau eine der Aussagen $\{\mathfrak{R}_1 \ni a_1, \cdots, a_{n+1}; \cdots; \mathfrak{R}_k \ni a_1, \cdots, a_{n+1}\}$.

Satz 62. Jeder lineare Raum lässt sich bis auf die Anordnung der Faktoren auf die einzige Weise in die direkte Summe von unzerlegbaren linearen Räumen zerlegen.

Beweis :

Es sei $\mathfrak{R} = \overline{\mathfrak{R}}_1 \overset{*}{+} \cdots \overset{*}{+} \overline{\mathfrak{R}}_k = \overline{\mathfrak{R}}_1' \overset{*}{+} \cdots \overset{*}{+} \overline{\mathfrak{R}}_m'$.

$\mathfrak{R} = \mathfrak{R}_1 \overset{*}{+} \cdots \overset{*}{+} \mathfrak{R}_k$, $\mathfrak{R} \geqq \overline{\mathfrak{R}}_1'$ —(Satz 61)$\rightarrow$ $\mathfrak{R}_1 \geqq \overline{\mathfrak{R}}_1'$ od. $\cdots$

od. $\mathfrak{R}_k \geqq \overline{\mathfrak{R}}_1'$ $\rightarrow$ z. B. $\mathfrak{R}_1 \geqq \overline{\mathfrak{R}}_1'$.

$\mathfrak{R} = \mathfrak{R}_1' \overset{*}{+} \cdots \overset{*}{+} \mathfrak{R}_k'$, $\mathfrak{R} \geqq \overline{\mathfrak{R}}_1$ —(Satz 61)$\rightarrow$ $\mathfrak{R}_1' \geqq \overline{\mathfrak{R}}_1$ od. $\cdots$

od. $\mathfrak{R}_k' \geqq \overline{\mathfrak{R}}_1$ $\rightarrow$ z. B. $\mathfrak{R}_i' \geqq \overline{\mathfrak{R}}_1$.

$\overline{\mathfrak{R}}_i' \geqq \overline{\mathfrak{R}}_1$, $\overline{\mathfrak{R}}_1 \geqq \overline{\mathfrak{R}}_1' \rightarrow \overline{\mathfrak{R}}_i' \geqq \overline{\mathfrak{R}}_1' \rightarrow \overline{\mathfrak{R}}_i' = \overline{\mathfrak{R}}_1'$.

$\overline{\mathfrak{R}}_i' \geqq \overline{\mathfrak{R}}_1 \geqq \overline{\mathfrak{R}}_1'$, $\overline{\mathfrak{R}}_i' = \overline{\mathfrak{R}}_1' \rightarrow \overline{\mathfrak{R}}_1 = \overline{\mathfrak{R}}_1'$.

$\therefore$ z. B. $\overline{\mathfrak{R}}_1 = \overline{\mathfrak{R}}_1'$.

$\therefore$ z. B. $\overline{\mathfrak{R}}_1 = \overline{\mathfrak{R}}_1', \cdots, \overline{\mathfrak{R}}_k = \overline{\mathfrak{R}}_k'$; $k = m$.

Daher lässt sich jeder lineare Raum bis auf die Anordnung der Faktoren auf eine und nur eine Weise in die direkte Summe von unzerlegbaren linearen Räumen zerlegen.

§ 5. Die Zerlegung des $\mathfrak{B}_2$-Raumes.

Definition XI. Falls für zwei Elemente a, b von $\mathfrak{B}_2$ die Relation ab oder $Ez, \overline{abz}$ besteht, so heisst, "*a trennt sich nicht von b*", und wird mit $a \sim b$ bezeichnet. Sonst heisst, "*a trennt sich von b*", und wird mit $a \nsim b$ bezeichnet.

Satz 63. Die Relation " $\sim$ " erfüllt das Äquivalenzgesetz.

Beweis: (i) Sie ist reflexiv, d. h.;

$aa \to a \sim a$.

(ii) Sie ist symmetrisch, d. h.;

$a \sim b \to ab$ od. $\{Ez, \overline{abz}\}$.

$ab \to ba \to b \sim a$.

$Ez, \overline{abz} \to Ez, \overline{baz} \to b \sim a$.

$\therefore\ a \sim b \to b \sim a$.

(iii) Sie ist transitiv, d. h.;

$a \sim b,\ b \sim c \to \{ab,\ bc\}$ od. $\{ab,\ Ey,\ \overline{bcy}\}$

od. $\{Ex,\ \overline{abx},\ bc\}$ od. $\{Ex,\ \overline{abx},\ Ey,\ \overline{bcy}\}$.

$ab,\ bc \to ac \to a \sim c$.

$ab,\ Ey,\ \overline{bcy} \to Ey,\ \overline{acy} \to a \sim c$.

$Ex,\ \overline{abx},\ bc \to Ex,\ \overline{acx} \to a \sim c$.

$Ex,\ \overline{abx},\ Ey,\ \overline{bcy} \to \overline{acxy}$ od. $\{abc,\ ab \neq 0,\ bc \neq 0\}$.

$\overline{acxy} \to Ez,\ \overline{acz} \to a \sim c$.

$abc,\ ab \neq 0,\ bc \neq 0 \to \overline{abc}$ od. ac.

$\overline{abc} \to Eb,\ \overline{acb} \to a \sim c$.

$ac \to a \sim c$.

$\therefore\ a \sim c,\ b \sim c \to a \sim c$, w. z. b. w.

Zusatz. Bezeichnet man die zwei sich mit einander nicht trennenden Elemente von $\mathfrak{B}_2$ als eine Klasse von Elementen, so zerlegt der $\mathfrak{B}_2$-Raum in eine Anzahl von Klassen, die untereinander keine gemeinsamen Elemente besitzen.

Definition XII. Eine solche Klasse von Elementen von $\mathfrak{B}_2$ heisst der *Primverknüpfungsraum*, in kurzen Worten, **$\mathfrak{P}$-Raum**, von $\mathfrak{B}_2$-Raum, und wird mit $\mathfrak{P}$ bezeichnet. Ferner heisst *der $\mathfrak{B}_2$-Raum sich in die direkte Summe von Primverknüpfungsräumen $\mathfrak{P}_1$, $\mathfrak{P}_2$, $\mathfrak{P}_3$, $\cdots$ zerlegen,* und wir bezeichnen in Zeichen :

$$\mathfrak{B}_2 = \mathfrak{P}_1 \overset{*}{+} \mathfrak{P}_2 \overset{*}{+} \mathfrak{P}_3 \overset{*}{+} \cdots .$$

Satz 64. $\mathfrak{P} \ni a_1, \cdots, a_n,$ und $a_1 \cdots a_n \neq 0 \rightarrow \mathfrak{P} \geqq \mathfrak{R}(a_1 \cdots a_n)$.

Beweis :

$\mathfrak{R}(a_1 \cdots a_n) \ni x \rightarrow a_1 \cdots a_n \neq 0,\ a_1 \cdots a_n x$.

$a_1 \cdots a_n \neq 0,\ a_1 \cdots a_n x \rightarrow \overline{a_1 \cdots a_n x}$ oder

$$\{ * a_2 \cdots a_n x \text{ od. } a_1 * \cdots a_n x \text{ od. } \cdots \text{ od. } a_1 \cdots a_{n-1} * x \}$$

$$\rightarrow \overline{a_1 \cdots a_n x} \text{ oder z. B. } a_2 \cdots a_n x .$$

$$\therefore \left.\begin{matrix} a_1 \cdots a_n \neq 0, \\ a_1 \cdots a_n x \end{matrix}\right\} \rightarrow \overline{a_1 \cdots a_n x} \text{ oder z. B. } a_2 \cdots a_n x .$$

$$\left.\begin{matrix} a_2 \cdots a_n \neq 0, \\ a_2 \cdots a_n x \end{matrix}\right\} \rightarrow \overline{a_2 \cdots a_n x} \text{ oder z. B. } a_3 \cdots a_n x .$$

$$\cdots \quad \cdots \quad \cdots \quad \cdots \quad \cdots \quad \cdots$$

$$\cdots \quad \cdots \quad \cdots \quad \cdots \quad \cdots \quad \cdots$$

$$\left.\begin{matrix} a_{n-1} a_n \neq 0, \\ a_{n-1} a_n x \end{matrix}\right\} \rightarrow \overline{a_{n-1} a_n x} \text{ oder z. B. } \overline{a_n x} .$$

$$\therefore \left.\begin{matrix} a_1 \cdots a_n \neq 0, \\ a_1 \cdots a_n x \end{matrix}\right\} \rightarrow \overline{a_1 \cdots a_n x} \text{ od. z. B. } \overline{a_2 \cdots a_n x} \text{ od. z. B. } \cdots \text{ od. z.B } \overline{a_n x} .$$

$$\therefore \left.\begin{matrix} a_1 \cdots a_n \neq 0, \\ a_1 \cdots a_n x \end{matrix}\right\} \rightarrow \text{z. B. } \overline{a_m \cdots a_n x} .$$

$\therefore\ \mathfrak{R}(a_1 \cdots a_n) \ni x \rightarrow$ z. B. $\overline{a_m \cdots a_n x}$.

$\overline{a_m \cdots a_n x}$ —(Satz 33)→ Ez, $\overline{a_n x z}$ oder $\overline{a_n x} \rightarrow a_n \sim x$.

$\therefore\ \mathfrak{R}(a_1 \cdots a_n) \ni x \rightarrow$ z. B. $a_n \sim x$.

$\mathfrak{P} \ni a_n,\ a_n \sim x \rightarrow \mathfrak{P} \ni x$.

$\therefore\ \mathfrak{R}(a_1 \cdots a_n) \ni x \rightarrow \mathfrak{P} \ni x$.

$\therefore\ \mathfrak{R}(a_1 \cdots a_n) \leqq \mathfrak{P}$, w. z. b. w.

Zusatz. $\mathfrak{P} \geqq M \rightarrow \mathfrak{P} \geqq \mathfrak{B}(M)$.

Satz 65. $\mathfrak{P} \ni \mathfrak{R} \rightarrow \overline{\mathfrak{R}}$.

Beweis : Falls der Rang von $\mathfrak{R}$ gleich 1 ist, ist es trival, so genügt es zu zeigen, im Fall der Rang von $\mathfrak{R}$ mehr als 2 ist.

Es sei nun

$\mathfrak{R} \geqq \mathfrak{R}^2(a_1a_2)$.

$\mathfrak{P} \ni a_1,\ a_2 \rightarrow a_1 \sim a_2$.

$a_1 \sim a_2,\ a_1a_2 \neq 0 \rightarrow Ez,\ \overline{a_1a_2z}$.

$Ez,\ \mathfrak{R}^2(a_1a_2) \ni z,\ \overline{a_1a_2z} \rightarrow \overline{\mathfrak{R}}^2(a_1a_2)$.

$\therefore$ (Satz 58)$\rightarrow \overline{\mathfrak{R}}$.

$\therefore$ $\mathfrak{P} \geqq \mathfrak{R} \rightarrow \overline{\mathfrak{R}}$, w. z. b. w.

Zusatz 1. $\mathfrak{P} \ni a_1, \cdots, a_n$, und $a_1 \cdots a_n \neq 0 \rightarrow Ez,\ \mathfrak{P} \ni z,\ \overline{a_1 \cdots a_n z}$.

Zusatz 2. $\mathfrak{P} \geqq \mathfrak{R}_1,\ \mathfrak{R}_2 \rightarrow$ Es ist nie $\mathfrak{R}_1 * \mathfrak{R}_2$.

Nach Satz 64, und Satz 65, ergibt sich,

Satz 66. Ist der Rang des $\mathfrak{P}$-Raumes endlich, so ist der $\mathfrak{P}$-Raum ein linearer Primraum.

Satz 67. $\left.\begin{array}{l}\mathfrak{P}_1 \ni a_1, \cdots, a_n, \text{ und } a_1 \cdots a_n \neq 0, \\ \mathfrak{P}_2 \ni b_1, \cdots, b_m, \text{ und } b_1 \cdots b_m \neq 0\end{array}\right\} \rightarrow a_1 \cdots a_n b_1 \cdots b_m \neq 0.$

Beweis: Wäre vorläufig $a_1 \cdots a_n b_1 \cdots b_m = 0$, so wäre,

$a_1 \cdots a_n b_1 \cdots b_m \rightarrow Ex,\ a_1 \cdots a_n x,\ b_1 \cdots b_m x$.

$\mathfrak{P}_1 \ni a_1, \cdots, a_n$, und $a_1 \cdots a_n \neq 0,\ a_1 \cdots a_n x$ —(Satz 64)$\rightarrow \mathfrak{P}_1 \ni x$.

$\mathfrak{P}_2 \ni b_1, \cdots, b_m$, und $b_1 \cdots b_m \neq 0,\ b_1 \cdots b_m x$ —(Satz 64)$\rightarrow \mathfrak{P}_2 \ni x$.

$Ex,\ \mathfrak{P}_1 \ni x,\ \mathfrak{P}_2 \ni x \rightarrow W$.

$\therefore$ $a_1 \cdots a_n b_1 \cdots b_m \rightarrow W$.

$\therefore$ $a_1 \cdots a_n b_1 \cdots b_m \neq 0$, w. z. b. w.

Satz 68. $\mathfrak{P}_1 \ni a_1, \cdots, \mathfrak{P}_k \ni a_k \rightarrow a_1 \cdots a_k \neq 0$.

Beweis: Wäre vorläufig $a_1 \cdots a_k = 0$, so wäre,

$a_1 \cdots a_k \rightarrow \overline{a_1 \cdots a_k}$ oder $a_1 \cdots * \cdots a_k$.

$\therefore$ $a_1 \cdots a_k \rightarrow \overline{a_1 \cdots a_k}$ oder z. B. $a_1 \cdots a_{k-1}$.

$a_1 \cdots a_{k-1} \rightarrow \overline{a_1 \cdots a_{k-1}}$ oder z. B. $a_1 \cdots a_{k-2}$.

... ...

$a_1a_2a_3 \rightarrow \overline{a_1a_2a_3}$ oder z. B. $\overline{a_1a_2}$.

$\therefore$ $a_1 \cdots a_k \rightarrow \overline{a_1 \cdots a_k}$ od. z. B. $\overline{a_1 \cdots a_{k-1}}$ od. z.B. $\cdots$ od. z.B. $\overline{a_1a_2}$.

$\therefore$ $a_1 \cdots a_k \rightarrow$ z. B. $\overline{a_1 \cdots a_m},\ 2 \leqq m \leqq k$.

$\overline{a_1 \cdots a_m} \rightarrow Ez,\ \overline{a_1a_2z}$ oder $\overline{a_1a_2} \rightarrow a_1 \sim a_2$.

$\therefore \quad a_1 \cdots a_k \rightarrow$ z. B. $a_1 \sim a_2$.

$\mathfrak{P}_1 \ni a_1, \ \mathfrak{P}_2 \ni a_2, \ a_1 \sim a_2 \rightarrow W.$

$\therefore \quad a_1 \cdots a_k \rightarrow W.$

$\therefore \quad a_1 \cdots a_k \neq 0,$ w. z. b. w.

Satz 69.

$$\left.\begin{array}{l} \mathfrak{P}_1 \ni a_1, \cdots, a_n, \text{ und } a_1 \cdots a_n \neq 0, \\ \mathfrak{P}_2 \ni b_1, \cdots, b_m, \text{ und } b_1 \cdots b_m \neq 0, \\ \cdots \quad \cdots \quad \cdots \quad \cdots \\ \mathfrak{P}_k \ni l_1, \cdots, l_s, \text{ und } l_1 \cdots l_s \neq 0 \end{array}\right\} \rightarrow a_1 \cdots a_n b_1 \cdots b_m \cdots l_1 \cdots l_s \neq 0.$$

Beweis: Wäre vorläufig $a_1 \cdots a_n b_1 \cdots b_m \cdots l_1 \cdots l_s = 0$, so wäre,

$a_1 \cdots a_n b_1 \cdots b_m \cdots l_1 \cdots l_s \rightarrow Ea, \ a_1 \cdots a_n a, \ ab_1 \cdots b_m \cdots l_1 \cdots l_s.$

$\mathfrak{P}_1 \ni a_1, \cdots, a_n,$ und $a_1 \cdots a_n \neq 0, \ a_1 \cdots a_n a$ –(Satz 64)→ $\mathfrak{P}_1 \ni a$.

$\therefore \quad a_1 \cdots a_n b_1 \cdots b_m \cdots l_1 \cdots l_s \rightarrow Ea, \ \mathfrak{P}_1 \ni a, \ ab_1 \cdots b_m \cdots l_1 \cdots l_s.$

$ab_1 \cdots b_m \cdots l_1 \cdots l_s \rightarrow Eb, \ \mathfrak{P}_2 \ni b, \ ab \cdots l_1 \cdots l_s.$

$\cdots \quad \cdots \quad \cdots \quad \cdots \quad \cdots \quad \cdots$

$ab \cdots l_1 \cdots l_s \rightarrow El, \ \mathfrak{P}_k \ni l, \ ab \cdots l.$

$$\therefore \quad a_1 \cdots a_n b_1 \cdots b_m \cdots l_1 \cdots l_s \rightarrow \left\{\begin{array}{l} Ea, \ Eb, \cdots, El, \ ab \cdots l, \\ \mathfrak{P}_1 \ni a, \ \mathfrak{P}_2 \ni b, \cdots, \mathfrak{P}_k \ni l. \end{array}\right.$$

$\mathfrak{P}_1 \ni a, \ \mathfrak{P}_2 \ni b, \cdots, \mathfrak{P}_k \ni l$ –(Satz 68)→ $ab \cdots l \neq 0$.

$ab \cdots l, \ ab \cdots l \neq 0 \rightarrow W.$

$\therefore \quad a_1 \cdots a_n b_1 \cdots b_m \cdots l_1 \cdots l_s \rightarrow W.$

$\therefore \quad a_1 \cdots a_n b_1 \cdots b_m \cdots l_1 \cdots l_s \neq 0,$ w. z. b. w.

Zusatz.

$$\left.\begin{array}{l} \mathfrak{P}_1 \ni a_1, \cdots, a_n, \\ \cdots \quad \cdots \quad \cdots, \\ \mathfrak{P}_k \ni l_1, \cdots, l_s, \\ a_1 \cdots a_n \cdots l_1 \cdots l_s \end{array}\right\} \rightarrow a_1 \cdots a_n \text{ od. } \cdots \text{. od. } l_1 \cdots l_s.$$

Satz 70.

$$\left.\begin{array}{l} \mathfrak{P}_1 \ni a_1, \cdots, a_n, \text{ und } a_1 \cdots a_n x \neq 0, \\ \mathfrak{P}_2 \ni b_1, \cdots, b_m, \text{ und } b_1 \cdots b_m x \neq 0, \\ \cdots \quad \cdots \quad \cdots \quad \cdots \quad \cdots, \\ \mathfrak{P}_k \ni l_1, \cdots, l_s, \text{ und } l_1 \cdots l_s x \neq 0 \end{array}\right\} \rightarrow a_1 \cdots a_n b_1 \cdots b_m \cdots l_1 \cdots l_s x \neq 0.$$

Beweis: Ist zum Beispiel $\mathfrak{P}_1 \ni x$, so ergibt sich,

$$\left.\begin{array}{l} \mathfrak{P}_1 \ni a_1, \cdots, a_n, x, \text{ und } a_1 \cdots a_n x \neq 0, \\ \mathfrak{P}_2 \ni b_1, \cdots, b_m, \text{ und } b_1 \cdots b_m \neq 0, \\ \cdots \quad \cdots \quad \cdots \quad \cdots \quad \cdots, \\ \mathfrak{P}_k \ni l_1, \cdots, l_s, \text{ und } l_1 \cdots l_s \neq 0 \end{array}\right\} -(\text{Satz } 69) \rightarrow a_1 \cdots a_n b_1 \cdots b_m \cdots l_1 \cdots l_s x \neq 0.$$

Ist $\mathfrak{P}_i \not\ni x$, $(i = 1, \cdots, k)$, so ergibt sich,

$$\left.\begin{array}{l} \mathfrak{P}_1 \ni a_1, \cdots, a_n, \text{ und } a_1 \cdots a_n \neq 0, \\ \mathfrak{P}_2 \ni b_1, \cdots, b_m, \text{ und } b_1 \cdots b_m \neq 0, \\ \cdots \quad \cdots \quad \cdots \quad \cdots \quad \cdots, \\ \mathfrak{P}_k \ni l_1, \cdots, l_s, \text{ und } l_1 \cdots l_s \neq 0, \\ \mathfrak{P} \ni x \qquad , \text{ und } x \neq 0 \end{array}\right\} -(\text{Satz } 69) \rightarrow a_1 \cdots a_n b_1 \cdots b_m \cdots l_1 \cdots l_s x \neq 0,$$

w. z. b. w.

Zusatz 1.

$$\left.\begin{array}{l} \mathfrak{P}_1 \ni a_1, \cdots, a_n, \\ \cdots \quad \cdots \quad \cdots, \\ \mathfrak{P}_k \ni l_1, \cdots, l_s, \\ a_1 \cdots a_n \cdots l_1 \cdots l_s x \end{array}\right\} \rightarrow a_1 \cdots a_n x \text{ od. } \cdots \text{ od. } l_1 \cdots l_s x.$$

Zusatz 2.

$$\left.\begin{array}{l} \mathfrak{P}_1 \ni a_1, \cdots, a_n, \text{ und } a_1 \cdots a_n \neq 0, \\ \cdots \quad \cdots \quad \cdots \quad \cdots \quad \cdots, \\ \mathfrak{P}_k \ni l_1, \cdots, l_s, \text{ und } l_1 \cdots l_s \neq 0, \\ a_1 \cdots a_n \cdots l_1 \cdots l_s x \end{array}\right\} \rightarrow$$

Es gilt genau eine der Aussagen

$$\left\{ a_1 \cdots a_n x, \cdots, \text{ und } l_1 \cdots l_s x \right\}.$$

Satz 71. $\mathfrak{P}_1 \geqq \mathfrak{R}_1, \cdots, \mathfrak{P}_k \geqq \mathfrak{R}_k \rightarrow \mathfrak{R}_1 * \cdots * \mathfrak{R}_k$.

Beweis: Seien die Basen von $\mathfrak{R}_1, \cdots$, und $\mathfrak{R}_k$ bzw. je $a_1 \cdots a_n, \cdots$, und $l_1 \cdots l_s$, so folgt,

$$\left.\begin{array}{l} \mathfrak{P}_1 \ni a_1, \cdots, a_n, \text{ und } a_1 \cdots a_n \neq 0, \\ \cdots \quad \cdots \quad \cdots \quad \cdots \quad \cdots, \\ \mathfrak{P}_k \ni l_1, \cdots, l_s, \text{ und } l_1 \cdots l_s \neq 0 \end{array}\right\} -(\text{Satz } 69) \rightarrow a_1 \cdots a_n \cdots l_1 \cdots l_s \neq 0.$$

$\therefore$ Rang $\mathfrak{R}(a_1 \cdots a_n \cdots l_1 \cdots l_s) =$ Rang $\mathfrak{R}_1(a_1 \cdots a_n) + \cdots +$ Rang $\mathfrak{R}_k(l_1 \cdots l_s)$.

Einerseits folgt,

$$\left.\begin{array}{l} \mathfrak{P}_1 \ni a_1, \cdots, a_n, \\ \cdots\cdots\cdots\cdots, \\ \mathfrak{P}_k \ni l_1, \cdots, l_s, \\ a_1 \cdots a_n \cdots l_1 \cdots l_s x \end{array}\right\} -(\text{Zusatz 1 zum Satz } 70) \rightarrow a_1 \cdots a_n x \text{ od. } \cdots \text{ od. } l_1 \cdots l_s x.$$

$\therefore\quad \mathfrak{R}(a_1\cdots a_n\cdots l_1\cdots l_s)\ni x \to \mathfrak{R}_1(a_1\cdots a_n)\ni x$ od. $\cdots$ od. $\mathfrak{R}_k(l_1\cdots l_s)\ni x$.

$\therefore\quad \mathfrak{R}(a_1\cdots a_n\cdots l_1\cdots l_s)\leqq \mathfrak{R}_1(a_1\cdots a_n)+\cdots+\mathfrak{R}_k(l_1\cdots l_s)$.

Anderseits folgt trivialerweise,

$$\mathfrak{R}(a_1\cdots a_n\cdots l_1\cdots l_s)\geqq \mathfrak{R}_1(a_1\cdots a_n)+\cdots+\mathfrak{R}_k(l_1\cdots l_s).$$

$$\therefore\quad \mathfrak{R}(a_1\cdots a_n\cdots l_1\cdots l_s)= \mathfrak{R}_1(a_1\cdots a_n)+\cdots+\mathfrak{R}_k(l_1\cdots l_s).$$

$$\therefore\quad \mathfrak{R}(a_1\cdots a_n\cdots l_1\cdots l_s)= \mathfrak{R}_1(a_1\cdots a_n)\overset{*}{+}\cdots\overset{*}{+}\mathfrak{R}_k(l_1\cdots l_s).$$

$$\therefore\quad \mathfrak{R}_1(a_1\cdots a_n) * \cdots * \mathfrak{R}_k(l_1\cdots l_s),\qquad \text{w. z. b. w.}$$

Zusatz. $\mathfrak{P}_1\ni\mathfrak{R}_1,\ \cdots,\ \mathfrak{P}_k\ni\mathfrak{R}_k \to$

$$\mathfrak{D}\left\{\mathfrak{B}(\mathfrak{R}_1\cdots\mathfrak{R}_i),\ \mathfrak{B}(\mathfrak{R}_{i+1}\cdots\mathfrak{R}_k)\right\}=\mathfrak{N},\quad 1\leqq i\leqq k-1.$$

Satz 72. $\mathfrak{R}=\overline{\mathfrak{D}}(\mathfrak{P}_1,\mathfrak{R})\overset{*}{+}\overline{\mathfrak{D}}(\mathfrak{P}_2,\mathfrak{R})\overset{*}{+}\overline{\mathfrak{D}}(\mathfrak{P}_3,\mathfrak{R})\overset{*}{+}\cdots.$

Beweis:

$$\left.\begin{array}{l}\mathfrak{B}_2=\mathfrak{P}_1+\mathfrak{P}_2+\mathfrak{P}_3+\cdots,\\ \mathfrak{B}_2\geqq\mathfrak{R}\end{array}\right\}\to \mathfrak{R}=\mathfrak{D}(\mathfrak{P}_1,\mathfrak{R})+\mathfrak{D}(\mathfrak{P}_2,\mathfrak{R})+\mathfrak{D}(\mathfrak{P}_3,\mathfrak{R})+\cdots.$$

Weiter ist $\mathfrak{D}(\mathfrak{P}_i,\mathfrak{R})$, $(i=1,2,3,\cdots)$ ein linearer Raun nach Zusatz zum Satz 64.

$$\left.\begin{array}{l}\mathfrak{P}_1\geqq\mathfrak{D}(\mathfrak{P}_1,\mathfrak{R}),\\ \mathfrak{P}_2\geqq\mathfrak{D}(\mathfrak{P}_2,\mathfrak{R}),\\ \mathfrak{P}_3\geqq\mathfrak{D}(\mathfrak{P}_3,\mathfrak{R}),\\ \cdots\quad\cdots\quad\cdots\end{array}\right\}\text{—(Satz 71)}\to \mathfrak{D}(\mathfrak{P}_1,\mathfrak{R}) * \mathfrak{D}(\mathfrak{P}_2,\mathfrak{R}) * \mathfrak{D}(\mathfrak{P}_3,\mathfrak{R}) * \cdots.$$

Ferner ergibt sich,

$$\mathfrak{P}_i\geqq\mathfrak{D}(\mathfrak{P}_i,\mathfrak{R}),\ (i=1,2,3,\cdots)\ \text{—(Satz 65)}\to \overline{\mathfrak{D}}(\mathfrak{P}_i,\mathfrak{R}).$$

Also ist,

$$\mathfrak{R}=\overline{\mathfrak{D}}(\mathfrak{P}_1,\mathfrak{R})\overset{*}{+}\overline{\mathfrak{D}}(\mathfrak{P}_2,\mathfrak{R})\overset{*}{+}\overline{\mathfrak{D}}(\mathfrak{P}_3,\mathfrak{R})\overset{*}{+}\cdots,\qquad \text{w. z. b. w.}$$

Zusatz 1. Für jeden linearen Primraum $\overline{\mathfrak{R}}$, gilt genau eine der Aussagen

$$\left\{\mathfrak{P}_1\geqq\overline{\mathfrak{R}},\ \mathfrak{P}_2\geqq\overline{\mathfrak{R}},\ \mathfrak{P}_3\geqq\overline{\mathfrak{R}},\ \cdots\right\}.$$

Zusatz 2. Für jeden linearen Primzyklus $\overline{a_1\cdots a_{n+1}}$, gilt genau eine der Aussagen

$\left\{\mathfrak{P}_1\ni a_1,\cdots,a_{n+1}\right.$, und $\mathfrak{P}_2\ni a_1,\cdots,a_{n+1}$, und $\mathfrak{P}_3\ni a_1,\cdots,a_{n+1}$, und $\left.\cdots\right\}$.

Satz 73. Eine beliebige Summenmenge der linearen Räume bzw. Primverknüpfungsräume von $\mathfrak{V}_2$ ist auch ein $\mathfrak{V}_2$-Raum.

Beweis: Der Beweis ist klar, weil die Grundannahme, sowie die Axiome 1*, 2, 3, 4, 5 in unserer Definition von $\mathfrak{V}_2$-Raums erfüllt sind.

Bemerkungen zum Schluss.

Es bleibt noch zu zeigen, dass wenn ferner die Endlichkeit des Ranges als Axiom zugefügt wird, so reduziert sich unser Primverknüpfungsraum zu einem projektiven Raum von endlicher Dimension, welcher auf Grund des Verknüpfungs- und des Dimensionsaxiom[16] aufgebaut wird. Dabei zerlegt sich also der $\mathfrak{V}_2$-Raum in die endliche direkte Summe von projektiven Räumen von endlicher Dimension.

(16) Z. B. vgl. Veblen-Young: Projective Geometry!

Zur Axiomatik der linearen Abhängigkeit. III.

Zur Axiomatik der linearen Abhängigkeit, III (Schluss).

Von Takeo NAKASAWA

(Eingegangen am 20 Juli, 1936)

§ 1. Einleitung.

In meinen früheren Arbeiten[1] habe ich eine Methode hergestellt und gezeigt, dass sich dieselbe für die Untersuchungen der linearen Abhängigkeit im projektiven Raumes mit Erfolg verwenden lässt. Die betrachtete Methode gehört zu einer algebraischen Symbolenrechnung, bei welcher sogar die einzige Relation d. h. "*gelten*" (natürlich auch die Verneinung "*nicht gelten*") für gewisse Reihen der Elementen aufgeprägt wird. Diese Rechnung habe ich deshalb nach G. Thomsen einen *Zyklenkalkül*[2] genannt.

In der vorliegenden Schrift handelt es sich

1. um den vollständigen Aufbau der projektiven Geometrie, indem die am Anfang meiner Arbeit aufgezählten Rechnungsprinzipien[3] als räumliche Axiome eines geometrischen Raumes gedeutet werden, und
2. um einige Anwendungen zum allgemeinen linearen Raum sowie Vergleichungen mit den verwandten Arbeiten von Herren G. Birkhoff und H. Whitney, und
3. um einen Unabhängigkeitsbeweis der dort angenommenen Axiomengruppe.

---

(1) T. Nakasawa, "*Zur Axiomatik der linearen Abhängigkeit. I*", Science Reports of the Tokyo Bunrika Daigaku, Sec. A, Vol. 2, No. 43, (235-255); *II*, ibid., Sec. A, Vol. 3, No. 51, (45-69). Wegen der Bequemlichkeit bezeichnen wir diese zwei Arbeiten bzw. kurz mit A1 und A2.

(2) Das Wort und die Idee des Zyklenkalküls geht wohl zum G. Thomsenschen Buch, "*Grundlagen der Elementargeometrie*", (Leipzig 1933) zurück. Unter Einfluss seiner Idee habe ich meine axiomatische Untersuchung der linearen Abhängigkeit angefangen.

(3) Vgl. S. 236 in A1, sowie S. 46 in A2!

Schreibt man nun für s Punkte $a_1, a_2, \cdots, a_s$ des klassischen n-dimensionalen projektiven Raumes[4] $a_1 \cdots a_s = 0$ (gewöhnlich in kurzen Zeichen $a_1 \cdots a_s$), bzw. $a_1 \cdots a_s \neq 0$, je nachdem sie linear abhängig oder linear unabhängig sind, so gelten bekanntlich die folgenden Relationen[5];

1° $a \neq 0$, aa.

2° $a_1 \cdots a_s \rightarrow a_1 \cdots a_s x$, $(s \geqq 2)$.

3° $a_1 \cdots a_i \cdots a_s \rightarrow a_i \cdots a_1 \cdots a_s$, $(s \geqq i \geqq 2)$.

4° $x a_1 \cdots a_s$, $a_1 \cdots a_s y \rightarrow a_1 \cdots a_s$ oder $x a_1 \cdots a_{s-1} y$, $(s \geqq 1)$.

5° $a_1 \cdots a_s x y \rightarrow E z$, $a_1 \cdots a_s z$, xyz, $(s \geqq 2)$.

Umgekehrt, wenn man diese fünf Sätze allein als Rechnungsprinzipien eines logischen Kalküls für den Elementen $a_1, a_2, \cdots, a_s, \cdots$ einer abstrakten Menge $\mathfrak{B}$ aufnimmt, dann lassen sich auf Grund der Axiome 1° bis 4° fast alle Eigenschaften der linearen Abhängigkeit herleiten, welche keine Existenzaussage enthalten. Wenn man ferner das letzte Axiom 5° dazu adjungiert, dann lässt sich der *Dualitätssatz* des linearen Raumes

$$\text{Rang}\,(A \frown B) + \text{Rang}\,(A \smile B) = \text{Rang}\,A + \text{Rang}\,B^{(6)}$$

beweisen, und sodann werden fast alle Sätze des projektiven Raumes beweisbar. Die Einzelheiten darüber habe ich bereits in meiner Arbeiten A1, A2 ausführlich betrachtet.

In der vorliegenden Schrift will ich die Unterschied klar herzustellen versuchen, welche Sätze der projektiven Geometrie aus den Axiomen 1°–5° herleitbar sind und welche nicht. Dafür scheint es mir zweckmässig das betrachtete 1°–5° Axiomensystem mit dem Veblen-Youngschen[7] zu vergleichen, da ich das letztere für eines der vollständigen und vorbildlichen Systeme halte.

---

(4) Vgl. z. B. Eugenio Bertini, "*Einführung in die projektive Geometrie mehrdimensionaler Räume*", Wien, 1906, (1–23)!

(5) Betreffs der hier benutzten Bezeichnungen vergleiche S. 235 in A1 oder S. 45 in A2!

(6) Satz 29 auf S. 48 in A2.

(7) O. Veblen and J. W. Young, "*Projective geometry, I*", Boston, 1910, (1–30).

§ 2. Die Vergleichung mit dem Veblen, Youngschen System.

Definition: Es sei $a_1, a_2, \cdots, a_n$ n linear unabhängige Elemente d. h. $a_1 \cdots a_n \neq 0$. Die Menge der allen Elementen x von $\mathfrak{B}$ derart, dass $a_1 \cdots a_n x = 0$ sind, nennen wir den von $a_1, a_2, \cdots, a_n$ erzeugten *linearen Raum* in $\mathfrak{B}$ vom Range n, in Zeichen $\mathfrak{R}^n(a_1 \cdots a_n)$, und den Zyklus $a_1 \cdots a_n$ nennen wir die Basis des $\mathfrak{R}^n(a_1 \cdots a_n)$. Insbesondere nennen wir den linearen Raum vom Range 1 bzw. 2 den *Punkt* bzw. die *Gerade*.

Dann bestehen die folgenden Sätze.

(i) Zwei Punkte $\mathfrak{R}(a)$, $\mathfrak{R}(b)$ stimmen dann und nur dann ein, wenn $ab = 0$.

(ii) Drei Punkte $\mathfrak{R}(a)$, $\mathfrak{R}(b)$, $\mathfrak{R}(c)$ liegen dann und nur dann auf einer einzigen Gerade, wenn $abc = 0$.

(iii) Zwei disjunkte Punkte $\mathfrak{R}(a)$, $\mathfrak{R}(b)$ bestimmen eine einzige Gerade $\mathfrak{R}(ab)$.

(iv) Zwei auf einer Gerade $\mathfrak{R}(ab)$ liegende disjunkte Punkte $\mathfrak{R}(x)$, $\mathfrak{R}(y)$ bestimmen dieselbe Gerade, d. h. $\mathfrak{R}(ab) = \mathfrak{R}(xy)$.

Die Beweise derselben finden sich im noch erweiterten Sinne bereits in A1.

(v) Wenn $\mathfrak{R}(a)$, $\mathfrak{R}(b)$, $\mathfrak{R}(c)$ nicht auf einer Gerade liegen, und $\mathfrak{R}(d)$ auf der Gerade $\mathfrak{R}(bc)$ liegt, und $\mathfrak{R}(e)$ auf der Gerade $\mathfrak{R}(ac)$ liegt, und noch $\mathfrak{R}(d) \neq \mathfrak{R}(e)$ ist, so existiert mindestens ein Durchschnittpunkt $\mathfrak{R}(f)$ von $\mathfrak{R}(ab)$ und $\mathfrak{R}(de)$.

Beweis: Nach der Voraussetzung und nach (i), (ii) folgt, dass $abc \neq 0$, $bcd = 0$, $ace = 0$, und $de \neq 0$.

$$\therefore \left.\begin{matrix} abc \neq 0\,, \\ bcd \quad\;\;, \\ ace \end{matrix}\right\} \rightarrow \left.\begin{matrix} abc \neq 0\,, \\ abcd\,, \\ abce \end{matrix}\right\} \rightarrow abde \rightarrow Ef\,,\ abf\,,\ def.$$

Da $ab \neq 0$, $de \neq 0$, $abf = 0$, und $def = 0$ sind, so ist der Punkt $\mathfrak{R}(f)$ der Durchschnitt der Geraden $\mathfrak{R}(ab)$ und $\mathfrak{R}(de)$, w. z. b. w.

(vi) Es seien $\mathfrak{R}(a_1 \cdots a_k)$ ein linearer Raum vom Range k und $\mathfrak{R}(a_{k+1})$ ein nicht darin liegender Punkt. So stimmen der linearen Raum $\mathfrak{R}(a_1 \cdots a_{k+1})$ vom Range $k+1$ ein mit der Menge $\sum_x \mathfrak{R}(a_{k+1}x)$, welche aus allen durch Verbindung des festen Punktes $\mathfrak{R}(a_{k+1})$ mit irgendeinem Punkt $\mathfrak{R}(x)$ von $\mathfrak{R}(a_1 \cdots a_k)$ entstehenden Geraden $\mathfrak{R}(a_{k+1}x)$ besteht.

Bewes: Sei $\mathfrak{R}(y)$ ein beliebiger in $\sum_x \mathfrak{R}(a_{k+1}x)$ liegender Punkt, so folgt, $a_1 \cdots a_k x$, $x a_{k+1} y \rightarrow^{(8)} a_1 \cdots a_{k+1} y - \mathfrak{R}(a_1 \cdots a_{k+1}) \ni \mathfrak{R}(y)$.

$\therefore \quad \sum_x \mathfrak{R}(a_{k+1} x) \leqq \mathfrak{R}(a_1 \cdots a_{k+1})$.

Sei $\mathfrak{R}(y)$ ein beliebiger in $\mathfrak{R}(a_1 \cdots a_{k+1})$ liegender Punkt, so folgt,

$a_1 \cdots a_{k+1} y \rightarrow Ex$, $a_1 \cdots a_k x$, $a_{k+1} y x$.

$\therefore \quad Ex$, $\mathfrak{R}(a_1 \cdots a_k) \ni \mathfrak{R}(x)$, $\mathfrak{R}(a_{k+1} x) \ni \mathfrak{R}(y)$.

$\therefore \quad \sum_x \mathfrak{R}(a_{k+1} x) \geqq \mathfrak{R}(a_1 \cdots a_{k+1})$.

Durch Vergleichung der beiden ergibt sich sodann

$$\sum_x \mathfrak{R}(a_{k+1} x) = \mathfrak{R}(a_1 \cdots a_{k+1}) \,, \quad \text{w. z. b. w.}$$

(vii) Bezeichnet man drei auf einer Gerade liegende Punkte als eine Klasse von Punkten, so zerfällt $\mathfrak{B}$ in eine Anzahl von Klassen von Elementen[9], die untereinander keine gemeinsame Elemente besitzen Sei $\mathfrak{P}_\nu$ eine solche Klasse von Elementen, so ist $\mathfrak{P}_\nu$ auch ein $\mathfrak{B}_2$-Raum[10] und $\mathfrak{B}$ zerfällt in die direkte Summe von $\mathfrak{P}_1$, $\mathfrak{P}_2$, $\mathfrak{P}_3$, $\cdots$[11]. Dann existieren mindestens drei Punkte auf jeder Gerade von $\mathfrak{P}_\nu$[12].

Der wesentliche Inhalt dieses Satzes findet sich bereits in A2, § 5, SS. 63–69.

Das Veblen-Youngsche Axiom ist folgendes[13]:

By a projective geometry is meant a set of elements which, for suggestiveness, are called points subject to the following four conditions:

I. If A and B are distinct points, there is one and only one line that contains A and B.

II. If A, B, C are non-collinear points and if a line l contains a point D of the line (BC) and a point E of the line (AC), where D and E are distinct points, then the line l contains a point F of the line (AB).

III. There are at least three points on every line.

IV. A k-space is defined by the following inductive definition. A point is a 1-space. If A_1, A_2, $\cdots$, A_{k+1} are points not all in the same k-space, the set of all points collinear with the point A_{k+1} and any point of the k-space $(A_1 A_2 \cdots A_k)$ is the $(k+1)$-space $(A_1 A_2 \cdots A_{k+1})$. Thus a line is a 2-space, and a plane is a 3-specе, and so on.

---

(8) Zusatz 2 zum Satz 24 auf S. 252 in A1.

(9) Satz 63 sowie Zusatz auf S. 63 in A2.

(10) Definition VI auf S. 47, sowie Satz 73 auf S. 69 in A2.

(11) Satz 71, Zusatz, Satz 72, und Zusatz 1 auf SS. 67–68 in A2.

(12) $\mathfrak{P}_\nu$ heisst der $\mathfrak{P}$-Raum. Vgl. Definition XII auf S. 63 in A2!

(13) O. Veblen and W. H. Bussy, "*Finite projective geometries*", Trans. Amer. Math. Soc., Vol. 7, 1906, (241–242).

Durch Vergleichung dieses Axiomensystems mit den oben bewiesenen Sätzen sehen wir offenbar, dass unsere Sätze (iii), (iv) zu I, (v) zu II, und (vi) zu IV inhaltlich entsprechen. Das Axiom III gilt aber nicht immer im $\mathfrak{B}_2$-Raum, da ein n-Simplex(14) sich als ein $\mathfrak{B}_2$-Raum auffassen lässt. Jedoch besteht dasselbe eben im $\mathfrak{P}$-Raum, folglich auch im linearen Primraum(15).

Also nach (vii) kann man folgendermassen behaupten.

Satz 74. *Jeder $\mathfrak{B}_2$-Raum lässt sich bis auf die Anordnung der Faktoren auf die einzige Weise in die direkte Summe von projektiven Räumen zerlegen. Jeder $\mathfrak{P}$-Raum oder jeder lineare Primraum ist ein projektiver Raum.*

Dem obigen Veblen-Youngschen Axiomensystem, welches eine Definition des, sozusagen, *dimensionsfreien* projektiven Raumes ist, wird gewöhnlich das folgende Dimensionsaxiom hinzugefügt.

V. There is a finite upper bound to the dimensions of the spaces. Let n be the upper bound, our set of points is called an n-dimensional projective space.

Dann besteht der folgende Satz.

Satz 75. *Jeder lineare Primraum vom Range n ist ein n-dimensionaler projektiver Raum. Jeder lineare Raum lässt sich bis auf die Anordnung der Faktoren auf die einzige Weise in die endliche direkte Summe von projektiven Räumen von endlichen Dimensionen zerlegen*(16).

§ 3. Anwendungen und verwandte Arbeiten.

1. Die Anwendung zum allgemeinen linearen Raum(17) und Hilbertschen Raum(18).

Die Abelsche Gruppe $\mathfrak{R}$ mit den linksseitigen Operatoren der reellen Zahlen (oder mit dem allgemeinen abstrakten Körper) heisst den allge-

(14) Unter ein n-simplex verstehen wir die Menge von n Elementen, in deren die *Gelten-Relation* folgendermassen definiert ist; d. h., der Zyklus, der mehr als zwei dieselbe Elementen enthält, setzen wir $=0$, sonst $\neq 0$.

(15) Definition VIII auf S. 62, sowie Zusatz 1 zum Satz 72 auf S. 68 in A2.

(16) Satz 62 auf S. 62, sowie Satz 57 auf S. 60 in A2.

(17) Vgl. z.B. S. Banach, "*Théorie des opérations linéaires*, Warszawa, 1932, Kap. II!

(18) D. Hilbert, "*Grundzüge einer allgemeinen Theorie der linearen Integralgleichungen*", 4. Mitteilung, Göttingen Nachrichten, 1906, (SS, 157-227).

meinenlinearen Raum[19]. Es seien nun $x_1, \cdots, x_s$ je s von Null verschiedene Elemente von $\mathfrak{R}$. Wenn es s reelle Zahlen $\alpha_1, \cdots, \alpha_s$ gibt, die nicht sämtlich gleich Null sind, und für die gilt: $\alpha_1 x_1 + \cdots + \alpha_s x_s = 0$ (d. h. $x_1, \cdots, x_s$ vektorabhängig !), bezeichnen wir dies mit $x_1 \cdots x_s = 0$, und sonst mit $x_1 \cdots x_s \neq 0$. Dann werden die fünf Axiome des $\mathfrak{B}_2$-Raum offenbar im $\mathfrak{R}$ erfüllt. Folglich ist folgendes zu behaupten :

Die Menge der allen von Null verschiedenen Elemente des allgemeinen linearen Raums ist ein $\mathfrak{B}_2$-Raum in Bezug auf die oben erklärte lineare Abhängigkeit.

Wenn man ferner die Bedingung $\sum_i \alpha_i = 0$ zur obigen Definition der linearen Abhängigkeit hinzufügt (d. h. $x_1, \cdots, x_s$ punktabhängig !), so sind die vier Axiome des $\mathfrak{B}_1$-Raums offenbar im $\mathfrak{R}$ erfüllt. Also kann man behaupten :

Der allgemeine lineare Raum ist ein $\mathfrak{B}_1$-Raum in Bezug auf die solchermasse definierte Abhängigkeit.

Da es bei den beiden Definitionen der linearen Abhängigkeit mindestens drei Punkte auf jeder Gerade gibt, so ist der allgemeine lineare Raum ein $\mathfrak{P}$-Raum. Der Hilbertsche Raum ist bekanntlich ein Sonderfall des allgemeinen linearen Raumes, daher ist unser Zyklenkalkül auch dazu anzuwenden.

2. Die Anwendung zur "*Lattice*"-Theorie[20], *Boolschen* Algebra[21], und *Verbändentheorie*[22].

Wir denken uns ein solches "Lattice C", welches kommutativ, assoziativ, reduktiv[23], und schwach distributiv[24] ist, und noch dem

(19) S. Banach, "*Sur les opérations dans les ensembles abtraits et leur application aux équations integrales*", Fund. Math., **3**, (1922), S. 135.
A. Tychonoff, "*Ein Fixpunktsatz*", Math. Ann., **111**, (1935), S. 767.

(20) G. Birkhoff, "*On the combination of subalgebras*", Proc. of the Cam. Phil. Soc., **29**, 1933, (441–464), sowie "*Applications of lattice algebra*", ibid., **30**, 1934, (115–122).

(21) E. V. Huntington, "*Set of independent postulates for the algebra of logic*", Trans. of the Amer. Math. Soc., Vol. **5**, 1904, (288–290).
G. Bergmann, "*Zur Axiomatik der Elementargeometrie*", Monat. für Math. und. Phys., **36**, 1929, (269–290).

(22) Fritz Klein, "*Zur Theorie der abstrakten Verknüpfungen*", Math. Ann., **105**, 1931, (308–323), sowie *Über einen Zerlegungssatz in der Theorie der abstrakten Verknüpfungen*", Math. Ann., **106**, 1932, (114–130).

(23) d. h. $a \frown (a \smile b) = a \smile (a \frown b) = a$ für beliebige a, b. L4 auf S. 743 in G. Birkhoff, "*Combinatorial relations in projective geometries*", Ann. of Math., Vol. **36**, 1935, (743–748). Wegen der Bequemlichkeit bezeichnen wir diese Arbeit kurz mit "G. Birkhoff C".

(24) d. h. $(a \smile c) \frown \{b \smile (a \frown c)\} = \{(a \smile c) \frown b\} \smile (a \frown c)$ für beliebige a, b, c. L5 auf S. 442 in G. Birkhoff, "*On the structure of abstract algebras*", Proc. of the Cam. Phil. Soc., **31**, 1935, (433–454).

Vielfachenkettensatz[25], genügt[26]. Dies ist nämlich nach G. Birkhoff sogenanntes "*modular Lattice*", *dessen Elemente lauter von endlichen Dimensionen sind*[27]. Die Menge der allen ein-dimensionalen Elemente, nämlich der allen Punkte von C, bezeichnen wir mit L, und s Punkte $a_1, \cdots, a_s$ derselben bezeichnen wir mit $a_1 \cdots a_s \neq 0$ bzw. $a_1 \cdots a_s = 0$, je nachdem, $\dim(a_1 \cup \cdots \cup a_s) = s$ oder $\dim(a_1 \cup \cdots \cup a_s) < s$ ist. So bestehen offenbar die fünf Axiome des $\mathfrak{B}_2$-Raums in L. Daher gilt;

Die Punktmenge von jedem "modular Lattice", das dem Vielfachenkettensatz genügt, ist ein $\mathfrak{B}_2$-Raum von endlichen Dimensionen.

Es sei hier bemerkt, dass in L aus $a_1 \neq a_2$ $a_1a_2 \neq 0$ folgt, und demnach besteht der Punkt $\mathfrak{R}(a)$ aus einem einzigen Element a. Setzt man nun noch das Gesetz des Komplements[28] in C voraus, so lässt sich jedes n-dimensionale Element a als $a = a_1 \cup \cdots \cup a_n$ bezüglich der n treffenden Punkte $a_1, \cdots, a_n$ darstellen[29], also kann man folgendermassen behaupten;

Die Punktmenge von jedem "complemented modular Lattice", das dem Vielfachenkettensatz genügt, ist ein $\mathfrak{B}_2$-Raum von endlichen Dimensionen, und jedes Element von Lattice entspricht eineindeutig zu einem linearen Raum von $\mathfrak{B}_2$-Raum.

Setzt man das distributive Gesetz $a \cap (b \cup c) = (a \cap b) \cup (a \cap c)$[30] an die Stelle des schwachen distributiven Gesetzes in "complemented modular Lattice", so ergibt sich die Boolsche Algebra; also ist jede Boolsche Algebra, die dem Vielfachenkettensatz genügt, auch selbstverständlich ein $\mathfrak{B}_2$-Raum. Da in der Boolschen Algebra können keine drei Punkte auf einer Gerade liegen, so zerfällt die Boolsche Algebra in die direkte Summe von endlichen Punkten; infolgedessen wird sie endlich. Es ist ohne weiteres einzusehen, dass sich der von Herrn F. Klein betrachtete sog. Verband auch als ein Lattice auffassen lässt.

---

(25) d. i. in jeder Reihe der Produkte a_1, $a_1 \cap a_2$, $a_1 \cap a_2 \cap a_3$, ... irgend zwei Absätze je gleich sind. L_4 auf S. 801 in G. Birkhoff, "*Abstract linear dependence and lattices*", Amer. Journ. of Math., Vol. **57**, 1935, (800–804).

(26) Das Gesetz des Komplements ist hier nicht immer notwendig.

(27) S. 745 in G. Birkhoff C.

(28) Gesetz des Komplements: Für jedes Element a gibt es wenigstens ein Element nämlich "Komplement" a' derart, dass $a \cap a' \cap x = a \cap a'$ und $a \cup a' \cup x = a \cup a'$ für jedes x sind. Vgl. L7 auf S. 743 in G. Birkhoff C!

(29) S. 746 in G. Birkhoff C.

(30) S. 705 in M. H. Stone, "*Postulates for Boolean algebras and generalized Boolean algebras*", Amer. Journ. of Math., Vol. **57**, 1935, (703-732).

3. Die Beziehung zur H. Whitneyschen Matroidentheorie(31).

Es seien M ein $\mathfrak{B}_1$-Raum(32) von endlichen vielen Elementen $e_1, e_2, \cdots, e_n$, und N seine Teilmenge. Sodann bestehen natürlich die folgende Sätze.

R_1 Rang $\mathfrak{N} = 0$(33).

R_2 Rang $(N+e) =$ Rang $N+k$, wo $k=1$ oder 0 ist.

R_3 Rang $(N+e_1) =$ Rang $(N+e_2) =$ Rang N

$\rightarrow$ Rang $(N+e_1+e_2) =$ Rang N.

Umgekehrt legt man diese drei Sätze der abstrakten aus $e_1, e_2, \cdots, e_n$ bestehenden Menge M als Axiome zugrunde, und schreibt man $e_1 \cdots e_n = 0$ bzw. $e_1 \cdots e_n \neq 0$, je nachdem Rang $(e_1 \cdots e_s) < s$ oder Rang $(e_1 \cdots e_s) = s$ ist, so wird M auch ein $\mathfrak{B}_1$-Raum. Diese R_1, R_2, und R_3 sind in der Tat die Axiome des H. Whitneyschen sogenannten Matroid(34), also kann man folgendermassen behaupten;

Jeder endliche $\mathfrak{B}_1$-Raum ist ein Matroid, und umgekehrt.

§ 4. Die Unabhängigkeit des Axiomensystems.

In diesem Paragraphen soll nun ein Axiomensystem, das zum 1°, 5°-Axiom äquivalent und gleichzeitig untereinander unabhängig ist, hergestellt werden. Zuerst sind die folgenden sieben Eigenschaften aus dem 1°, 5°-Axiom leicht herzuleiten.

1* $a \neq 0$.

2* aba.

3* $ac,\ bc \rightarrow ab$.

4* $abc,\ abd \rightarrow ab$ oder bcd.

5* $abcd \rightarrow acbd$.

6* $a_1 \cdots a_s z,\ z b_1 b_2 \rightarrow a_1 \cdots a_s b_1 b_2,\ s \geqq 1$.

7* $a_1 \cdots a_s b_1 b_2 \rightarrow Ez,\ a_1 \cdots a_s z,\ z b_1 b_2,\ s \geqq 1$.

---

(31) H. Whitney, "*On the abstract properties of linear dependence*", Amer. Journ. of Math., Vol. **57**, 1935, (509-533).

S. Maclane, "*Some interpretations of abstract linear dependence in terms of projective geometry*", Amer. Journ. of Math., Vol. **58**, 1936, (236-240).

(32) S. 236 in A1.

(33) $\mathfrak{N}$ bedeutet die leere Menge.

(34) S. 510 in der obigen H. Whitneyschen Arbeit.

Um die Äquivalenz der beiden Axiomensysteme zu zeigen, wollen wir behaupten ;

(1) Es gilt aa für jedes a.

Beweis :
$$\begin{array}{lll} aaa & & \\ az, & zaa & \\ az, & az, & \to aa, \end{array} \qquad \text{w. z. b. w.}$$

(2) $ab \to ba$.

Beweis : bb, $ab \to ba$, w. z. b. w.

(3) Es gilt baa für beliebige a, b.

Beweis :
$$\begin{array}{l} \left.\begin{array}{l} aba, \\ aba \end{array}\right\} \longrightarrow baa, \\ \qquad\quad \downarrow \\ \qquad\quad ab \\ \qquad\quad ba, \; aaa \to baa, \qquad \text{w. z. b. w.} \end{array}$$

(4) Es gilt aab für beliebige a, b.

Beweis :
$$\begin{array}{l} \left.\begin{array}{l} baa, \\ bab \end{array}\right\} \longrightarrow aab, \\ \qquad\quad \downarrow \\ \qquad\quad ba \\ \qquad\quad ab, \; bab \to aab, \qquad \text{w. z. b. w.} \end{array}$$

(5) $abc \to bac$.

Beweis :
$$\begin{array}{l} \left.\begin{array}{l} aba, \\ abc \end{array}\right\} \longrightarrow bac, \\ \qquad\quad \downarrow \\ \qquad\quad ab \\ \qquad\quad ba, \; aac \to bac, \qquad \text{w. z. b. w.} \end{array}$$

(6) $abc \to bca$.

Beweis :
$$\begin{array}{l} \left.\begin{array}{l} abc, \\ aba \end{array}\right\} \longrightarrow bca, \\ \qquad\quad \downarrow \\ \qquad\quad ab \\ \qquad\quad ba, \; aca \to bca, \qquad \text{w. z. b. w.} \end{array}$$

(7) $a_1 \cdots a_s \to a_1 \cdots a_s x$ für jedes x, $s \geqq 2$.

Beweis: $a_1 \cdots a_s$

$a_1 \cdots a_s$, $a_s a_s x$

$a_1 \cdots a_s x$, w. z. b. w.

(8) $a_1 \cdots a_m z$, $z b_1 \cdots b_n \to a_1 \cdots a_m b_1 \cdots b_n$, $m \geqq 1$, $n \geqq 2$.

Beweis: (Vollständige Induktion in Bezug auf n)

$a_1 \cdots a_m z$, $z b_1 \cdots b_n$

$a_1 \cdots a_m z$, $z b_1 \cdots b_{n-2} z_1$, $z_1 b_{n-1} b_n$

$a_1 \cdots a_m b_1 \cdots b_{n-2} z_1$, $z_1 b_{n-1} b_n$

$a_1 \cdots a_m b_1 \cdots b_n$, w. z. b. w.

(9) $a_1 \cdots a_m b_1 \cdots b_n \to Ez$, $a_1 \cdots a_m z$, $z b_1 \cdots b_n$, $m \geqq 1$, $n \geqq 2$.

Beweis: (Vollständige Induktion in Bezug auf n)

$a_1 \cdots a_m b_1 \cdots b_n$

$a_1 \cdots a_m b_1 \cdots b_{n-2} z_1$, $z_1 b_{n-1} b_n$

$a_1 \cdots a_m z$, $z b_1 \cdots b_{n-2} z_1$, $z_1 b_{n-1} b_n$

$a_1 \cdots a_m z$, $z b_1 \cdots b_n$, w. z. b. w.

(10) $a_1 \cdots a_i \cdots a_s \to a_i \cdots a_1 \cdots a_s$, $s = 4$, $4 \geqq i \geqq 2$.

Beweis:	$abcd$	$abcd$	$abcd$	
	abz, zcd	$bacd$	$badc$	
	baz, zcd	$bcad$	$bdac$	
	$bacd$,	$cbad$,	$dbca$,	w. z. b. w.

(11) $a_1 \cdots a_i \cdots a_s \to a_i \cdots a_1 \cdots a_s$, $s \geqq 5$, $s \geqq i \geqq 2$.

Beweis: (Vollständige Induktion in Bezug auf s)

(i) $i \leqq s - 2$.

$a_1 \cdots a_i \cdots a_s$

$a_1 \cdots a_i \cdots a_{s-2} z$, $z a_{s-1} a_s$

$a_i \cdots a_1 \cdots a_{s-2} z$, $z a_{s-1} a_s$

$a_i \cdots a_1 \cdots a_s$, w. z. b. w.

(ii) $i = s - 1$ (iii) $i = s$

$$\begin{gathered} a_1 \cdots a_{s-3} a_{s-2} a_{s-1} a_s \\ a_{s-2} \cdots a_{s-3} a_1 a_{s-1} a_s \\ a_{s-2} \cdots a_{s-3} z, \quad z a_1 a_{s-1} a_s \end{gathered}$$

$$\begin{array}{l|l} a_{s-2} \cdots a_{s-3} z, \quad z a_{s-1} a_1 a_s & a_{s-2} \cdots a_{s-3} z, \; z a_s a_{s-1} a_1 \\ a_{s-2} \cdots a_{s-3} a_{s-1} a_1 a_s & a_{s-2} \cdots a_{s-3} a_s a_{s-1} a_1 \\ a_{s-1} \cdots a_{s-3} a_{s-2} a_1 a_s, & a_s \cdots a_{s-3} a_{s-2} a_{s-1} a_1, \end{array}$$

w. z. b. w.

(12) $$\left.\begin{array}{l} a b x_1 \cdots x_m, \\ a b y \end{array}\right\} \longrightarrow ab \text{ od. } b x_1 \cdots x_m y, \quad m \geqq 1.$$

Beweis: $\left.\begin{array}{l} a b x_1 \cdots x_m, \\ a b y \end{array}\right\}$

$\therefore$ $\left.\begin{array}{l} a b z, \; z x_1 \cdots x_m, \\ a b y \end{array}\right\}$

$\therefore$ $(ab \text{ od. } bzy), \; z x_1 \cdots x_m$

$\therefore$ $ab \text{ od. } (byz, \; z x_1 \cdots x_m)$

$\therefore$ $ab \text{ od. } b x_1 \cdots x_m y,$ w. z. b. w.

(13) $$\left.\begin{array}{l} a_1 \cdots a_s x_1 \cdots x_m, \\ a_1 \cdots a_s y \end{array}\right\} \longrightarrow a_1 \cdots a_s \text{ od. } a_2 \cdots a_s x_1 \cdots x_m y, \quad s \geqq 2, \; m \geqq 1.$$

Beweis: (Vollständige Induktion in Bezug auf s)

$\left.\begin{array}{l} a_1 \cdots a_s x_1 \cdots x_m, \\ a_1 \cdots a_s y \end{array}\right\}$

$\therefore$ $\left.\begin{array}{l} a_1 \cdots a_s x_1 \cdots x_m, \\ a_1 \cdots a_{s-1} z, \; z a_s y \end{array}\right\}$

$\therefore$ $(a_1 \cdots a_{s-1} \text{ od. } a_2 \cdots a_s x_1 \cdots x_m z), \; z a_s y$

$\therefore$ $a_1 \cdots a_{s-1} \text{ od. } (a_2 \cdots a_s x_1 \cdots x_m z, \; z a_s y)$

$\therefore$ $a_1 \cdots a_{s-1} \text{ od. } a_s z \text{ od. } a_2 \cdots a_s x_1 \cdots x_m y$

$\therefore$ $a_1 \cdots a_{s-1} \text{ od. } (a_1 \cdots a_{s-1} z, \; z a_s) \text{ od. } a_2 \cdots a_s x_1 \cdots x_m y$

$\therefore$ $a_1 \cdots a_{s-1} \text{ od. } a_1 \cdots a_s \text{ od. } a_2 \cdots a_s x_1 \cdots x_m y$

$\therefore$ $a_1 \cdots a_s \text{ od. } a_2 \cdots a_s x_1 \cdots x_m y,$ w. z. b. w.

Folglich gilt ;

Satz 76. *Das 1^*, 7^*-Axiom und 1°, 5°-Axiom sind miteinander äquivalent.*

Wir wollen nun die Unabhängigkeit des 1^*, 7^*-Axiomensystems zeigen.

(1) 1^* ist von übrigen Axiomen unabhängig.

Beweis: Dafür brauchen wir uns nur eine aus einzigen Element bestehende Menge zu denken, indem wir allen Zyklen als " $=0$ " setzen.

(2) 2^* ist von übrigen unabhängig.

Beweis: Dafür brauchen wir uns nur eine aus einzigen Elemente bestehende Menge zu denken, indem wir dann allen Zyklen als " $\neq 0$ " setzen.

(3) 3^* ist von übrigen unabhängig.

Beweis: Wir denken uns eine aus zwei Elementen a, b bestehende Menge, und deuten dann die Abhängigkeit der Zyklen folgendermassen :

Anzahl der Elemente des Zyklus	1	2	$3 \leqq$
Festsetzung der Abhängigkeit des Zyklus	$\neq 0$	$ab \neq 0$ sonst, $=0$	$=0$

(4) 4^* ist von übrigen unabhängig.

Beweis: Um ein konkretes Model zu geben denken wir uns eine aus drei Elementen a, b, c bestehende Menge, und definieren dann die Abhängigkeit der Zyklen durch die folgende Festsetzung :

Anzahl der Elemente des Zyklus	1	2	3	$4 \leqq$
Festsetzung der Abhängigkeit des Zyklus	$\neq 0$	$A \neq 0$ $B = 0$	$abc \neq 0$ sonst, $=0$	$=0$

wobei :

A : der aus lauter verschiedenen Elementen bestehende Zyklus und

B : der mindestens ein Paar von denselben Elementen enthaltende Zyklus.

(5) 5* ist von übrigen unabhängig.

Beweis: Als ein konkretes Model denken wir uns eine aus fünf Elementen a, b, c, d, e bestehende Menge, und definieren die Abhängigkeit der Zyklen durch folgende Festsetzung:

Anzahl der Elemente des Zyklus	1	2	3	4	5 $\leqq$
Festsetzung der Abhängigkeit des Zyklus	$\neq 0$	$A \neq 0$ $B = 0$	$B = 0$ $(abe) = 0$ $(cde) = 0$ sonst, $\neq 0$	$B = 0$ $(e) = 0$ $((ab)(cd)) = 0$ sonst, $\neq 0$	$= 0$

wobei:

A: der aus lauter verschiedenen Elementen bestehende Zyklus,

B: der mindestens ein Paar von denselben Elementen enthaltende Zyklus,

(abe): der in beliebiger Reihenfolge aus a, b, e bestehende Zyklus, und dasselbe gilt für (cde),

(e): der wenigstens einmal e enthaltende Zyklus, und

$((ab)(cd))$: einer der $abcd$, $bacd$, $abdc$, und $badc$ ist.

(6) 6_1^* ist von übrigen unabhängig.

Beweis: Als ein Model denken wir uns eine aus drei Elementen a, b, c bestehende Menge, indem wir die Abhängigkeit der Zyklen folgendermassen festsetzen:

Anzahl der Elemente des Zyklus	1	2	3	4 $\leqq$
Festsetzung der Abhängigkeit des Zyklus	$\neq 0$	$= 0$	$aab \neq 0$ sonst, $= 0$	$= 0$

(7) 6_s^*, $(s \geqq 2)$ ist von übrigen unabhängig.

Beweis: Wir betrachten eine aus einem einzigen Element bestehende Menge, inwelcher die Abhängigkeit der Zyklen folgendermassen definiert ist:

Anzahl der Elemente des Zyklus	1	$2 \leqq, \leqq s+1$	$s+2 \leqq$
Festsetzung der Abhängigkeit des Zyklus	[illegible]	$=0$	$\neq 0$

(8) 7_1^* ist von übrigen unabhängig.

Beweis : Wir denken uns eine aus einem einzigen Elemente bestehende Menge. Die Festsetzung der Abhängigkeit der Zyklen ist hierbei die folgende :

Anzahl der Elemente des Zyklus	$1 \leqq, \leqq 2$	$3 \leqq$
Festsetzung der Abhängigkeit des Zyklus	$\neq 0$	$=0$

(9) 7_s^*, $(s \geqq 2)$ ist von übrigen unabhängig.

Beweis : Wir denken uns eine aus $s+2$ Elementen bestehende Menge mit der folgenden Festsetzung der Abhängigkeit der Zyklen :

Anzahl der Elemente des Zyklus	$1 \leqq, \leqq s+1$	$s+2 \leqq$
Festsetzung der Abhängigkeit des Zyklus	$A \neq 0$ sonst, $=0$	$=0$

wobei A der aus lauter verschiedenen Elementen bestehende Zyklus ist.

Schliesslich kann man behaupten ;

Satz 77. *Das 1^*, 7^*-Axiom ist untereinander unabhängig.*

Das hier betrachtete Axiomensystem ist selbstverständlich widerspruchsfrei, aber kategorisch ist es nicht.

Bei Verlegung dieser Arbeit hat mein Lehrer Prof. K. Nakamura sich der grossen Mühe unterzogen, eine vollständige Korrektur mitzulesen, und mich vielfach mit kritischen Bemerkungen und wertvollen Ratschläge und herzlichen Anregungen unterstützt. Auch an dieser Stelle sei ihm der herzlichste Dank ausgesprochen.

Über die Abbildungskette vom Projektionsspektrum

Über die Abbildungskette vom Projektionsspektrum.

Von Takeo NAKASAWA.

(Eingegangen am 13. Dezember, 1937)

Einleitung.

In vorliegender Arbeit wollen wir zwei neue Begriffe, d. h. *Abbildungskette* und *Projektionspunktfolge* einführen, und dadurch wird dasjenige untersucht, welches man etwa die Analyse und die Verfeinerung eines *Homöomorphiesatzes*[1] *von Alexandroff* nennen darf. Durch die Abbildungskette wollen wir nämlich denselben Satz in die einseitige Homomorphiesätze zerlegen, welche die topologischen Eigenschaften der Abbildungsketten zum deutlichen Ausdruck bringen, und durch die Projektionspunktfolge den Beweis derselben Sätze einigermassen vereinfachen. Als nächstes wollen wir unter Verwendung des Abbildungskettenbegriffs einen Beweis des Satzes von Borsuk[2] angeben. Und zuletzt betrachten wir eine Erweiterung des Abbildungskettenbegriffs und einige daraus verfolgte Ergebnisse.

Bezeichnungen.

Zuerst schicken wir die ausführlichen Definitionen für die häufig vorkommenden Buchstaben bzw. Bezeichnungen voraus: Damit sparen wir die mühsame Wiederholung derselben Definitionen.

A bzw. B bedeuten je ein Kompaktum, d. i. einen kompakten metrisierbaren Raum.

$\{A_m\}$ bzw. $\{B_m\}$ bedeuten je eine *geometrisch realisierte* Spektrenentwicklung von A oder von B, d. i. ein bzw. von A oder von B konstruiertes und bzw. in A oder in B geometrisch realisiertes Projektionsspektrum.

$\{\mathfrak{A}_m\}$ bzw. $\{\mathfrak{B}_m\}$ bedeuten je eine Unterteilungsfolge von A oder von B, deren Nerfenfolge bzw. $\{A_m\}$ oder $\{B_m\}$ ist, d. i. Nerf von $\mathfrak{A}_m = A_m$ und Nerf von $\mathfrak{B}_m = B_m$ sind.

---

(1) P. Alexandroff, *Gestalt und Lage*, Ann. of Math. **30** (1929), S. 134.

(2) K. Borsuk, *Sur les rétracts*, Fund. Math. **17** (1931), S. 165.

α_m bzw. β_m bedeuten je ein Durchmesser der Überdeckung $\mathfrak{A}_m$ oder der Überdeckung $\mathfrak{B}_m$, d. i. $\mathfrak{A}_m = \alpha_m$-Überkeckung und $\mathfrak{B}_m = \beta_m$-Überdeckung sind.

$d(\mathfrak{A}_m)$ bzw. $d(\mathfrak{B}_m)$ bedeuten je eine *Lebesguesche Konstante*, welche in Bezug auf die Überdeckung $\mathfrak{A}_m$ oder in Bezug auf die Überdeckung $\mathfrak{B}_m$ bestimmt wird, (Vgl. Gestalt und Lage, S. 115, Hilfssatz).

a_m (bzw. $a_{i_1\cdots i_m}$) bedeutet ein Eckpunkt von A_m, und b_m (bzw. $b_{j_1\cdots j_m}$) bedeutet ein Eckpunkt von B_m.

Eine Folge, welche gleichzeitig eine Projektionsfolge und eine Eckpunktfolge ist, nennen wir die *Projektionspunktfolge*.

Was die Bedeutung der anderen Bezeichnungen betrifft, vgl. die a. a. O.[1] zitierte Alexandroffsche Arbeit!

1.

Satz 1. *Wenn für jedes m eine simpliziale Abbildung f_{ν_m} von A_{ν_m} in[8] B_m derart existiert, dass für beliebige p und q, $p \leqq q$, und einen beliebigen Eckpunkt a_{ν_q} von A_{ν_q} die Relation*

$$(A) \qquad f_{\nu_p}\pi_p^q(a_{\nu_q}) \cdot\!-\!\cdot\ \pi_p^q f_{\nu_q}(a_{\nu_q})$$

in B_p stattfindet, dann bestimmt die Folge $\{f_{\nu_m}\}$ der simplizialen Abbildungen f_{ν_m} eine eindeutige Abbildung von A in B.

Beweis: Es sei $\{a_{i_1\cdots i_m}\}$ die einen Punkt x von A definierende Projektionspunktfolge. Wegen der Ungleichung $\rho\big(f_{\nu_p}(a_{i_1\cdots i_{\nu_p}}), f_{\nu_q}(a_{i_1\cdots i_{\nu_q}})\big) \leqq 2\beta_p$, konvergiert die Folge $\big\{f_{\nu_m}(a_{i_1\cdots i_{\nu_m}})\big\}$. Den Häufungspunkt bezeichnen wir mit y, und wir definieren die Abbildung f von A in B durch $f(x) = y$. Da x definierende Projektionspunktfolgen untereinander benachbart sind, so sind ihre Abbildungen durch f_{ν_m} auch benachbart. Daher ist f eine eindeutige Abbildung von A in B.

Es sei nun $\{x^n\} \to x$ in A, und $f(x^n) = y^n$ in B. Sei $\{a^n_{i_1\cdots i_m}\}$ die x^n definierende Projektionspunktfolge, dann kann für beliebiges p ein hinreihend grosses n so gewählt werden, dass die Relation $a_{i_1\cdots i_{\nu_p}} \cdot\!-\!\cdot\ a^n_{i_1\cdots i_{\nu_p}}$ stattfindet; daher ist $\rho\big(f_{\nu_p}(a_{i_1\cdots i_{\nu_p}}), f_{\nu_p}(a^n_{i_1\cdots i_{\nu_p}})\big) < 2\beta_p$. Aus $\rho\big(f_{\nu_p}(a_{i_1\cdots i_{\nu_p}}), f_{\nu_q}(a_{i_1\cdots i_{\nu_q}})\big) < 2\beta_p$ ist anderseits $\rho\big(f_{\nu_p}(a_{i_1\cdots i_{\nu_p}}), y\big) \leqq 2\beta_p$. Ebenfalls ist $\rho\big(f_{\nu_p}(a^n_{i_1\cdots i_{\nu_p}}), y^n\big) \leqq 2\beta_p$. Daher ist $\rho(y, y^n) < 6\beta_p$; infolge dessen ist f stetig, w. z. b. w.

Satz 1'. *Wenn für jedes m eine simpliziale Abbildung f_{ν_m} von A_{ν_m} auf*[4] *B_m derart existiert, dass für beliebige p und q, $p \leqq q$, und einen beliebigen Eckpunkt a_{ν_q} von A_{ν_q} die Relation*

$$(A) \qquad f_{\nu_p}\pi_p^q(a_{\nu_q}) \cdot - \cdot \pi_p^q f_{\nu_q}(a_{\nu_q})$$

in B_p stattfindet, dann bestimmt die Folge $\{f_{\nu_m}\}$ der simplizialen Abbildungen f_{ν_m} eine eindeutige stetige Abbildung von A auf B.

Beweis: Nach Satz 1 genügt es zu beweisen, dass wenigstens ein Punkt von A auf einen beliebigen Punkt y von B sich abbildet. Es sei $\{a^m_{i_1\cdots i_{\nu_m}}\}$ eins der Urbilder der y definierenden Projektionspunktfolgen $\{b_{j_1\cdots j_m}\}$ in Bezug auf $\{f_{\nu_m}\}$, und sei ferner x^m der Häufungspunkt einer Projektionspunktfolge $\{a^m_{i_1\cdots i_s}\}$, derer ν_m^{te} Eckpunkt $a^m_{i_1\cdots i_{\nu_m}}$ ist. Da A kompakt ist, so konvergiert eine passend gewählte Teilfolge $\{x^{m_\mu}\}$ der Folge $\{x^m\}$ gegen einen Punkt x von A. Es sei ferner $f(x^m) = y^m$, $f(x) = y^*$, so folgt,

$$\lim \rho(y, b_{j_1\cdots j_{m_\mu}}) = 0\,,$$

$$\lim \rho(b_{j_1\cdots j_{m_\mu}}, y^{m_\mu}) \leqq \lim 2\beta_{m_\mu} = 0\,,$$

$$\lim \rho(y^{m_\mu}, y^*) = 0\,.$$

Daher ist $\rho(y, y^*) = 0$.

Also ist $f(x) = y$, w. z. b. w.

Satz 2. *Wenn eine eindeutige stetige Abbildung f von A in B existiert, dann können eine Teilfolge $\{A_{\nu_m}\}$ von $\{A_m\}$ und eine simpliziale Abbildung f_{ν_m} von A_{ν_m} in B_m für jedes m so gefunden werden, dass für beliebige p und q, $p \leqq q$, und einen beliebigen Eckpunkt a_{ν_q} von A_{ν_q} die Relation*

$$(A) \qquad f_{\nu_p}\pi_p^q(a_{\nu_q}) \cdot - \cdot \pi_p^q f_{\nu_q}(a_{\nu_q})$$

in B_p stattfindet, und dass die Folge $\{f_{\nu_m}\}$ die Abbildung f definiert.

Beweis: Setzt man $f(A) = A'$ in B, so ist die Folge $\{f(\mathfrak{A}_m)\}$ eine Unterteilungsfolge von A'. Dann lässt sich für beliebiges m ein

(3), (4) Hierbei bedeutet "*in*" "*auf einen echten oder unechten Teilkomplex* (*bzw. Teilmenge*) *von*", und ähnlich "*auf*" "*auf einen unechten Teilkomplex* (*bwz. Teilmenge*) *von*", und ein Teilkomplex Q eines gegebenen Komplexes P soll ein *unechter Teilkomplex* heissen, wenn Q alle Eckpunkte von P enthält.

hinreihend grosses ν_m so finden, dass der Durchmesser von $f(\mathfrak{A}_{\nu_m})$ kleiner als die Lebesguesche Konstante $d(\mathfrak{B}_m)$ von $\mathfrak{B}_m$ ist. Dann lässt sich der Komplex $f(A_{\nu_m})$ in B durch eine simpliziale Abbildung φ_{ν_m} in B_m abbilden. Setzt man $\varphi_{\nu_m} f = f_{\nu_m}$, so ist f_{ν_m} eine simpliziale Abbildung von A_{ν_m} in B_m. Wegen der Ungleichung $\rho\big(f(a_{i_1\cdots i_{\nu_p}}), f(a_{i_1\cdots i_{\nu_q}})\big) < d(\mathfrak{B}_p)$ gilt

$$f_{\nu_p}(a_{i_1\cdots i_{\nu_p}}) \cdot\!-\!\cdot\; \pi^q_p f_{\nu_q}(a_{i_1\cdots i_{\nu_q}}) \quad \text{in} \quad B_p\,,$$

was die Bedingung (A) ist. Wegen der Ungleichung $\rho\big(f(a_{i_1\cdots i_{\nu_m}}), f_{\nu_m}(a_{i_1\cdots i_{\nu_m}})\big) < \beta_m$ gilt auch

$$\lim f_{\nu_m}(a_{i_1\cdots i_{\nu_m}}) = \lim f(a_{i_1\cdots i_{\nu_m}}) = f(x)\,.$$

Also definiert die Folge $\{f_{\nu_m}\}$ die Abbildung f, w. z. b. w.

Satz 2'. *Wenn eine eindeutige stetige Abbildung f von A auf B existiert, dann können eine Teilfolge $\{A_{\nu_m}\}$ von $\{A_m\}$ und eine simpliziale Abbildung f_{ν_m} von A_{ν_m} auf B_m für jedes m so gefunden werden, dass für beliebige p und q, $p \leq q$, und einen beliebigen Eckpunkt a_{ν_q} von A_{ν_q} die Relation*

$$(A) \qquad f_{\nu_p}\pi^q_p(a_{\nu_q}) \cdot\!-\!\cdot\; \pi^q_p f_{\nu_q}(a_{\nu_q})$$

in B_p stattfindet, und dass die Folge $\{f_{\nu_m}\}$ die Abbildung f definiert.

Beweis: Nehmen wir die Lebesguesche Konstante, welche bei dem Beweis des Satzes 2 auftritt, hinreichend klein, so wird wegen der Gleichung $f(A) = B$ $f(A_{\nu_m})$ durch eine simpliziale Abbildung *auf* B_m abgebildet. Was übrig bleibt läuft ganz analog wie bei dem Satz 2.

Nun wollen wir die Definition der Abbildungskette ausführlich aussprechen :

Definition 1. Die Folge $\{f_{\nu_m}\}$ von simplizialen Abbildungen, welche die Bedingung (A) erfüllt, soll nun eine *Abbildungskette* von $\{A_m\}$ *in* (oder *auf*) $\{B_m\}$ heissen.

Durch diesen Begriff können wir die oben bewiesenen Sätze in der folgenden Form aussprechen :

Satz I. *Die eindeutige stetige Abbildung von A in B lässt sich vollständig durch die Abbildungskette von $\{A_m\}$ in $\{B_m\}$ charakterisieren.*

Satz I'. *Die eindeutige stetige Abbildung von A auf B lässt sich vollständig durch die Abbildungskette von $\{A_m\}$ auf $\{B_m\}$ charakterisieren.*

2.

Satz II[5]. *A und B sind dann und nur dann ineinander homöomorph, wenn sich eine Teilfolge $\{A_{\lambda_{2m-1}}\}$ von $\{A_m\}$, eine Teilfolge $\{B_{\lambda_{2m}}\}$ von $\{B_m\}$, und für jedes m eine simpliziale Abbildung $f_{\lambda_{2m+1}}$ von $A_{\lambda_{2m+1}}$ in $B_{\lambda_{2m}}$ und eine simpliziale Abbildung $g_{\lambda_{2m}}$ von $B_{\lambda_{2m}}$ in $A_{\lambda_{2m-1}}$ so finden lassen, dass für beliebige p und q, $p \leqq q$, und einen beliebigen Eckpunkt $a_{\lambda_{2q+1}}$ von $A_{\lambda_{2q+1}}$ bzw. $b_{\lambda_{2q}}$ von $B_{\lambda_{2q}}$ die Relationen*

$$(A) \qquad f_{\lambda_{2p+1}}\pi^{2q+1}_{2p+1}(a_{\lambda_{2q+1}}) \cdot\!-\!\cdot\ \pi^{2q}_{2p}f_{\lambda_{2q+1}}(a_{\lambda_{2q+1}}),$$

$$g_{\lambda_{2p}}\pi^{2q}_{2p}(b_{\lambda_{2q}}) \cdot\!-\!\cdot\ \pi^{2q-1}_{2p-1}g_{\lambda_{2q}}(b_{\lambda_{2q}}),$$

$$(B) \qquad g_{\lambda_{2q}}f_{\lambda_{2q+1}}(a_{\lambda_{2q+1}}) \cdot\!-\!\cdot\ \pi^{2q+1}_{2q-1}(a_{\lambda_{2q+1}}),$$

$$f_{\lambda_{2q-1}}g_{\lambda_{2q}}(b_{\lambda_{2q}}) \cdot\!-\!\cdot\ \pi^{2q}_{2q-2}(b_{\lambda_{2q}})$$

bzw. in $B_{\lambda_{2p}}$, in $A_{\lambda_{2p-1}}$, in $A_{\lambda_{2q-1}}$, in $B_{\lambda_{2q-2}}$ stattfinden.

Beweis: Seien A und B ineinander homöomorph. Dann können nach Satz 2 eine Teilfolge $\{A_{\nu_m}\}$ von $\{A_m\}$, eine Teilfolge $\{B_{\mu_m}\}$ von $\{B_m\}$, und für jedes m eine simpliziale Abbildung f_{ν_m} von A_{ν_m} in B_m und eine simpliziale Abbildung g_{μ_m} von B_{μ_m} in A_m so gefunden werden, dass zwei Relationen von (A) erfüllt werden. Wählen wir weiter die Teilfolge $\{A_{\lambda_{2m-1}}\}$ bzw. $\{B_{\lambda_{2m}}\}$ von $\{A_{\nu_m}\}$ bzw. $\{B_{\mu_m}\}$ nach dem Regel, wie in der folgenden Figur angegeben wird;

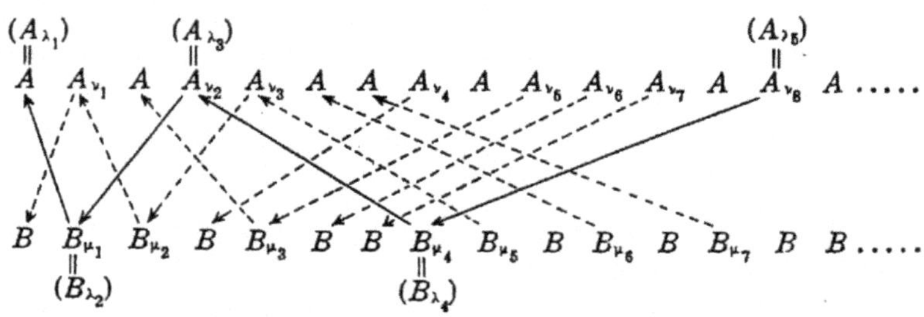

---

(5) Dies ist wörtlich der sog. *Homöomorphiesatz* von Alexandroff, (Ann. of Math., 30 (1929), S. 134).

Dann ist $\rho\left(gf_{\lambda_{2q+1}}(a_{i_1\cdots i_{\lambda_{2q+1}}}),\ gf(a_{i_1\cdots i_{\lambda_{2q+1}}})\right) < d(\mathfrak{A}_{\lambda_{2q-1}})^{(6)}$..

Also ist $g_{\lambda_{2q}} f_{\lambda_{2q+1}}(a_{i_1\cdots i_{\lambda_{2q+1}}}) \cdot - \cdot\ a_{i_1\cdots i_{\lambda_{2q-1}}}$.

Das ist die erste Relation von (B), und auf dieselbe Weise ist die zweite zu beweisen. Damit ist die Bedingung notwendig.

Um die Hinreichendheit ze beweisen genügt es nach Satz 1 zu zeigen, dass $gf(x) = x$ in A und $fg(y) = y$ in B sind. Sei $\{a_{i_1\cdots i_m}\}$ die einen Punkt x von A definierende Projektionspunktfolge, so konvergiert die Folge $\left\{f_{\lambda_{2m+1}}(a_{i_1\cdots i_{\lambda_{2m+1}}})\right\}$, welche man auch $\{b^{2m+1}_{j_1\cdots j_{\lambda_{2m}}}\}$ schreibt, zu dem Punkt $f(x)$ in B, welcher auch y heissen möge. Es sei ferner $\{b_{j_1\cdots j_{\lambda_{2m}}}\}$ die y definierende Projektionspunktfolge. Dann existiert für beliebiges k ein hinreichend grosses n, so dass die Relation $b_{j_1\cdots j_{\lambda_{2k}}} \cdot - \cdot$ $b^{2n+1}_{j_1\cdots j_{\lambda_{2k}}}$ gilt. Daher ist

$$g_{\lambda_{2k}}(b_{j_1\cdots j_{\lambda_{2k}}}) \cdot - \cdot\ g_{\lambda_{2k}}\pi^{2n}_{2k}(b^{2n+1}_{j_1\cdots j_{\lambda_{2n}}}) \cdot - \cdot\ \pi^{2n-1}_{2k-1} g_{\lambda_{2n}}(b^{2n+1}_{j_1\cdots j_{\lambda_{2n}}})$$

$$= \pi^{2n-1}_{2k-1} g_{\lambda_{2n}} f_{\lambda_{2n+1}}(a_{i_1\cdots i_{\lambda_{2n+1}}}) \cdot - \cdot\ \pi^{2n+1}_{2k-1}(a_{i_1\cdots i_{\lambda_{2n+1}}}) = a_{i_1\cdots i_{\lambda_{2k-1}}}.$$

Also gilt $g_{\lambda_{2k}}(b_{j_1\cdots j_{\lambda_{2k}}}) \cdot - \cdot - \cdot - \cdot\ a_{i_1\cdots i_{\lambda_{2k-1}}}$.

Daher ist $gf(x) = \lim g_{\lambda_{2k}}(b_{j_1\cdots j_{\lambda_{2k}}}) = \lim a_{i_1\cdots i_{\lambda_{2k-1}}} = x$.

Daher ist $gf(x) = x$.

Gleicherweise ist die Relation $fg(y) = y$ zu beweisen, w. z. b. w.

Man kann die Bedingung des Satzes II ein wenig modifizieren, indem man die Bedingung von *in*-Abbildbarkeit mit der stärkeren, d. h. Abbildbarkeit von $A_{\lambda_{2m+1}}$ *auf* einen unechten Teilkomplex von $B_{\lambda_{2m}}$ ersetzt, und die zweite Relation von (B) auslässt. Die so veränderte Bedingung ist wieder notwendig und hinreichend. Denn dieselbe Bedingung ist tatsächlich notwendig nach Satz 2′ und dem Notwendigkeitsbeweis des Satzes II. Anderseits folgt nach Satz 1′ $f(A) = B$, und nach dem Hinreichendheitsbeweis des Satzes II gilt $gf =$ *Identität*; dabei wird die zweite Formel von (B) keineswegs gebraucht, womit die Bedingung hinreichend ist. Infolgedessen kann man den folgenden Satz aussprechen:

Satz II′. *A und B sind dann und nur dann ineinander homöomorph, wenn sich eine Teilfolge* $\{A_{\lambda_{2m-1}}\}$ *von* $\{A_m\}$, *eine Teilfolge* $\{B_{\lambda_{2m}}\}$ *von* $\{B_m\}$, *und für jedes m eine simpliziale Abbildung* $f_{\lambda_{2m+1}}$

(6) g bedeutet hierbei bloss f^{-1}.

von $A_{\lambda_{2m+1}}$ auf $B_{\lambda_{2m}}$ und eine simpliziale Abbildung $g_{\lambda_{2m}}$ von $B_{\lambda_{2m}}$ in $A_{\lambda_{2m-1}}$ so finden lassen, dass für beliebige p und q, $p \leqq q$, und einen beliebigen Eckpunkt $a_{\lambda_{2q+1}}$ von $A_{\lambda_{2q+1}}$ bzw. $b_{\lambda_{2q}}$ von $B_{\lambda_{2q}}$ die Relationen

$$(A) \qquad f_{\lambda_{2p+1}}\pi^{2q+1}_{2p+1}(a_{\lambda_{2p+1}}) \cdot - \cdot \pi^{2q}_{2p} f_{\lambda_{2q+1}}(a_{\lambda_{2q+1}}) \,,$$

$$g_{\lambda_{2p}}\pi^{2q}_{2p}(b_{\lambda_{2q}}) \cdot - \cdot \pi^{2q-1}_{2p-1} g_{\lambda_{2q}}(b_{\lambda_{2q}}) \,,$$

$$(B) \qquad g_{\lambda_{2q}} f_{\lambda_{2q+1}}(a_{\lambda_{2q+1}}) \cdot - \cdot \pi^{2q+1}_{2q-1}(a_{\lambda_{2q+1}})$$

bzw. in $B_{\lambda_{2p}}$, in $A_{\lambda_{2p-1}}$, in $A_{\lambda_{2q-1}}$ stattfinden.

3.

Satz 3. *Es seien $\{f_{\lambda_m}\}$ und $\{g_{\mu_m}\}$ Abbildungsketten von $\{A_m\}$ in $\{B_m\}$, und es seien f bzw. g die eindeutige stetige Abbildungen, welche $\{f_{\lambda_m}\}$ bzw. $\{g_{\mu_m}\}$ definieren. Wenn für gegebenes ε ein hinreichend grosses s existiert, so dass für irgendeinen Eckpunkt $a_{i_1 \cdots i_{\lambda_s}}$ von A_{λ_s} bzw. $a_{i_1 \cdots i_{\mu_s}}$ von A_{μ_s} die Relation*(7)

$$(C) \qquad f_{\lambda_s}(a_{i_1 \cdots i_{\lambda_s}}) \cdot - \cdot - \cdot - \cdot g_{\mu_s}(a_{i_1 \cdots i_{\mu_s}})$$

in B_s stattfindet, so gilt die Ungleichung $\rho\big(f(x), g(x)\big) < \varepsilon$ in B für jeden Punkt x von A.

Beweis: Man nehme s so gross, dass die Ungleichung $\varepsilon > 8\beta_s$ gilt. Sei $\{a_{i_1 \cdots i_m}\}$ die einen Punkt x von A definierende Projektionspunktfolge, so sind

$$\rho\big(f_{\lambda_s}(a_{i_1 \cdots i_{\lambda_s}}), f(x)\big) \leqq 2\beta_s \,,$$

$$\rho\big(g_{\mu_s}(a_{i_1 \cdots i_{\mu_s}}), g(x)\big) \leqq 2\beta_s \,,$$

Anderseits ist nach (C)

$$\rho\big(f_{\lambda_s}(a_{i_1 \cdots i_{\lambda_s}}), g_{\mu_s}(a_{i_1 \cdots i_{\mu_s}})\big) < 4\beta_s \,.$$

---

(7) Es wäre notwendig zu bemerken, dass die Indexe von $a_{i_1 \cdots i_{\lambda_s}}$ und $a_{i_1 \cdots i_{\mu_s}}$, d. h. die zwei Eckpunkte in ein und derselben Projektionspunktfolge umgefasst werden.

Also ist $\rho\big(f(x),\, g(x)\big) < 8\beta_s < \varepsilon$ in B, w. z. b. w.

Satz 4. *Es seien* $\{f_{\lambda_m}\}$ *und* $\{g_{\mu_m}\}$ *Abbildungsketten von* $\{A_m\}$ *in* $\{B_m\}$, *und es seien* f *bzw.* g *die eindeutige Abbildungen, welche* $\{f_{\lambda_m}\}$ *bzw.* $\{g_{\mu_m}\}$ *definieren. Wenn es für gegebenes* s *ein hinreichend kleines* ε *existiert, so dass für jeden Punkt* x *von* A *die Ungleichung* $\rho\big(f(x),\, g(x)\big) < \varepsilon$ *in* B *gilt, so gilt für beliebigen Eckpunkt* $a_{i_1\cdots i_{\lambda_s}}$ *von* A_{λ_s} *bzw.* $a_{i_1\cdots i_{\mu_s}}$ *von* A_{μ_s} *die folgende Relation in* B_s:

$$(C) \qquad f_{\lambda_s}(a_{i_1\cdots i_{\lambda_s}}) \cdot\!-\!\cdot\!-\!\cdot\!-\!\cdot\; g_{\mu_s}(a_{i_1\cdots i_{\mu_s}})\,.$$

Beweis: Man nehme ε so klein, dass die Ungleichung $d(\mathfrak{B}_s) > \varepsilon$ gilt, und weiter n so gross, dass die Ungleichung $d(\mathfrak{B}_s) - \varepsilon > 4\beta_n$ gilt. Dann gelten

$$\rho\big(f_{\lambda_n}(a_{i_1\cdots i_{\lambda_n}}),\, f(x)\big) \leqq 2\beta_n\,,$$

$$\rho\big(g_{\mu_n}(a_{i_1\cdots i_{\lambda_n}}),\, g(x)\big) \leqq 2\beta_n\,.$$

Anderseits gilt nach der Voraussetzung $\rho\big(f(x),\, g(x)\big) < \varepsilon$.

Also ist $\rho\big(f_{\lambda_n}(a_{i_1\cdots i_{\lambda_n}}),\, g_{\nu_n}(a_{i_1\cdots i_{\mu_n}})\big) < 4\beta_n + \varepsilon < d(\mathfrak{B}_s)$.

Also ist $\pi_s^n f_{\lambda_n}(a_{i_1\cdots i_{\lambda_n}}) \cdot\!-\!\cdot\; \pi_s^n g_{\mu_n}(a_{i_1\cdots i_{\mu_n}})$.

Anderseits gelten nach der Bedingung (A)

$$f_{\lambda_s}(a_{i_1\cdots i_{\lambda_s}}) \cdot\!-\!\cdot\; \pi_s^n f_{\lambda_n}(a_{i_1\cdots i_{\lambda_n}})\,,$$

$$g_{\nu_s}(a_{i_1\cdots i_{\mu_s}}) \cdot\!-\!\cdot\; \pi_s^n g_{\nu_n}(a_{i_1\cdots i_{\mu_n}})\,.$$

Also ist $f_{\lambda_s}(a_{i_1\cdots i_{\lambda_s}}) \cdot\!-\!\cdot\!-\!\cdot\!-\!\cdot\; g_{\mu_s}(a_{i_1\cdots i_{\nu_s}})$, w. z. b. w.

Daraus lassen sich die folgenden Sätze aussprechen:

Satz III. *Zwei eindeutige stetige Abbildungen von* A *in* B *haben dann und nur dann die sog. "kleine Entfernung", wenn deren zugehörigen Abbildungsketten von* $\{A_m\}$ *in* $\{B_m\}$ *für hinreichend grosses* s *die Bedingung* (C) *erfüllen.*

Satz III′. *Zwei Abbildungsketten* $\{f_{\lambda_m}\}$, $\{g_{\mu_m}\}$ *von* $\{A_m\}$ *in* $\{B_m\}$ *definieren dann und nur dann dieselbe eindeutige stetige Abbildung von* A *in* B, *wenn für jedes* m *die Bedingung* (C) *erfüllt wird.*

Satz III''. *Dass die Bedingung (C) für jedes m erfüllt wird, ist ein Äquivalenzrelation in der Menge der Abbildungsketten.*

Unter Verwendung des Abbildungskettenbegriffs will ich einen neuen Beweis eines Satzes von Borsuk[(8)] angeben.

Satz von Borsuk. *Der Abbildungsraum*[(9)] *ist separabel.*

Beweis: Für irgendeine feste Zahl λ_m können wir die Menge der Abbildungsketten, deren m-te Glied A_{λ_m} in B_m abbildet, in die Klassen einteilen, indem wir Abbildungsketten als gleich ansehen, wenn deren zugehörigen simplizialen Abbildungen von A_{λ_m} in B_m dieselbe Eckpunktzuordnung bewirken. Da die Anzahl der simplizialen Abbildungen von A_{λ_m} in B_m offenbar endlich ist, so ist die Anzahl der Klassen auch endlich. Aus jeder Klasse ziehen wir je eine Abbildungskette als Vertreter der Klasse heraus; es seien $f^m_{1\lambda_m}$, $f^m_{2\lambda_m}$, ..., $f^m_{l(\lambda_m)\lambda_m}$ diese Vertreter. Dann ist die Menge der allen Vertreter $M = \sum_{m=1}^{\infty} \sum_{\lambda_m=m}^{\infty} \sum_{i=1}^{l(\lambda_m)} f^m_{i\lambda_m}$ offenbar eine abzählbare Menge. Nun denken wir uns eine beliebige Abbildungskette $g = \{g_{\nu_m}\}$, und nehmen wir sodann eine endliche Teilmenge $f^s_{1\mu_s}$, $f^s_{2\mu_s}$, ..., $f^s_{l(\mu_s)\nu_s}$ von M in Bezug auf μ_m und für beliebiges festes s, so soll g, gemäss der obigen Klasseneinteilung in Bezug auf die simpliziale Abbildung von A_{μ_s} in B_s, mit irgendeiner Abbildungskette von $\{f^s_{1\mu_s}, \ldots, f^s_{l(\mu_s)\mu_s}\}$ zu derselben Klasse angehören. Nun gehört g, z. B., mit $f^s_{i\mu_s}$ zu derselben Klasse, und sei $f^s_{i\mu_s} = \{f_{\nu_1}, f_{\nu_2}, \ldots, f_{\nu_{s-1}}, f_{\mu_s}, f_{\nu_{s+1}}, \ldots\}$, so gilt für jeden Eckpunkt $a_{i_1 \cdots i_{\mu_s}}$ von A_{μ_s} $f_{\mu_s}(a_{i_1 \cdots i_{\mu_s}}) = g_{\mu_s}(a_{i_1 \cdots i_{\mu_s}})$; dies ist die Bedingung (C). Also gilt nach Satz 3 die Ungleichung $\rho(f^s_i, g) < \varepsilon$; damit liegt die abzählbare Menge M in dem Abbildungsraum überalldicht. Also ist der Abbildungsraum separabel, w. z. b. w.

4.

Definition 2. In diesen Paragraphen bedeuten $k, k_1, \ldots, k_6$ gegebene, voneinander unabhängige Konstante. Und "$a \cdot - (k) - \cdot b$"

(8), (9) Wir definieren eine Metrik in der Menge der eindeutigen stetigen Abbildungen von A in B durch $\rho(f, g) = \underset{x \in A}{\mathrm{Max}}\, \rho(f(x), g(x))$ in B. Die so metrisierte Menge der Abbildungen nennen wir den *Abbildungsraum* von A in B. Der Satz, dass der Abbildungsraum beschränkt, vollständig und separabel ist, ist von *K. Borsuk* (K. Borsuk, *Sur les rétracts*, Fund. Math. 17 (1931), S. 164–165) bewiesen. Was hier von mir betrachtet handelt sich nur um den Beweis der Separabilität.

bedeutet, dass man a und b mit einer aus höchstens k Kanten bestehenden Kette verbinden kann, und a heisst von b *k-benachbart*. Besonders ist a von b gewöhnlich benachbart im Sinne von Alexandroff, wenn $k=1$ ist.

Definition 3. Eine Eckpunktabbildung f von A_{ν_m} in B_m heisst eine *pseudosimpliziale Abbildung*, wenn aus $a \cdot\!-\!\cdot\, b$ in A_{ν_m} die Relation $f(a) \cdot - (k) - \cdot f(b)$ in B_m folgt. Für $k=1$ ist die pseudosimpliziale Abbildung offenber eine gewöhnliche simpliziale Abbildung.

Definition 4. Eine Folge $\{f_{\nu_m}\}$ der pseudosimplizialen Abbildungen f_{ν_m}, $(m = 1, 2, \ldots)$ von A_{ν_m} in (auf) B_m heisst die *Abbildungskette im weiteren Sinne* von $\{A_m\}$ in (auf) $\{B_m\}$, wenn für beliebige p und q, $p \leqq q$, und für einen beliebigen Eckpunkt a_{ν_q} von A_{ν_q} die Relation

$$(A') \qquad f_{\nu_p} \pi_p^q (a_{\nu_q}) \cdot - (k_2) - \cdot\, \pi_p^q f_{\nu_q} (a_{\nu_q}) \quad \text{in} \quad B_p$$

stattfindet. Und für $k_1 = k_2 = 1$ wird die Abbildungskette im weiteren Sinne ein gewöhnliche Abbildungskette.

Satz 5. *Die Abbildungskette im weiteren Sinne von $\{A_m\}$ in (auf) $\{B_m\}$ definiert eine eindeutige stetige Abbildung von A in (auf) B.*

Der Beweis läuft ganz ähnlich mit demselben vom Satz 1 (Satz 1').

Satz IV. *A und B sind ineinander homöomorph, wenn sich eine Teilfolge $\{A_{\lambda_{2m-1}}\}$ von $\{A_m\}$, eine Teilfolge $\{B_{\lambda_{2m}}\}$ von $\{B_m\}$, und für jedes m eine pseudosimpliziale Abbildung $f_{\lambda_{2m+1}}$ von $A_{\lambda_{2m+1}}$ in $B_{\lambda_{2m}}$ und eine pseudosimpliziale Abbildung $g_{\lambda_{2m}}$ von $B_{\lambda_{2m}}$ in $A_{\lambda_{2m-1}}$ so finden lassen, dass für beliebige p und q, $p \leqq q$, und einen beliebigen Eckpunkt $a_{\lambda_{2q+1}}$ von $A_{\lambda_{2q+1}}$ bzw. $b_{\lambda_{2q}}$ von $B_{\lambda_{2q}}$ die Relationen*

$$(A') \qquad f_{\lambda_{2p+1}} \pi_{2p+1}^{2q+1} (a_{\lambda_{2q+1}}) \cdot - (k_3) - \cdot\, \pi_{2p}^{2q} f_{\lambda_{2q+1}} (a_{\lambda_{2q+1}}),$$

$$g_{\lambda_{2p}} \pi_{2p}^{2q} (b_{\lambda_{2q}}) \cdot - (k_4) - \cdot\, \pi_{2p-1}^{2q-1} g_{\lambda_{2q}} (b_{\lambda_{2q}}),$$

$$(B') \qquad g_{\lambda_{2q}} f_{\lambda_{2q+1}} (a_{\lambda_{2q+1}}) \cdot - (k_5) - \cdot\, \pi_{2q-1}^{2q+1} (a_{\lambda_{2q+1}}),$$

$$f_{\lambda_{2q-1}} g_{\lambda_{2q}} (b_{\lambda_{2q}}) \cdot - (k_6) - \cdot\, \pi_{2q-2}^{2q} (b_{\lambda_{2q}})$$

bzw. in $B_{\lambda_{2p}}$, in $A_{\lambda_{2p-1}}$, in $A_{\lambda_{2q-1}}$, in $B_{\lambda_{2q-2}}$ stattfinden.

Der Beweis läuft ganz ähnlich mit dem Hinreichendheitsbeweis des Satzes II.

Satz IV'. *A und B sind miteinander homöomorph, wenn sich eine Teilfolge* $\{f_{\lambda_{2m-1}}\}$ *von* $\{A_m\}$, *eine Teilfolge* $\{B_{\lambda_{2m}}\}$ *von* $\{B_m\}$, *und für jedes m eine pseudosimpliziale Abbildung* $f_{\lambda_{2m+1}}$ *von* $A_{\lambda_{2m+1}}$ *auf* $B_{\lambda_{2m}}$ *und eine pseudosimpliziale Abbildung* $g_{\lambda_{2m}}$ *von* $B_{\lambda_{2m}}$ *in* $A_{\lambda_{2m-1}}$ *so finden lassen, dass für beliebige p und q,* $p \leq q$, *und einen beliebigen Eckpunkt* $a_{\lambda_{2q+1}}$ *von* $A_{\lambda_{2q+1}}$ *bzw.* $b_{\lambda_{2q}}$ *von* $B_{\lambda_{2q}}$ *die Relationen*

$$(A') \qquad f_{\lambda_{2p+1}}\pi^{2q+1}_{2p+1}(a_{\lambda_{2q+1}}) \cdot\!-\!(k_3)\!-\!\cdot\ \pi^{2q}_{2p} f_{\lambda_{2q+1}}(a_{\lambda_{2q+1}}) \,,$$

$$g_{\lambda_{2p}}\pi^{2q}_{2p}(b_{\lambda_{2q}}) \cdot\!-\!(k_4)\!-\!\cdot\ \pi^{2q-1}_{2p-1} g_{\lambda_{2q}}(b_{\lambda_{2q}}) \,,$$

$$(B') \qquad g_{\lambda_{2q}} f_{\lambda_{2q+1}}(a_{\lambda_{2q+1}}) \cdot\!-\!(k_5)\!-\!\cdot\ \pi^{2q+1}_{2q-1}(a_{\lambda_{2q+1}})$$

bzw. in $B_{\lambda_{2p}}$, *in* $A_{\lambda_{2p-1}}$, *in* $A_{\lambda_{2q-1}}$ *stattfinden.*

Der Beweis läuft ganz ähnlich mit dem Hinreichendheitsbeweis des Satzes II'.

Satz 6. *Es seien* $\{f_{\lambda_m}\}$ *und* $\{g_{\mu_m}\}$ *Abbildungsketten im weiteren Sinne von* $\{A_m\}$ *in* $\{B_m\}$, *und es seien f bzw. g die eindeutige stetige Abbildungen, welche* $\{f_{\lambda_m}\}$ *bzw.* $\{g_{\mu_m}\}$ *definieren. Wenn für gegebenes* ε *ein hinreichend grosses s existiert, so dass für irgendeinen Eckpunkt* $a_{i_1\cdots i_{\lambda_s}}$ *von* A_{λ_s} *bzw.* $a_{i_1\cdots i_{\mu_s}}$ *von* A_{μ_s} *die Relation*

$$(C') \qquad f_{\lambda_s}(a_{i_1\cdots i_{\lambda_s}}) \cdot\!-\!(k_6)\!-\!\cdot\ g_{\mu_s}(a_{i_1\cdots i_{\mu_s}})$$

in B_s *stattfindet, so gilt die Ungleichung* $\rho\big(f(x), g(x)\big) < \varepsilon$ *in B für jeden punkt x von A.*

Der Beweis läuft ganz ähnlich nach demselben vom Satz 3.

Satz 7. *Es seien* $\{f_{\lambda_m}\}$ *und* $\{g_{\mu_m}\}$ *Abbildungsketten im weiteren Sinne von* $\{A_m\}$ *in* $\{B_m\}$, *und es seien f bzw. g die eindeutige stetige Abbildungen, welche* $\{f_{\lambda_m}\}$ *bzw.* $\{g_{\mu_m}\}$ *definieren. Wenn es für gegebenes s ein hinreichend kleines* ε *existiert, so dass für jeden Punkt x von A die Ungleichung* $\rho\big(f(x), g(x)\big) < \varepsilon$ *in B gilt, so gilt für beliebigen Eckpunkt* $a_{i_1\cdots i_{\lambda_s}}$ *von* A_{λ_s} *bzw.* $a_{i_1\cdots i_{\mu_s}}$ *von* A_{μ_s} *die folgende Relation in* B_s:

$$(C') \qquad f_{\lambda_s}(a_{i_1\cdots i_{\lambda_s}}) \cdot\!-\!(k_6)\!-\!\cdot\ g_{\mu_s}(a_{i_1\cdots i_{\mu_s}}) \,.$$

Der Beweis läuft ganz ähnlich nach demselben vom Satz 4. Es sei hier bemerkt, dass $k_6 \geq k_3 + k_4 + 1$ sein soll.

Satz V. *Zwei Abbildungsketten im weiteren Sinne* $\{f_{\lambda_m}\}$, $\{g_{\mu_m}\}$ *von* $\{A_m\}$ *in* $\{B_m\}$ *definieren dann und nur dann dieselbe eindeutige Abbildung von* A *in* B, *wenn für jedes* m *die Bedingung* (C') *erfüllt wird.*

Satz V′. *Wenn für zwei Abbildungsketten* $\{f_{\lambda_m}\}$, $\{g_{\mu_m}\}$ *von* $\{A_m\}$ *in* $\{B_m\}$ *die Relation* $f_{\lambda_m}(a_{i_1\cdots i_{\lambda_m}}) \cdot - (k) - \cdot g_{\mu_m}(a_{i_1\cdots i_{\mu_m}})$ *für jedes* m *gilt, so gilt auch für jedes* m *die Relation* $f_{\lambda_m}(a_{i_1\cdots i_{\lambda_m}}) \cdot - \cdot - \cdot - \cdot g_{\mu_m}(a_{i_1\cdots i_{\mu_m}})$.

On Axiomatics of Linear Dependence I: The $\mathfrak{B}_1$-Space

Takeo Nakasawa 30 June, 1935 (received)

Introduction

In this investigation, an axiom system for the notion of linear dependence in an n-dimensional projective space is to be presented with the aid of *cycle calculation*, which Mr. G. Thomson introduced[1] by his relatively abstract treatment of elementary geometry. First, we will deal with the geometry of an abstract space of linear dependence of the first type, whose exact definition will follow shortly.

Preliminaries

In this article, we need the following notation:

1. The notation $A \longrightarrow B$ means that B follows from A.
2. The notation $A \longleftrightarrow B$ means that $A \longrightarrow B$ and $B \longrightarrow A$.
3. The notation $A \longrightarrow \perp$ means that A turns into contradiction.
4. The notation A, B means that A and B hold.

---

[1] G. Thomsen: Grundlagen der Elementargeometrie, (Leipzig 1933), (p. 67–p. 70).

5. The notation $A \vee B$ means that at least one of A and B holds.
6. The notation $A \underset{\longrightarrow}{(S)} B$ means that B follows from A on the basis of the statement S.
7. The notation

$$\left.\begin{matrix} A_1 \\ A_2 \\ \vdots \\ A_k \end{matrix}\right\} \longrightarrow B$$

means that B follows from $A_1, A_2, \cdots$ and A_k.

8. The notation

$$A \longrightarrow \left\{\begin{matrix} B_1 \\ B_2 \\ \vdots \\ B_k \end{matrix}\right.$$

means that the k statements $B_1, B_2, \cdots$ and B_k follow from A at the same time.

9. The notation A_i $(i = 1, \cdots, m)$ means $A_1, A_2, \cdots$ and A_m.
10. We also need the following notation of set theory: $\supseteqq$, $\not\supseteqq$, $\supset$, $=$, $\neq$, $\ni$, $\notin$, $\cup$, etc.

Axioms

Fundamental Assumptions: We are provided with a certain set of elements; $\mathfrak{B}_1 \ni a_1, a_2, \cdots, a_s \cdots$. For finite series of elements, which we want to call *cycles*, we consider the relations "*(linearly) dependent*", in notation: $a_1 \cdots a_s = 0$[2], and "*(linearly) independent*", in notation: $a_1 \cdots a_s \neq 0$. These relations should abide by the following axioms:

Axiom 1 *(reflexivity)*

$$aa$$

Axiom 2 *(conclusion)*

$$a_1 \cdots a_s \to a_1 \cdots a_s x \ \ (s = 1, 2, \cdots)$$

---

[2] From now on, usually, instead of "$a_1 \cdots a_s = 0$", we will use the abbreviation "$a_1 \cdots a_s$", as Thomsen uses in his cycle calculation (cf. l.c.[(1)]).

Axiom 3 *(exchange)*

$$a_1 \cdots a_i \cdots a_s \to a_i \cdots a_1 \cdots a_s, (s = 2, 3, \cdots ; i = 2, \cdots , s)$$

Axiom 4 *(transitivity)*

$$a_1 \cdots a_s \neq 0,\ xa_1 \cdots a_s,\ a_1 \cdots a_s y \to xa_1 \cdots a_{s-1} y,\ (s = 1, 2, \cdots)$$

Definition 5 *Such a set $\mathfrak{B}_1$ is called an abstract space of linear dependence of the first type, shortly, $\mathfrak{B}_1$-**space**.*

Claim 6 *Let us take the set of all points of the classical n-dimensional projective space as a $\mathfrak{B}_1$-space. We use the expressions "linearly dependent" and "linearly independent" in the usual sense. Then the fundamental assumptions and Axioms 1–4 are fulfilled, as is easy to see*[3].

Claim 7 *As the above axiom system contains no existence statement, any subset of a $\mathfrak{B}_1$-space is also a $\mathfrak{B}_1$-space.*

Furthermore, we get the following readily from Axioms 1, 2, 3 and 4.

Claim 8 *A cycle in which the same element occurs twice or more is linearly dependent; consequently, the elements in a linearly independent cycle are all distinct.*

Linear Spaces

Definition 9 *Let $a_1, \cdots , a_n \in \mathfrak{B}_1$ with $a_1 \cdots a_n \neq 0$. The set of all elements x of $\mathfrak{B}_1$ with $a_1 \cdots a_n x$ is called the linear space in $\mathfrak{B}_1$ of dimension n with $a_1, \cdots , a_n$ as its basis, in notation: $\mathfrak{R}^n(a_1 \cdots a_n)$.*

Definition 10 *The set of all elements x of $\mathfrak{B}_1$ with $x = 0$ is called the root of $\mathfrak{B}_1$, in notation: $\mathfrak{N}$*[4].

Then, because $x = 0 \xrightarrow{(Axioms\ 2,\ 3)} a_1 \cdots a_n x$, we remark that

Claim 11 *For any linear space $\mathfrak{R}$, we have $\mathfrak{R} \supseteqq \mathfrak{N}$. In words, any linear space always contains the root in itself.*

Because $\xrightarrow{(Axiom\ 1)} a_i a_i \xrightarrow{(Axioms\ 2,\ 3)} a_1 \cdots a_n a_i$ $(i = 1, \cdots , n)$, we have

3 Therefore, from now on, we will use often the word "point" instead of "element".

4 The same set is, on the other hand, a linear space of order 0. i.e., $\mathfrak{R}^0$ unless we assume the empty cycle to be linearly dependent.

Claim 12 $\mathfrak{R}^n(a_1 \cdots a_n) \ni a_1, \cdots, a_n$. *In words, the linear space contains its bases in itself.*

Furthermore, because

$$a_1 \cdots a_m x,\ m \leqq n \xrightarrow{(Axiom\ 2,\ 3)} a_1 \cdots a_m \cdots a_n x,$$

we have

Claim 13 $n \geqq m \to \mathfrak{R}^n(a_1 \cdots a_n) \supseteqq \mathfrak{R}^m(a_1 \cdots a_m)$.

Proposition 14 *For $n \geqq 1$, we have*

$$\left.\begin{array}{l} a_1 \cdots a_n \neq 0, \\ a_1 \cdots a_n x_1, \\ a_1 \cdots a_n x_2 \end{array}\right\} \to a_2 \cdots a_n x_1 x_2$$

Proof. We have

$$\left.\begin{array}{l} a_1 \cdots a_n \neq 0, \\ a_1 \cdots a_n x_1, \\ a_1 \cdots a_n x_2 \end{array}\right\} \xrightarrow{(Axiom\ 3)} \left\{\begin{array}{l} a_2 \cdots a_n a_1 \neq 0, \\ x_1 a_2 \cdots a_n a_1, \\ a_2 \cdots a_n a_1 x_2 \end{array}\right\}$$
$$\xrightarrow{(Axiom\ 4)} x_1 a_2 \cdots a_n x_2 \xrightarrow{(Axiom\ 3)} a_2 \cdots a_n x_1 x_2$$

Therefore we have

$$\left.\begin{array}{l} a_1 \cdots a_n \neq 0, \\ a_1 \cdots a_n x_1, \\ a_1 \cdots a_n x_2 \end{array}\right\} \to a_2 \cdots a_n x_1 x_2$$

□

Proposition 15 *For $n \geqq 2$, we have*

$$\left.\begin{array}{l} a_1 \cdots a_n \neq 0, \\ a_1 \cdots a_n x_1, \\ a_1 \cdots a_n x_2, \\ a_1 \cdots a_n x_3 \end{array}\right\} \to a_3 \cdots a_n x_1 x_2 x_3$$

Proof.

1. For $a_2 \cdots a_n x_1 \neq 0$, we have

$$\left.\begin{array}{l} a_1 \cdots a_n \neq 0, \\ a_1 \cdots a_n x_1, \\ a_1 \cdots a_n x_2, \\ a_1 \cdots a_n x_3 \end{array}\right\} \xrightarrow{(Proposition\ 14)} \left\{\begin{array}{l} a_2 \cdots a_n x_1 \neq 0, \\ a_2 \cdots a_n x_1 x_2, \\ a_2 \cdots a_n x_2 x_3 \end{array}\right\}$$
$$\xrightarrow{(Proposition\ 14)} a_3 \cdots a_n x_1 x_2 x_3$$

2. For $a_2 \cdots a_n x_2 \neq 0$, we have

$$\left.\begin{array}{l} a_1 \cdots a_n \neq 0, \\ a_1 \cdots a_n x_1, \\ a_1 \cdots a_n x_2, \\ a_1 \cdots a_n x_3 \end{array}\right\} \xrightarrow{(Proposition\ 14)} \left\{\begin{array}{l} a_2 \cdots a_n x_2 \neq 0, \\ a_2 \cdots a_n x_2 x_1, \\ a_2 \cdots a_n x_2 x_3 \end{array}\right\}$$

$$\xrightarrow{(Proposition\ 14)} a_3 \cdots a_n x_2 x_1 x_3$$

$$\xrightarrow{(Axiom\ 3)} a_3 \cdots a_n x_1 x_2 x_3$$

3. If $a_2 \cdots a_n x_1$ and $a_2 \cdots a_n x_2$, we have

$$a_1 \cdots a_n \neq 0 \xrightarrow{(Axiom\ 2,\ 3)} \left.\begin{array}{l} a_2 \cdots a_n \neq 0, \\ a_2 \cdots a_n x_1, \\ a_2 \cdots a_n x_2 \end{array}\right\}$$

$$\xrightarrow{(Proposition\ 14)} a_3 \cdots a_n x_1 x_2 \xrightarrow{(Axiom\ 2)} a_2 \cdots a_n x_1 x_2 x_3$$

Therefore we have

$$\left.\begin{array}{l} a_1 \cdots a_n \neq 0, \\ a_1 \cdots a_n x_1, \\ a_1 \cdots a_n x_2, \\ a_1 \cdots a_n x_3 \end{array}\right\} \rightarrow a_3 \cdots a_n x_1 x_2 x_3$$

□

Proposition 16 *For $n \geqq m - 1$, we have*

$$\left.\begin{array}{l} a_1 \cdots a_n \neq 0, \\ a_1 \cdots a_n x_1, \\ a_1 \cdots a_n x_2, \\ \cdots\cdots\cdots\cdots, \\ a_1 \cdots a_n x_m \end{array}\right\} \rightarrow a_m \cdots a_n x_1 \cdots x_m$$

Proof. We prove this by induction on m. If m equals 1, it is trivial. If m equals 2 or 3, it is already dealt with in Proposition 14 or 15 respectively. Therefore it suffices to show the case of $[m]$ on the inductive hypothesis in case of $[m-1]$.

1. If at least one of the relations $a_2 \cdots a_n x_1$, $a_2 \cdots a_n x_2$, ..., $a_2 \cdots a_n x_m$ fails, say, $a_2 \cdots a_n x_1 \neq 0$, then we have

$$\left.\begin{array}{l} a_1 \cdots a_n \neq 0, \\ a_1 \cdots a_n x_1, \\ a_1 \cdots a_n x_2, \\ \cdots\cdots\cdots\cdots\cdots, \\ a_1 \cdots a_n x_m \end{array}\right\} \xrightarrow{(Proposition\ 14)} \left\{\begin{array}{l} a_2 \cdots a_n x_1 \neq 0, \\ a_2 \cdots a_n x_1 x_2, \\ a_2 \cdots a_n x_1 x_3, \\ \cdots\cdots\cdots\cdots, \\ a_2 \cdots a_n x_1 x_m \end{array}\right\}$$

$$\xrightarrow{[m-1]} a_m \cdots a_n x_1 \cdots x_m$$

2. If all of the relations $a_2 \cdots a_n x_1$, $a_2 \cdots a_n x_2$, ..., $a_2 \cdots a_n x_m$ hold, then we have

$$a_1 \cdots a_n \neq 0 \underrightarrow{(Axiom\ 2,\ 3)} \left.\begin{array}{l} a_2 \cdots a_n \neq 0, \\ a_2 \cdots a_n x_1, \\ a_2 \cdots a_n x_2, \\ \cdots\cdots\cdots\cdots, \\ a_2 \cdots a_n x_{m-1} \end{array}\right\}$$

$$\underrightarrow{[m-1]}\, a_m \cdots a_n x_1 \cdots x_{m-1} \underrightarrow{(Axiom\ 2)} a_m \cdots a_n x_1 \cdots x_m$$

Therefore we have

$$\left.\begin{array}{l} a_1 \cdots a_n \neq 0, \\ a_1 \cdots a_n x_1, \\ a_1 \cdots a_n x_2, \\ \cdots\cdots\cdots\cdots, \\ a_1 \cdots a_n x_m \end{array}\right\} \to a_m \cdots a_n x_1 \cdots x_m \text{ [5]}$$

□

The above proposition in case of $n = m - 1$ implies

$$\left.\begin{array}{l} a_1 \cdots a_n \neq 0, \\ a_1 \cdots a_n x_1, \\ a_1 \cdots a_n x_2, \\ \cdots\cdots\cdots\cdots, \\ a_1 \cdots a_n x_{n+1} \end{array}\right\} \to x_1 x_2 \cdots x_{n+1}$$

which is summarized as follows.

Proposition 17 *We have*

$$\mathfrak{R}^n(a_1 \cdots a_n) \ni x_1, x_2, \cdots, x_{n+1} \to x_1 x_2 \cdots x_{n+1}$$

In words, $n+1$ points contained in the linear space of dimension n are dependent.

Corollary 18 $\mathfrak{R}^n(a_1 \cdots a_n) \ni x_1, \cdots, x_m$ *and* $n < m \to x_1 \cdots x_m$. *In words, more than n points contained in the linear space of dimension n are dependent.*

Proof. By the above proposition we have $x_1 \cdots x_{n+1}$. Therefore we have

$$x_1 \cdots x_{n+1} \underrightarrow{(Axiom\ 2)} x_1 \cdots x_m$$

□

---

[5] It should be noted, that we do not use Axiom 1 in this proof at all.

Proposition 19 *We have*

$$\mathfrak{R}_1^n \supseteqq \mathfrak{R}_2^n \to \mathfrak{R}_1^n \subseteqq \mathfrak{R}_2^n$$

In words, if a linear space $\mathfrak{R}_1^n$ of dimension n contains another linear space $\mathfrak{R}_2^n$ of dimension n, then they agree[6].

Proof. Let $a_1 \cdots a_n$ and $b_1 \cdots b_n$ be the bases of $\mathfrak{R}_1$ and $\mathfrak{R}_2$ respectively. Then we have

$$\mathfrak{R}_1 \supseteqq \mathfrak{R}_2 \ni b_1, \cdots, b_n \to \mathfrak{R}_1 \ni b_1, \cdots, b_n$$

Therefore we have

$$\left.\begin{array}{l} a_1 \cdots a_n \neq 0, \\ a_1 \cdots a_n b_1, \\ \cdots\cdots\cdots\cdots, \\ a_1 \cdots a_n b_n, \end{array}\right\} \xrightarrow{(Proposition\ 17)} b_1 \cdots b_n x \to \mathfrak{R}_2 \ni x$$

On the other hand we have

$$\mathfrak{R}_1 \ni x \to a_1 \cdots a_n x$$

Therefore we have

$$\mathfrak{R}_1 \ni x \to \mathfrak{R}_2 \ni x$$

which is tantamount to saying that

$$\mathfrak{R}_1 \subseteqq \mathfrak{R}_2$$

□

Corollary 20 *It is not the case that $\mathfrak{R}_2^n$ is properly contained by $\mathfrak{R}_1^n$.*

Proposition 21 *We have*

$$n < m \to \mathfrak{R}_2^m \not\subseteqq \mathfrak{R}_1^n$$

In words, any linear space does not contain a linear space of greater dimension.

Proof. Let $a_1 \cdots a_n$ and $b_1 \cdots b_m$ be the bases of $\mathfrak{R}_1$ and $\mathfrak{R}_2$ respectively. Suppose for contradiction that $\mathfrak{R}_1 \supseteqq \mathfrak{R}_2$, then we have

$$\mathfrak{R}_1^n(a_1 \cdots a_n) \supseteqq \mathfrak{R}_2^m(b_1 \cdots b_m) \ni b_1, \cdots, b_m \to \mathfrak{R}_1^n(a_1 \cdots a_n) \ni b_1, \cdots, b_m$$

6 If two linear spaces $\mathfrak{R}_1$, $\mathfrak{R}_2$ agree with each other, we write $\mathfrak{R}_1 = \mathfrak{R}_2$.

so that

$$\mathfrak{R}_1^n(a_1 \cdots a_n) \ni b_1, \cdots, b_m \text{ and } n < m \xrightarrow{\text{(Corollary 18 to Proposition 17)}} b_1 \cdots b_m \rightarrow \perp$$

Thus we have reached the desired contradiction. □

Corollary 22 *We have*

$$n \neq m \rightarrow \mathfrak{R}_1^n \neq \mathfrak{R}_2^m$$

In words, linear spaces of distinct dimensions can not agree.

Proposition 23 *We have*

$$\mathfrak{R}_1^n(a_1 \cdots a_n) = \mathfrak{R}_2^m(b_1 \cdots b_m) \leftrightarrow n = m,\ a_1 \cdots a_n b_i, (i = 1, \cdots m)$$

In words, two linear spaces $\mathfrak{R}_1$ and $\mathfrak{R}_2$ agree only if their dimensions are equal and one of the linear spaces, say, $\mathfrak{R}_1$ contains the base points of the other, namely, $\mathfrak{R}_2$.

Proof.

1. The condition is necessary: From Corollary 22 to Proposition 21, we have $n = m$. Furthermore we have

$$\mathfrak{R}_1 = \mathfrak{R}_2 \ni b_1, \cdots, b_m \longrightarrow \mathfrak{R}_1 \ni b_1, \cdots, b_m \longrightarrow a_1, \cdots, a_n b_i, (i = 1, \cdots, m)$$

2. The condition is sufficient:We have

$$\left.\begin{array}{rl} & a_1 \cdots a_n \neq 0, \\ & a_1 \cdots a_n b_1, \\ & \cdots\cdots\cdots\cdots, \\ & a_1 \cdots a_n b_n, \\ \mathfrak{R}_1 \ni x \longrightarrow & a_1 \cdots a_n x \end{array}\right\} \xrightarrow{(Proposition\ 17)} b_1 \cdots b_n x \rightarrow \mathfrak{R}_2 \ni x$$

Therefore we have $\mathfrak{R}_1 \subseteqq \mathfrak{R}_2$, so that

$$\mathfrak{R}_1^n \subseteqq \mathfrak{R}_2^n \xrightarrow{(Proposition\ 19)} \mathfrak{R}_1 = \mathfrak{R}_2$$

□

Corollary 24 *We have*

$$\mathfrak{R}(a_1 \cdots a_n) \ni b_1, \cdots, b_n \text{ and } b_1 \cdots b_n \neq 0 \to \mathfrak{R}(a_1 \cdots a_n) = \mathfrak{R}(b_1 \cdots b_n)$$

In words, any independent sequence of n points in a linear space of dimension n can be taken as its basis.

Corollary 25 *If $a_1 \cdots a_n \neq 0$ and $b_1 \cdots b_n \neq 0$, then*

$$\left.\begin{array}{l} a_1 \cdots a_n b_1, \\ \cdots\cdots\cdots\cdots, \\ a_1 \cdots a_n b_n \end{array}\right\} \longleftrightarrow \left\{\begin{array}{l} b_1 \cdots b_n a_1, \\ \cdots\cdots\cdots\cdots, \\ b_1 \cdots b_n a_n \end{array}\right\}$$

Proposition 26 *We have*

$$\begin{aligned} &\mathfrak{R}^n(a_1 \cdots a_n) \ni b_1, \cdots, b_m \text{ with } b_1 \cdots b_m \neq 0,\ n > m \\ &\qquad \to \text{There exists a subsequence } a_{i_{m+1}}, \cdots, a_{i_n} \\ &\qquad \text{in the sequence} \\ &\qquad a_1, \cdot\cdot, a_n \text{ with } b_1 \cdots b_m a_{i_{m+1}} \cdots a_{i_n} \neq 0 \end{aligned}$$

In words, the linear space $\mathfrak{R}^n(a_1 \cdots a_n)$ has a basis containing the given independent sequence $b_1, \cdots, b_m$ together with a subsequence $a_{i_{m+1}}, \cdots, a_{i_n}$ of $a_1, \cdot\cdot, a_n$.

Proof. It suffices to show that there is an element $a_{i_{m+1}}$ in the sequence $a_1, \cdot\cdot, a_n$ in such a way that $b_1 \cdots b_m a_{i_{m+1}} \neq 0$, which will provide the desired result by induction on m. Suppose for contradiction that $b_1 \cdots b_m a_i = 0$ $(i = 1, \cdots, n)$, then we have

$$\left.\begin{array}{l} b_1 \cdots b_m \neq 0, \\ b_1 \cdots b_m a_1, \\ \cdots\cdots\cdots\cdots, \\ b_1 \cdots b_m a_n, \\ m < n \end{array}\right\} \xrightarrow{\text{(Corollary 18)}} a_1 \cdots a_n \to \perp$$

Hence, $b_1 \cdots b_m a_i \neq 0$ must for at least one a_i. □

Proposition 27 *We have*

$$\begin{aligned}\mathfrak{R}_1(a_1 \cdots a_n) \ni b_1, \cdots, b_m \, and b_1 \cdots b_m \neq 0 \\ \to \mathfrak{R}_1(a_1 \cdots a_n) \supseteqq \mathfrak{R}_2(b_1 \cdots b_m)\end{aligned}$$

In words, the linear space which contains the basis of another linear space contains the other linear space itself.

Proof. By Proposition 17 we have $n \geqq m$. By Proposition 26 there exists a subsequence $a_{i_{m+1}}, \cdots, a_{i_n}$ of the sequence $a_1, \cdots, a_n$ with $b_1 \cdots b_m a_{i_{m+1}} \cdots a_{i_n} \neq 0$. By Corollary to Proposition 23 we have

$$\mathfrak{R}_1(a_1 \cdots a_n) = \mathfrak{R}(b_1 \cdots b_m a_{i_{m+1}} \cdots a_{i_n})$$

Therefore we have

$$\begin{aligned}\mathfrak{R}_2(b_1 \cdots b_m) \ni x \to b_1 \cdots b_m x \to b_1 \cdots b_m a_{i_{m+1}} \cdots a_{i_n} x \\ \to \mathfrak{R}(b_1 \cdots b_m a_{i_{m+1}} \cdots a_{i_n}) \ni x \to \mathfrak{R}_1(a_1 \cdots a_n) \ni x\end{aligned}$$

which implies that

$$\mathfrak{R}_2 \ni x \to \mathfrak{R}_1 \ni x$$

Therefore $\mathfrak{R}_1(a_1 \cdots a_n) \supseteqq \mathfrak{R}_2(b_1 \cdots b_m)$. □

Proposition 28 *We have*

$$\mathfrak{R}_1^n(a_1 \cdots a_n) \supset \mathfrak{R}_2^m(b_1 \cdots b_m) \leftrightarrow n > m, \; a_1 \cdots a_n b_i, \; (i = 1, \cdots, m)$$

In words, the linear space $\mathfrak{R}_1$ properly contains another linear space $\mathfrak{R}_2$ only if the dimension of $\mathfrak{R}_1$ is larger than that of $\mathfrak{R}_2$ and $\mathfrak{R}_1$ contains the basis of $\mathfrak{R}_2$.

Proof.

1. The condition is necessary:

   $$\left.\begin{array}{l}\text{By Corollary 20 of Proposition 19, we have } n \neq m, \\ \text{Proposition 21 shows that } n < m \text{ does not hold}\end{array}\right\} \longrightarrow n > m$$

   Furthermore we have

   $$\begin{aligned}\mathfrak{R}_1 \supset \mathfrak{R}_2 \ni b_1, \cdots, b_m \longrightarrow \mathfrak{R}_1 \ni b_1, \cdots, b_m \longrightarrow a_1 \cdots a_n b_i, \\ (i = 1, \cdots, m)\end{aligned}$$

2. The condition is sufficient:We have

$$a_1 \cdots a_n b_i,\ (i = 1, \cdots, m) \longrightarrow \mathfrak{R}_1(a_1 \cdots a_n) \ni b_1, \cdots, b_m \quad \text{and}$$
$$b_1 \cdots b_m \neq 0 \ \underrightarrow{(Proposition\ 27)}\ \mathfrak{R}_1(a_1 \cdots a_n) \supseteqq \mathfrak{R}_2(b_1 \cdots b_m)$$

$$\left.\begin{array}{c} \mathfrak{R}_1(a_1 \cdots a_n) \supseteqq \mathfrak{R}_2(b_1 \cdots b_m), \\ n > m \end{array}\right\} \underrightarrow{(Proposition\ 23)}\ \mathfrak{R}_1 \supset \mathfrak{R}_2$$

Therefore we conclude that

$$\mathfrak{R}_1^n(a_1 \cdots a_n) \supset \mathfrak{R}_2^m(b_1 \cdots b_m)$$

□

Dimension

Definition 29 *Given a set M, the largest number of elements in independent cycles in M is called the dimension of M, in notation:* $\dim M$[7]. *An independent cycle in M whose number of elements equals* $\dim M$ *is called a basis of M.*[8]

It is easy to see the following:

Claim 30 *If one regards the linear space as either a linear space or a set, then the meanings of dimension in both cases agree. The same is also true for the meaning of the basis.*

Claim 31 *We have*

$$M_1 \supseteqq M_2 \longrightarrow \dim M_1 \geqq \dim M_2$$

Proposition 32 *Let $a_1 \cdots a_n$ be a basis of M. Then we have*

$$\mathfrak{R}_1(a_1 \cdots a_n) \supseteqq M$$

Proof. Let x be any element of M. By the very definition, we have

$$a_1 \cdots a_n \neq 0, a_1 \cdots a_n x$$

so that $\mathfrak{R}(a_1 \cdots a_m) \ni x$. Therefore $\mathfrak{R}(a_1 \cdots a_m) \supseteqq M$. □

---

[7] One would understand the meaning analogously that $\dim M$ is equal to ∞ or 0

[8] If $M \ni a_1, a_2, \cdots$, we write "dim $(a_1 a_2 \cdots)$" instead of "dim M".

Definition 33 *We denote the intersection of k sets $M_1, \cdots, M_k$ by $\mathfrak{D}(M_1 \cdots M_k)$, where commas are occasionally inserted between the consecutive M_i and M_{i+1}.*

It is easy to see that

Claim 34 *We have*

$$\mathfrak{D}(M_1 M_2) = \mathfrak{D}(M_2 M_1)$$
$$\mathfrak{D}(\mathfrak{D}(M_1 \cdots M_{k-1}),\ M_k) = \mathfrak{D}(M_1 \cdots M_k)$$
$$\mathfrak{D}(M_1,\ \mathfrak{D}(M_2 \cdots M_k)) = \mathfrak{D}(M_1 \cdots M_k)$$

Proposition 35 *The space $\mathfrak{D}(\mathfrak{R}_1 \mathfrak{R}_2)$ is a linear space.*

Proof. Let $a_1 \cdots a_n$, $b_1 \cdots b_m$, and $c_1 \cdots c_k$ be bases of $\mathfrak{R}_1$, $\mathfrak{R}_2$ and $\mathfrak{D}(\mathfrak{R}_1 \mathfrak{R}_2)$ respectively. Then, by dint of Proposition 32, we have

$$\mathfrak{R}(c_1 \cdots c_k) \supseteqq \mathfrak{D}(\mathfrak{R}_1 \mathfrak{R}_2)$$

On the other hand, we have

$$\mathfrak{R}_1,\ \mathfrak{R}_2 \supseteq \mathfrak{D}(\mathfrak{R}_1 \mathfrak{R}_2) \ni c_1,\ \cdots,\ c_k$$

so that we have

$$\mathfrak{R}_1 \ni c_1,\ \cdots,\ c_k \text{ and } c_1 \cdots c_k \neq 0 \ \underrightarrow{(Proposition\ 27)}\ \mathfrak{R}_1 \supseteqq \mathfrak{R}(c_1 \cdots c_k)$$
$$\mathfrak{R}_2 \ni c_1,\ \cdots,\ c_k \text{ and } c_1 \cdots c_k \neq 0 \ \underrightarrow{(Proposition\ 27)}\ \mathfrak{R}_2 \supseteqq \mathfrak{R}(c_1 \cdots c_k)$$

Therefore we have

$$\mathfrak{D}(\mathfrak{R}_1 \mathfrak{R}_2) = \mathfrak{R}(c_1 \cdots c_k)$$

so that $\mathfrak{D}(\mathfrak{R}_1 \mathfrak{R}_2)$ is a linear space. □

From Proposition 35 and Claim 34, we have

Corollary 36 *The space $\mathfrak{D}(\mathfrak{R}_1 \cdots \mathfrak{R}_k)$ is a linear space.*

From Claim 11 we have

Claim 37 $\mathfrak{D}(\mathfrak{R}_1 \cdots \mathfrak{R}_k) \supseteqq \mathfrak{R}$

Definition 38 *Given k sets $M_1, \cdots, M_k$, the minimal linear space containing all of $M_1, \cdots, M_k$ is called the linear space generated by $M_1, \cdots, M_k$ and is denoted by $\mathfrak{B}(M_1 \cdots M_k)$*[9].

---

[9] If $M \ni a_1, a_2, \cdots$, we will often write also "$\mathfrak{B}(a_1 a_2 \cdots)$" instead of "$\mathfrak{B}(M)$".

Now we have

Claim 39 $\mathfrak{B}(M_1M_2) = \mathfrak{B}(M_2M_1)$

Claim 40 $\mathfrak{B}(M_1 \cup \cdots \cup M_k) = \mathfrak{B}(M_1 \cdots M_k)$

Proposition 41 *Let $a_1 \cdots a_n$ be a basis of M. Then we have*

$$\mathfrak{B}(M) = \mathfrak{R}(a_1 \cdots a_n)$$

In words, the linear space generated by a finite set M is no other than the linear space whose basis is that of M.

Proof. By Proposition 32, we have

$$\mathfrak{R}(a_1 \cdots a_n) \supseteqq M$$

On the other hand, by Definition 38, we have

$$\mathfrak{B}(M) \supseteqq M \ni a_1, \cdots, a_n \longrightarrow \mathfrak{B}(M) \ni a_1, \cdots, a_n \xrightarrow{(Proposition\ 27)} \mathfrak{B}(M) \supseteqq \mathfrak{R}(a_1 \cdots a_n)$$

Therefore we have

$$\mathfrak{B}(M) \supseteqq \mathfrak{R}(a_1 \cdots a_n) \supseteqq M$$

so that

$$\mathfrak{B}(M) = \mathfrak{R}(a_1 \cdots a_n)$$

□

Corollary 42 $\dim \mathfrak{B}(M) = \dim M$

Proposition 43 *We have*

$$\mathfrak{R} \supseteqq M \longrightarrow \mathfrak{R} \supseteqq \mathfrak{B}(M)$$

In words, any linear space containing a set M in itself contains the linear space generated by M.

Proof. Let $a_1 \cdots a_n$ be a basis of M. Then we have

$$\mathfrak{R} \supseteqq M \ni a_1, \cdots, a_n \longrightarrow \mathfrak{R} \ni a_1, \cdots, a_n \xrightarrow{(Proposition\ 27)} \mathfrak{R} \supseteqq \mathfrak{R}(a_1 \cdots a_n)$$

Since $\mathfrak{R}(a_1 \cdots a_n) = \mathfrak{B}(M)$, we have $\mathfrak{R} \supseteqq \mathfrak{B}(M)$. □

Proposition 44 *We have*

$$\mathfrak{R} \supseteqq M_1, \ \cdots, \ M_k \longrightarrow \mathfrak{R} \supseteqq \mathfrak{B}(M_1 \cdots M_k)$$

In words, any linear space containing k sets $M_1, \ \cdots, \ M_k$ in itself contains the linear space generated by $M_1, \ \cdots, \ M_k$.

Proof. We have

$$\mathfrak{R} \supseteqq M_1, \ \cdots, \ M_k \longrightarrow \mathfrak{R} \supseteqq M_1 \cup \cdots \cup M_k \xrightarrow{(Proposition\ 43)}$$
$$\mathfrak{R} \supseteqq \mathfrak{B}(M_1 \cup \cdots \cup M_k) \xrightarrow{(Claim\ 40)} \mathfrak{R} \supseteqq \mathfrak{B}(M_1 \cdots M_k)$$

□

It is easy to see that

Corollary 45 *We have*

$$\mathfrak{B}(\mathfrak{B}(M_1 \cdots M_{k-1}), \ M_k) = \mathfrak{B}(M_1 \cdots M_k)$$
$$\mathfrak{B}((M_1, \mathfrak{B}(M_2 \cdots M_k)) = \mathfrak{B}(M_1 \cdots M_k)$$

Proposition 46 *We have*

$$\mathfrak{B}_1(\mathfrak{R}_1(a_1 \cdots a_n), \mathfrak{R}_2(b_1 \cdots b_m)) = \mathfrak{B}_2(a_1 \cdots a_n b_1 \cdots b_m)$$

Proof. We have

$$\left.\begin{array}{l} \mathfrak{B}_1 \supseteqq \mathfrak{R}_1 \ni a_1, \ \cdots, \ a_n \text{ and} \\ \mathfrak{B}_1 \supseteqq \mathfrak{R}_2 \ni b_1, \ \cdots, \ b_m \end{array}\right\} \longrightarrow \mathfrak{B}_1 \ni a_1, \cdots, a_n, b_1, \cdots, b_m$$
$$\xrightarrow{(Proposition\ 43)} \mathfrak{B}_1 \supseteqq \mathfrak{R}_2(a_1 \cdots a_n b_1 \cdots b_m)$$

On the other hand, we have

$$\left.\begin{array}{l} \mathfrak{B}_2 \ni a_1, \cdots, a_n \text{ and } a_1 \cdots a_n \neq 0 \xrightarrow{(Proposition\ 27)} \mathfrak{B}_2 \supseteqq \mathfrak{R}_1(a_1 \cdots a_n), \\ \mathfrak{B}_2 \ni b_1, \cdots, b_m \text{ and } b_1 \cdots b_m \neq 0 \xrightarrow{(Proposition\ 27)} \mathfrak{B}_2 \supseteqq \mathfrak{R}_2(b_1 \cdots b_m) \end{array}\right\}$$

$$\xrightarrow{(Proposition\ 44)} \mathfrak{B}_2 \supseteqq \mathfrak{B}_1(\mathfrak{R}_1 \mathfrak{R}_2)$$

Therefore we conclude that

$$\mathfrak{B}_1(\mathfrak{R}_1(a_1 \cdots a_n), \mathfrak{R}_2(b_1 \cdots b_m)) = \mathfrak{B}_2(a_1 \cdots a_n b_1 \cdots b_m)$$

□

The following proposition can be established by the same token.

Proposition 47 *We have*

$$\mathfrak{B}_1(\mathfrak{R}_1(a_1 \cdots a_n), \cdots, \mathfrak{R}_k(l_1 \cdots l_m)) = \mathfrak{B}_2(a_1 \cdots a_n \cdots l_1 \cdots l_m)$$

Proposition 48 *We have*

$$\dim \mathfrak{D}(\mathfrak{R}_1^n \mathfrak{R}_2^m) + \dim \mathfrak{B}(\mathfrak{R}_1^n \mathfrak{R}_2^m) \leqq n + m$$

In words, the sum of the dimension of the intersection of two linear spaces of dimensions n and m respectively and the dimension of the linear space generated by the same two linear spaces is equal to or less than $n + m$.[10]

Proof. Let $a_1 \cdots a_n, b_1 \cdots b_m$ and $c_1 \cdots c_k$ be bases of $\mathfrak{R}_1, \mathfrak{R}_2$ and $\mathfrak{D}(\mathfrak{R}_1\mathfrak{R}_2)$ respectively. Then, by Proposition 26 and Corollary 24, we have

$$\mathfrak{R}_1(a_1 \cdots a_n) = \mathfrak{R}_1(c_1 \cdots c_k a_{k+1} \cdots a_n)$$
$$\mathfrak{R}_2(b_1 \cdots b_m) = \mathfrak{R}_2(c_1 \cdots c_k b_{k+1} \cdots b_m)$$

so that

$$\begin{aligned}\mathfrak{B}(\mathfrak{R}_1\mathfrak{R}_2) &= \mathfrak{B}(c_1 \cdots c_k a_{k+1} \cdots a_n c_1 \cdots c_k b_{k+1} \cdots b_m)\\ &= \mathfrak{B}(c_1 \cdots c_k a_{k+1} \cdots a_n b_{k+1} \cdots b_m)\end{aligned}$$

holds by Proposition 46. Therefore we have

$$\begin{aligned}\dim \mathfrak{B}(\mathfrak{R}_1\mathfrak{R}_2) &= \dim \mathfrak{B}(c_1 \cdots c_k a_{k+1} \cdots a_n b_{k+1} \cdots b_m)\\ &= \dim (c_1 \cdots c_k a_{k+1} \cdots a_n b_{k+1} \cdots b_m)\\ &\leqq n + m - k\end{aligned}$$

By Proposition 35 we have

$$\dim \mathfrak{D}(\mathfrak{R}_1\mathfrak{R}_2) = \dim \mathfrak{R}(c_1 \cdots c_k) = k$$

so that

$$\dim \mathfrak{D}(\mathfrak{R}_1\mathfrak{R}_2) + \dim \mathfrak{B}(\mathfrak{R}_1\mathfrak{R}_2) \leqq n + m$$

□

Proposition 49 *We have*

$$\dim \mathfrak{D}(M_1 M_2) + \dim \mathfrak{B}(M_1 M_2) \leqq \dim M_1 + \dim M_2$$ [11]

---

[10] It should be pointed out here that the equality does not always hold.

[11] Propositions 48 and 49 are equivalent in this context, because, as already noted in Claim 7, one can regard the sum set $M_1 \cup M_2$ of M_1 and M_2 of Proposition 49 as our $\mathfrak{B}_1$-space.

Proof. If the space generated by a given set is a linear space, we have the following inequality by Proposition 48:

$$\begin{aligned}&\dim \mathfrak{D}(\mathfrak{B}(M_1), \mathfrak{B}(M_2)) + \dim \mathfrak{B}(\mathfrak{B}(M_1), \mathfrak{B}(M_2))\\ &\leqq \dim \mathfrak{B}(M_1) + \dim \mathfrak{B}(M_2)\end{aligned}$$

By Corollary 42 of Proposition 41 we have

$$\begin{aligned}\dim \mathfrak{B}(M_1) &= \dim M_1\\ \dim \mathfrak{B}(M_2) &= \dim M_2\end{aligned}$$

By Corollary 45 of Proposition 44 we have

$$\dim \mathfrak{B}(\mathfrak{B}(M_1)\mathfrak{B}(M_2)) = \dim \mathfrak{B}(M_1 M_2)$$

Furthermore we have

$$\left.\begin{aligned}\mathfrak{B}(M_1) \supseteqq M_1,\\ \mathfrak{B}(M_2) \supseteqq M_2,\end{aligned}\right\} \longrightarrow \mathfrak{D}(\mathfrak{B}(M_1),\ \mathfrak{B}(M_2)) \supseteqq \mathfrak{D}(M_1 M_2)$$

Therefore we have

$$\dim \mathfrak{D}(\mathfrak{B}(M_1), \mathfrak{B}(M_2)) \geqq \dim \mathfrak{D}(M_1 M_2)$$

so that

$$\dim \mathfrak{D}(M_1 M_2) + \dim \mathfrak{B}(M_1 M_2) \leqq \dim M_1 + \dim M_2$$

□

Proposition 50 *If the dimensions of k sets $M_1, \cdots, M_k$ are all finite, then we have*

$$\dim (M_1 \cup \cdots \cup M_k) \leqq \dim M_1 + \cdots + \dim M_k$$

Proof. By Proposition 49 we have

$$\begin{aligned}&\dim \mathfrak{B}(M_1, \mathfrak{B}(M_2 \cdots M_k))\\ &\leqq \dim M_1 + \dim \mathfrak{B}(M_2 \cdots M_k)\\ &\leqq \dim M_1 + \dim M_2 + \dim \mathfrak{B}(M_3 \cdots M_k)\\ &\leqq \cdots\\ &\leqq \dim M_1 + \dim M_2 + \cdots + \dim M_k\end{aligned}$$

Since we have

$$\begin{aligned}\dim (M_1 \cup \cdots \cup M_k) &= \dim \mathfrak{B}(M_1 \cdots M_k)\\ &= \dim \mathfrak{B}(M_1, \mathfrak{B}(M_2 \cdots M_k))\end{aligned}$$

we conclude that

$$\dim (M_1 \cup \cdots \cup M_k) \leqq \dim M_1 + \cdots + \dim M_k$$

□

The Reduction Method

With our cycle calculation, we need often, as one of the most effective proof methods, a so-called *reduction method.* In this paragraph, I want to explain this, and show some of the propositions which can be proved by using this method.

Proposition 51 *We have*

$$a_1 \cdots a_n x,\ xy \longrightarrow a_1 \cdots a_n y \vee x$$

Proof. We have

$$xy \longrightarrow \left.\begin{array}{l} a_2 \cdots a_n x \neq 0, \\ a_1 a_2 \cdots a_n x, \\ a_2 \cdots a_n xy \end{array}\right\} \xrightarrow{(Axiom\ 4)} a_1 a_2 \cdots a_n y$$

Therefore

$$a_1 \cdots a_n x,\ \ xy \longrightarrow a_1 \cdots a_n y \vee a_2 \cdots a_n x$$

In other words,

$$a_1 \cdots a_n x,\ \ xy \longrightarrow a_1 \cdots a_n y \vee \{a_2 \cdots a_n x, xy\}$$

which is depicted more diagrammatically as follows:

$$\left.\begin{array}{r} a_1 \cdots a_n x, \\ xy \end{array}\right\} \quad \begin{array}{c} \rightarrow \\ \downarrow \\ a_1 \cdots a_n y. \end{array} \quad \left.\begin{array}{r} a_2 \cdots a_n x, \\ xy \end{array}\right\}$$

By repeating this argument, we have

$$\left.\begin{array}{r} a_1 \cdots a_n x, \\ xy \end{array}\right\} \quad \begin{array}{c} \rightarrow \\ \downarrow \\ a_1 \cdots a_n y. \end{array} \quad \left.\begin{array}{r} a_2 \cdots a_n x, \\ xy \end{array}\right\} \quad \begin{array}{c} \rightarrow \\ \downarrow \\ a_2 \cdots a_n y. \end{array} \quad \left.\begin{array}{r} a_3 \cdots a_n x, \\ xy \end{array}\right\}$$

$$\begin{array}{c} \rightarrow \cdots \rightarrow \\ \downarrow \\ a_3 \cdots a_n y. \end{array} \quad \left.\begin{array}{r} a_n x, \\ xy \end{array}\right\} \begin{array}{c} \rightarrow x. \\ \downarrow \\ a_n y. \end{array}$$

Since we have

$$a_1 \cdots a_n y \leftarrow a_2 \cdots a_n y \leftarrow a_3 \cdots a_n y \leftarrow \cdots \leftarrow a_n y$$

the above diagram can be elaborated into

$$\begin{array}{ccccc}
\left.\begin{array}{r} a_1\cdots a_n x, \\ xy \end{array}\right\} & \begin{array}{c}\rightarrow \\ \downarrow \end{array} & \left.\begin{array}{r} a_2\cdots a_n x, \\ xy \end{array}\right\} & \begin{array}{c}\rightarrow \\ \downarrow \end{array} \\
 & a_1\cdots a_n y. & \leftarrow & a_2\cdots a_n y
\end{array}$$

$$\begin{array}{ccccc}
\left.\begin{array}{r} a_3\cdots a_n x, \\ xy \end{array}\right\} & \begin{array}{l}\rightarrow\cdots\rightarrow \\ \downarrow \end{array} & \left.\begin{array}{r} a_n x, \\ xy \end{array}\right\} & \begin{array}{l}\rightarrow x. \\ \downarrow \end{array} \\
\leftarrow & a_3\cdots a_n y & \leftarrow\cdots\leftarrow & a_n y
\end{array}$$

Thus we finally have

$$\begin{array}{cc}
\left.\begin{array}{r} a_1\cdots a_n x, \\ xy \end{array}\right\} & \begin{array}{l}\rightarrow x, \\ \downarrow \end{array} \\
 & a_1\cdots a_n y
\end{array}$$

□

The above proof method is called the *reduction method*. The following proposition should be taken as a methodological example of the propositions which can be established by this token.

Proposition 52 *We have*

$$a_1\cdots a_n x_1\cdots x_k,\ x_1\cdots x_k y \rightarrow a_1\cdots a_n x_1\cdots x_{k-1} y \vee x_1\cdots x_k$$

Proof.

$$\begin{array}{c}
\left.\begin{array}{r} a_2\cdots a_n x_1\cdots x_k \neq 0 \\ a_1\cdots a_n x_1\cdots x_k, \\ x_1\cdots x_k y \end{array}\right\} \longrightarrow \left.\begin{array}{r} a_2\cdots a_n x_1\cdots x_k \neq 0 \\ a_1 a_2\cdots a_n x_1\cdots x_k, \\ a_2\cdots a_n x_1\cdots x_k y \end{array}\right\} \xrightarrow{(Axiom\ 4)} \\
a_1\cdots a_n x_1\cdots x_{k-1} y
\end{array}$$

Therefore

$$\begin{array}{cc}
\left.\begin{array}{r} a_1\cdots a_n x_1\cdots x_k, \\ x_1\cdots x_k y \end{array}\right\} & \begin{array}{c}\rightarrow a_2\cdots a_n x_1\cdots x_k \\ \downarrow \\ a_1\cdots a_n x_1\cdots x_{k-1} y \end{array}
\end{array}$$

Continuing this argument, we have

$$\begin{array}{ccccc}
\left.\begin{array}{r} a_1\cdots a_n x_1\cdots x_k, \\ x_1\cdots x_k y \end{array}\right\} & \begin{array}{c}\rightarrow \\ \downarrow \end{array} & \left.\begin{array}{r} a_2\cdots a_n x_1\cdots x_k, \\ x_1\cdots x_k y \end{array}\right\} & \begin{array}{l}\rightarrow\cdots\rightarrow \\ \downarrow \end{array} \\
 & a_1\cdots a_n x_1\cdots x_{k-1} y. & \leftarrow & a_2\cdots a_n x_1\cdots x_{k-1} y
\end{array}$$

$$\begin{array}{cc}
\left.\begin{array}{r} a_n x_1\cdots x_k, \\ x_1\cdots x_k y \end{array}\right\} & \begin{array}{l}\rightarrow x_1\cdots x_k. \\ \downarrow \end{array} \\
\leftarrow\cdots\leftarrow & a_n x_1\cdots x_{k-1} y
\end{array}$$

Therefore we have

$$\left.\begin{array}{r} a_1\cdots a_n x_1\cdots x_k, \\ x_1\cdots x_k y \end{array}\right\} \begin{array}{c} \to x_1\cdots x_k, \\ \downarrow \\ a_1\cdots a_n x_1\cdots x_{k-1} y \end{array}$$ [12]

□

Proposition 53 *We have*

$$a_1\cdots a_n x,\ b_1\cdots b_m x \longrightarrow a_1\cdots a_n b_1\cdots b_m \vee x$$

Proof. We have

$$\left.\begin{array}{r} a_1\cdots a_n x, \\ b_1\cdots b_m x \end{array}\right\} \to \left.\begin{array}{r} a_1\cdots a_n b_2\cdots b_m x, \\ b_1 b_2\cdots b_m x \end{array}\right\} \xrightarrow{(Proposition\ 52)} \begin{array}{c} b_2\cdots b_m x. \\ \downarrow \\ a_1\cdots a_n b_1\cdots b_m. \end{array}$$

Therefore we have

$$\left.\begin{array}{r} a_1\cdots a_n x, \\ b_1\cdots b_m x \end{array}\right\} \begin{array}{c} \to \\ \downarrow \\ a_1\cdots a_n b_1\cdots b_m. \end{array} \left.\begin{array}{r} a_1\cdots a_n x, \\ b_2\cdots b_m x \end{array}\right\}.$$

so that

$$\left.\begin{array}{r} a_1\cdots a_n x, \\ b_1\cdots b_m x \end{array}\right\} \begin{array}{c} \to \\ \downarrow \\ a_1\cdots a_n b_1\cdots b_m. \end{array} \begin{array}{c} \\ \\ \leftarrow \end{array} \left.\begin{array}{r} a_1\cdots a_n x, \\ b_2\cdots b_m x \end{array}\right\} \begin{array}{c} \to\cdots\to \\ \downarrow \\ a_1\cdots a_n b_2\cdots b_m \end{array}$$

$$\left.\begin{array}{r} a_1\cdots a_n x, \\ b_m x \\ \leftarrow\cdots\leftarrow \end{array}\right\} \begin{array}{c} \to x. \\ \downarrow \\ a_1\cdots a_n b_m \end{array}$$

Therefore we have

$$\left.\begin{array}{r} a_1\cdots a_n x, \\ b_1\cdots b_m x \end{array}\right\} \begin{array}{c} \to x \\ \downarrow \\ a_1\cdots a_n b_1\cdots b_m \end{array}$$

□

[12] The proof of Proposition 52 is a typical example of the reduction method.

Proposition 54 *We have*

$$a_1\cdots a_n x_1\cdots x_k,\ b_1\cdots b_m x_1\cdots x_k \to a_1\cdots a_n b_1\cdots b_m x_1\cdots x_{k-1}\vee x_1\cdots x_k$$

Proof. We have

$$\left.\begin{array}{r} a_1\cdots a_n x_1\cdots x_k,\\ b_1\cdots b_m x_1\cdots x_k\end{array}\right\}\to \left.\begin{array}{r} a_1\cdots a_n b_2\cdots b_m x_1\cdots x_k,\\ b_1 b_2\cdots b_m x_1\cdots x_k\end{array}\right\}$$

$$\xrightarrow{(Proposition\ 52)} \begin{array}{c} b_2\cdots b_m x_1\cdots x_k.\\ \downarrow\\ a_1\cdots a_n b_1\cdots b_m x_1\cdots x_{k-1}.\end{array}$$

Therefore

$$\left.\begin{array}{r} a_1\cdots a_n x_1\cdots x_k,\\ b_1\cdots b_m x_1\cdots x_k\end{array}\right\} \begin{array}{c} \to\\ \downarrow\\ a_1\cdots a_n b_1\cdots b_m x_1\cdots x_{k-1}.\end{array} \left.\begin{array}{r} a_1\cdots a_n x_1\cdots x_k,\\ b_2\cdots b_m x_1\cdots x_k\end{array}\right\}.$$

so that

$$\left.\begin{array}{r} a_1\cdots a_n x_1\cdots x_k,\\ b_1\cdots b_m x_1\cdots x_k\end{array}\right\} \begin{array}{c} \to\\ \downarrow\\ a_1\cdots a_n b_1\cdots b_m x_1\cdots x_{k-1}.\end{array} \begin{array}{c} \left.\begin{array}{r} a_1\cdots a_n x_1\cdots x_k,\\ b_2\cdots b_m x_1\cdots x_k\end{array}\right\}\\ \leftarrow \end{array}$$

$$\begin{array}{c} \to\cdots\to\\ \downarrow\\ a_1\cdots a_n b_2\cdots b_m x_1\cdots x_{k-1}\end{array}$$

$$\begin{array}{c}\left.\begin{array}{r} a_1\cdots a_n x_1\cdots x_k,\\ b_m x_1\cdots x_k\end{array}\right\}\\ \leftarrow\cdots\leftarrow\end{array} \begin{array}{c} \to\\ \downarrow\\ a_1\cdots a_n b_m x_1\cdots x_{k-1}\end{array} \quad x_1\cdots x_k.$$

Therefore we have

$$\left.\begin{array}{r} a_1\cdots a_n x_1\cdots x_k,\\ b_1\cdots b_m x_1\cdots x_k\end{array}\right\} \begin{array}{c} \to x_1\cdots x_k\\ \downarrow\\ a_1\cdots a_n b_1\cdots b_m x_1\cdots x_{k-1}\end{array}$$

□

Corollary 55 *We have*

$$a_1\cdots a_n x,\ bx,\ x\neq 0\to a_1\cdots a_n b$$

Corollary 56 *We have*

$$a_1 \cdots a_n x,\ b_1 b_2 x,\ x \neq 0 \to a_1 \cdots a_n b_1 b_2$$

Corollary 57 *We have*

$$a_1 \cdots a_n x,\ b_1 \cdots b_m x,\ x \neq 0 \to a_1 \cdots a_n b_1 \cdots b_m.$$

Corollary 58 *We have*

$$a_1 \cdots a_n b_1 \cdots b_m \neq 0 \to \mathfrak{D}(\mathfrak{R}(a_1 \cdots a_n),\ \mathfrak{R}(b_1 \cdots b_m)) = \mathfrak{R}$$

Corollary 59 *We have*

$$\begin{gathered} a_1 \cdots a_n x_1 \cdots x_k,\ b_1 \cdots b_m x_1 \cdots x_k,\ x_1 \cdots x_k \neq 0 \to \\ a_1 \cdots a_n b_1 \cdots b_m x_1 \cdots x_{i-1} x_{i+1} \cdots x_k,\ (i = 1, \cdots, k) \end{gathered}$$

Corollary 60 *We have*

$$\begin{gathered} a_1 \cdots a_n b_1 \cdots b_m d_1 \cdots d_k \neq 0 \to \\ \mathfrak{D}(\mathfrak{R}(a_1 \cdots a_n d_1 \cdots d_k),\ \mathfrak{R}(b_1 \cdots b_m d_1 \cdots d_k)) = \mathfrak{R}(d_1 \cdots d_k) \end{gathered}$$

Remark 61 *If the premises of a proposition, as above, consist of two cycles, then there is still, besides the above reduction method, an always useful proof method which traces back to Proposition* 48. *We will reprove Proposition* 53 *by this new method.*

Proposition 62 *(the same as Proposition* 53*) We have*

$$a_1 \cdots a_n x,\ b_1 \cdots b_m x \longrightarrow a_1 \cdots a_n b_1 \cdots b_m \vee x$$

Proof. It is trivial to see that

$$\begin{aligned} a_1 \cdots a_n &\to a_1 \cdots a_n b_1 \cdots b_m \\ b_1 \cdots b_m &\to a_1 \cdots a_n b_1 \cdots b_m \end{aligned}$$

If $a_1 \cdots a_n \neq 0$ and $b_1 \cdots b_m \neq 0$, then

$$\begin{aligned} &\dim \mathfrak{D}(\mathfrak{R}(a_1 \cdots a_n), \mathfrak{R}(b_1 \cdots b_m)) + \dim \mathfrak{B}(\mathfrak{R}(a_1 \cdots a_n), \mathfrak{R}(b_1 \cdots b_m)) \\ &\leqq n + m \end{aligned}$$

by Proposition 48. At the same time we have

$$a_1 \cdots a_n x, b_1 \cdots b_m x \to \mathfrak{D}(\mathfrak{R}(a_1 \cdots a_n), \mathfrak{R}(b_1 \cdots b_m)) \ni x$$

Furthermore we have

$$\mathfrak{B}(\mathfrak{R}(a_1 \cdots a_n), \mathfrak{R}(b_1 \cdots b_m)) \ni a_1, \cdots, a_n, b_1, \cdots, b_m$$

Therefore we have

$$\dim \mathfrak{D}(\mathfrak{R}(a_1 \cdots a_n), \mathfrak{R}(b_1 \cdots b_m)) \geqq \dim (x)$$
$$\dim \mathfrak{B}(\mathfrak{R}(a_1 \cdots a_n), \mathfrak{R}(b_1 \cdots b_m)) \geqq \dim (a_1 \cdots a_n b_1 \cdots b_m)$$

so that

$$\dim (x) + \dim (a_1 \cdots a_n b_1 \cdots b_m) \leqq n + m$$

which means that

$$x \neq 0 \ \rightarrow a_1 \cdots a_n b_1 \cdots b_m$$

or equivalently

$$a_1 \cdots a_n b_1 \cdots b_m \neq 0 \rightarrow x$$

or equivalently again

$$a_1 \cdots a_n b_1 \cdots b_m \vee x$$

We conclude that

$$a_1 \cdots a_n x, \ b_1 \cdots b_m x \rightarrow a_1 \cdots a_n b_1 \cdots b_m \vee x$$

□

Proposition 63 *We have*

$$a_1 \cdots a_n, \ a_i b_i, \ a_i \neq 0 \ (i = 1, \cdots, n) \longrightarrow b_1 \cdots b_n$$

Proof. We have

$$a_1 \cdots a_n, \ a_i b_i \ \underrightarrow{(Proposition\ 51)} \ a_1 \cdots a_{i-1} b_i a_{i+1} \cdots a_n \vee a_i$$

so that

$$\left.\begin{array}{r} a_1 \cdots a_n, \\ a_1 b_1 \end{array}\right\} \begin{array}{c} \rightarrow \\ \downarrow \\ a_1 \rightarrow \perp . \end{array} \quad \left.\begin{array}{r} b_1 a_2 \cdots a_n, \\ a_2 b_2 \end{array}\right\} \begin{array}{c} \rightarrow \\ \downarrow \\ a_2 \rightarrow \perp \end{array}$$

$$\left.\begin{array}{r} b_1 b_2 a_3 \cdots a_n, \\ a_3 b_3 \end{array}\right\} \begin{array}{l} \rightarrow \cdots \rightarrow \\ \downarrow \\ a_3 \rightarrow \perp \end{array} \quad \left.\begin{array}{r} b_1 \cdots b_{n-1} a_n, \\ a_n b_n \end{array}\right\} \begin{array}{l} \rightarrow b_1 \cdots b_n . \\ \downarrow \\ a_n \rightarrow \perp \end{array}$$

Therefore we have

$$a_1 \cdots a_n, \ a_i b_i, \ a_i \neq 0 \ \ (i = 1, \ \cdots, \ n) \longrightarrow b_1 \cdots b_n$$

□

Corollary 64 *We have*

$$a_1 \cdots a_n, \neq 0,\ a_i b_i,\ b_i \neq 0,\ (i = 1, \cdots, n) \longrightarrow b_1 \cdots b_n \neq 0$$

Proof. We have

$$b_1 \cdots b_n,\ a_i b_i,\ b_i \neq 0,\ (i = 1, \cdots, n) \underrightarrow{(Proposition\ 63)} a_1 \cdots a_n \longrightarrow \perp$$

Therefore

$$a_1 \cdots a_n, \neq 0,\ a_i b_i,\ b_i \neq 0,\ (i = 1, \cdots, n) \longrightarrow b_1 \cdots b_n \neq 0$$

□

Proposition 65 *We have*

$$\left.\begin{array}{l} \widehat{a_1} a_2 \cdots a_n x, \\ a_1 \widehat{a_2} \cdots a_n x, \\ \cdots\cdots\cdots\cdots, \\ a_1 a_2 \cdots \widehat{a_n} x \end{array}\right\} \to a_1 a_2 \cdots a_n \vee x$$

Proof. We have

$$\left.\begin{array}{l} \widehat{a_1} a_2 \cdots a_n x, \\ a_1 \widehat{a_2} \cdots a_n x \end{array}\right\} \begin{array}{c} \underrightarrow{(Proposition\ 52)} \\ \downarrow \\ a_1 a_2 \cdots a_n \end{array} \widehat{a_1}\widehat{a_2} a_3 \cdots a_n x$$

so that

$$\left.\begin{array}{l} \widehat{a_1} a_2 \cdots a_n x, \\ a_1 \widehat{a_2} \cdots a_n x \end{array}\right\} \begin{array}{c} \to \\ \downarrow \\ a_1 a_2 \cdots a_n. \end{array} \left.\begin{array}{l} \widehat{a_1}\widehat{a_2} a_3 \cdots a_n x, \\ a_1 a_2 \widehat{a_3} \cdots a_n x \end{array}\right\}$$

$$\begin{array}{c} \to \\ \downarrow \\ a_1 a_2 \cdots a_n. \end{array} \left.\begin{array}{l} \widehat{a_1}\widehat{a_2}\widehat{a_3} a_4 \cdots a_n x, \\ a_1 a_2 a_3 \widehat{a_4} \cdots a_n x \end{array}\right\} \begin{array}{c} \to \cdots \\ \downarrow \\ a_1 a_2 \cdots a_n. \end{array}$$

$$\cdots \to \left.\begin{array}{l} \widehat{a_1} \cdots \widehat{a_{n-1}} a_n x, \\ a_1 \cdots a_{n-1} \widehat{a_n} x \end{array}\right\} \begin{array}{c} \to x. \\ \downarrow \\ a_1 a_2 \cdots a_n. \end{array}$$

Therefore

$$\left.\begin{array}{l} \widehat{a_1} a_2 \cdots a_n x, \\ a_1 \widehat{a_2} \cdots a_n x, \\ \cdots\cdots\cdots\cdots, \\ a_1 a_2 \cdots \widehat{a_n} x \end{array}\right\} \begin{array}{l} \to x, \\ \downarrow \\ a_1 a_2 \cdots a_n, \end{array}$$

□

Corollary 66 *We have*

$$a_1 \cdots a_n \neq 0 \to \mathfrak{D}(\mathfrak{R}_1(\widehat{a_1} a_2 \cdots a_n), \mathfrak{R}_2(a_1 \widehat{a_2} \cdots a_n), \cdots, \\ \mathfrak{R}_n(a_1 \cdots a_{n-1} \widehat{a_n})) = \mathfrak{N}$$

In words, there is no point which is contained in all of $(n-1)$-sided simplices of an n-simplex at the same time.

Proposition 67 *We have*

$$\left.\begin{array}{l} \widehat{a_1} a_2 \cdots a_n x_1 \cdots x_k, \\ a_1 \widehat{a_2} \cdots a_n x_1 \cdots x_k, \\ \cdots\cdots\cdots\cdots\cdots\cdots, \\ a_1 a_2 \cdots \widehat{a_n} x \cdots x_k \end{array}\right\} \to a_1 a_2 \cdots a_n x_1 \cdots x_{k-1} \vee x_1 \cdots x_k$$

Proof. We have

$$\left.\begin{array}{l} \widehat{a_1} a_2 \cdots a_i \cdots a_n x_1 \cdots x_k, \\ a_1 a_2 \cdots \widehat{a_i} \cdots a_n x_1 \cdots x_k \end{array}\right\} \begin{array}{c} \xrightarrow{(Proposition\ 52)} \\ \downarrow \\ a_1 a_2 \cdots a_n x_1 \cdots x_{k-1} \end{array} \widehat{a_1} a_2 \cdots \widehat{a_i} \cdots a_n x_1 \cdots x_k$$

so that

$$\left.\begin{array}{l} \widehat{a_1} a_2 \cdots a_n x_1 \cdots x_k, \\ a_1 \widehat{a_2} \cdots a_n x_1 \cdots x_k, \\ \cdots\cdots\cdots\cdots\cdots\cdots, \\ a_1 a_2 \cdots \widehat{a_n} x_1 \cdots x_k \end{array}\right\} \begin{array}{c} \to \\ \downarrow \\ a_1 a_2 \cdots a_n x_1 \cdots x_{k-1}. \end{array} \begin{array}{c} \left.\begin{array}{l} \widehat{a_2} a_3 \cdots a_n x_1 \cdots x_k, \\ a_2 \widehat{a_3} \cdots a_n x_1 \cdots x_k, \\ \cdots\cdots\cdots\cdots\cdots\cdots, \\ a_2 a_3 \cdots \widehat{a_n} x_1 \cdots x_k \end{array}\right\} \\ \leftarrow \end{array}$$

$$\begin{array}{c} \to \cdots \\ \downarrow \\ a_2 a_3 \cdots a_n x_1 \cdots x_{k-1} \end{array} \begin{array}{c} \\ \cdots \to \\ \leftarrow \cdots \end{array} \begin{array}{c} \left.\begin{array}{l} \widehat{a_{n-1}} a_n x_1 \cdots x_k, \\ a_{n-1} \widehat{a_n} x_1 \cdots x_k \end{array}\right\} \\ \cdots \leftarrow \end{array}$$

$$\begin{array}{c} \to x_1 \cdots x_k. \\ \downarrow \\ a_{n-1} a_n x_1 \cdots x_{k-1} \end{array}$$

Therefore

$$\left.\begin{array}{l} \widehat{a_1} a_2 \cdots a_n x_1 \cdots x_k, \\ a_1 \widehat{a_2} \cdots a_n x_1 \cdots x_k, \\ \cdots\cdots\cdots\cdots\cdots\cdots, \\ a_1 a_2 \cdots \widehat{a_n} x_1 \cdots x_k \end{array}\right\} \begin{array}{c} \to \quad x_1 \cdots x_k, \\ \downarrow \\ a_1 a_2 \cdots a_n x_1 \cdots x_{k-1}, \end{array}$$

□

Corollary 68 *We have*

$$a_1 \cdots a_n \neq 0, m \leqq n \rightarrow \mathfrak{D}(\mathfrak{R}_1(\widehat{a_1} a_2 \cdots a_m a_{m+1} \cdots a_n), \mathfrak{R}_2(a_1 \widehat{a_2} \cdots a_m a_{m+1} \cdots a_n), \ldots, \mathfrak{R}_m(a_1 a_2 \cdots \widehat{a_m} a_{m+1} \cdots a_n)) = \mathfrak{R}(a_{m+1} \cdots a_n)$$

Concluding Remarks

Concerning §5, it should be stressed that none of the propositions there depend on Axiom 1 at all. This is very important, for we intend to build geometries based upon only Axioms 2,3,4 in the future. Concerning this entire article, we should remark that no proposition here contains an existential statement at all, which will be crucial in the sequel to this article. There we will shortly develop the theory of an abstract space of linear dependence of the second type.

On Axiomatics of Linear Dependence. II. The $\mathfrak{B}_2$-Space

Takeo Nakasawa
Received November, 28, 1935

Introduction

In the present paper, which is a sequel to my earlier work[13], the geometry of the second abstract space of linear dependence $\mathfrak{B}_2$ (Definition 6) is to be presented precisely, after giving such auxiliary notions as circuit (Definition 15), principal linear space (Definition 25), separation of linear spaces (Definition 29) and separation of elements (Definition 58), we want to present our main theorems concerning the decomposition of a linear space into principal linear spaces (Proposition 56) and the decomposition of our $\mathfrak{B}_2$-space into principal spaces (Proposition 59).

The present axiom system is distinguished from that in my earlier work[1] in that firstly, the condition $a \neq 0$ should be satisfied in the first axiom, and secondly, as the last axiom, the existence of the intersecting element is added. Therefore, the $\mathfrak{B}_2$-space is a strengthening of the $\mathfrak{B}_1$-space earlier investigated.

---

[13] T. NAKASAWA, "On axiomatics of linear dependence. I", Science Reports of the Tokyo Bunrika Daigaku, Section A, Volume 2, No. 43, (p 235–255).

Notation

In addition to the notations which we use in my earlier work, we want to add some new ones, specifically 1, 7, 8, 11:

1. $\therefore$	means "therefore".
2. $A \to B$	means that B follows from A.
3. $A \leftrightarrow B$	means that $A \to B$ and $B \to A$.
4. A, B	means that A and B.
5. A_i $(i = 1, \cdots, k)$	means that A_1, A_2, $\cdots$, and A_k.
6. $A \vee B$	means that at least one of A and B.
7. $A_1 \vee \cdots \vee A_k$	means that at least one of A_1, $A_2, \cdots$, and A_k.
8. for example A_1	means that A_1 or A_2 or $\cdots$ or A_k.
9. $A \longrightarrow \perp$.	means that A gives rise to contradiction.
10. $A\ \underset{\longrightarrow}{(S)}\ B$	means that B follows from A on the basis of S.
11. $(\exists z)\overrightarrow{A}(z)$	means that an element z satisfying $A(z)$ exists.
12. $\left.\begin{matrix} A_1, \\ \vdots, \\ A_k \end{matrix}\right\} \to B$	means that B follows from the simultaneous *existence* of k statements A_1, A_2, $\cdots$, A_k.
13. $A \to \left\{\begin{matrix} B_1, \\ \vdots, \\ B_k \end{matrix}\right.$	means that k statements B_1, B_2, $\cdots$, B_k follow from A simultaneously.

14. The following notations of set theory are also used in their usual meaning; $\supseteqq$, $\not\supseteqq$, $\supset$, $=$, $\neq$, $\ni$, $\notin$, $\cup$, etc.

Axioms

We are provided with a certain set of elements; $\mathfrak{B}_2 \ni a_1, a_2, \cdots, a_s, \cdots$. For certain series of elements, which we will call *cycles*, we are supplied with the relations "*linearly dependent*", in notation $a_1 \cdots a_s = 0$, usually abbreviated to $a_1 \cdots a_s$, or "*linearly independent*", in notation $a_1 \cdots a_s \neq 0$. These relations should now satisfy the following axioms:

Axiom 1* *(reflexivity)*

$$a \neq 0, aa$$

Axiom 2 *(addition)*

$$a_1 \cdots a_s \to a_1 \cdots a_n x, \ (s = 1, 2, \cdots)$$

Axiom 3 *(exchange)*

$$a_1 \cdots a_i \cdots a_s \to a_i a_1 \cdots \widehat{a_i} \cdots a_s, \ (s = 2, 3, \cdots; i = 2, \cdots, s)$$

Axiom 4 *(transitivity)*

$$xa_1 \cdots a_s, a_1 \cdots a_s y \rightarrow xa_1 \cdots a_{s-1} y, \ (s = 1, 2, \cdots)$$

Axiom 5 *(Intersection)*

$$a_1 \cdots a_s xy \rightarrow (\exists z)(a_1 \cdots a_s z, xyz) \ (s = 2, 3, \cdots)$$

Definition 6 *Such a set $\mathfrak{B}_2$ is called a second abstract space of linear dependence, in short, a $\mathfrak{B}_2$-space.*

Remark 7 *We take the set of all points of the classic n-dimensional projective space as a $\mathfrak{B}_2$-space, and we call certain linear series of points dependent or lineally independent in the usual sense, so that our basic framework for a $\mathfrak{B}_2$-space applies (in particular, the Axioms 1*–5 are satisfied). Therefore the projective space appears as a typical representation of a $\mathfrak{B}_2$-space with respect to the linear independence of point sequences. From now on we will often use the word "point" instead of "element".*

Remark 8 *As mentioned above, any $\mathfrak{B}_2$-space is also a $\mathfrak{B}_1$-space. Therefore all propositions holding in any $\mathfrak{B}_1$-space are also valid in any $\mathfrak{B}_2$-space.*

Remark 9 *Further, it follows easily from Axiom 1* that the root[14] of $\mathfrak{B}_2$ is an empty set. From now on, we will also denote the empty set with the character $\mathfrak{N}$.*

The Intersecting Point

Proposition 10 *We have*

$$a_1 \cdots a_n b_1 \cdots b_m \rightarrow (\exists z)(a_1 \cdots a_n z, b_1 \cdots b_m z)$$

Proof. We proceed by induction on m.

1. If m is equal to 1, then we have

$$a_1 \cdots a_n b_1 \xrightarrow{\text{(Axiom 1*)}} (\exists z)(a_1 \cdots a_n z, b_1 z)$$

2. If m is equal to 2, then we have

$$a_1 \cdots a_n b_1 b_2 \xrightarrow{\text{(Axiom 5)}} (\exists z)(a_1 \cdots a_n z, b_1 b_2 z)$$

[14] cf. loc. cit. p. 237!

3. Now it suffices to show that the case $[m]$ follows from the case $[m-1]$. We have

$$a_1 \cdots a_n b_1 \cdots b_m \to a_1 \cdots a_n b_1 \cdots b_{m-2} b_{m-1} b_m \xrightarrow{\text{(Axiom 5)}} (\exists z)(a_1 \cdots a_n b_1 \cdots b_{m-2} y, b_{m-1} b_m y)$$

$$a_1 \cdots a_n b_1 \cdots b_{m-2} y \xrightarrow{[m-1]} (\exists z)(a_1 \cdots a_n z, b_1 \cdots b_{m-2} yz)$$

Then we have

$$b_1 \cdots b_{m-2} yz, b_{m-1} b_m y \xrightarrow{\text{(Corollary 56)}} b_1 \cdots b_{m-2} b_m z$$

Therefore we have

$$a_1 \cdots a_n b_1 \cdots b_m \to (\exists z)(a_1 \cdots a_n z, b_1 \cdots b_m z)$$

□

Proposition 11 *We have*

$$\dim \mathfrak{D}(\mathfrak{R}_1 \mathfrak{R}_2) + \dim \mathfrak{B}(\mathfrak{R}_1 \mathfrak{R}_2) = \dim \mathfrak{R}_1 + \dim \mathfrak{R}_2$$

In other words, the sum of the dimensions of the intersection and the sum of two given linear spaces is equal to the sum of the dimensions of the two given linear spaces.

Proof. Let the basis of $\mathfrak{D}(\mathfrak{R}_1 \mathfrak{R}_2)$ be $c_1 \cdots c_k$, and let the bases of $\mathfrak{R}_1$ and $\mathfrak{R}_2$ be $c_1 \cdots c_k a_{k+1} \cdots a_n$ and $c_1 \cdots c_k b_{k+1} \cdots b_m$ respectively; then

$$\mathfrak{B}(\mathfrak{R}_1 \mathfrak{R}_2) = \mathfrak{B}(c_1 \cdots c_k a_{k+1} \cdots a_n b_{k+1} \cdots b_m)$$

follows by Proposition 48. If $c_1 \cdots c_k a_{k+1} \cdots a_n b_{k+1} \cdots b_m = 0$ were to obtain, then we would have

$$c_1 \cdots c_k a_{k+1} \cdots a_n b_{k+1} \cdots b_m \xrightarrow{\text{(Proposition 10)}} (\exists z)(c_1 \cdots c_k a_{k+1} \cdots a_n z, b_{k+1} \cdots b_m z)$$

Then we have

$$\left.\begin{array}{l} c_1 \cdots c_k a_{k+1} \cdots a_n z \to \mathfrak{R}_1(c_1 \cdots c_k a_{k+1} \cdots a_n) \ni z \\ b_{k+1} \cdots b_m z \to \mathfrak{R}_2(c_1 \cdots c_k b_{k+1} \cdots b_m) \ni z \end{array}\right\} \to \mathfrak{D}(\mathfrak{R}_1 \mathfrak{R}_2) \ni z \to c_1 \cdots c_k z$$

$$c_1 \cdots c_k z, b_{k+1} \cdots b_m z \xrightarrow{\text{(Corollary I.57)}^{15}} c_1 \cdots c_k b_{k+1} \cdots b_m \to \perp$$

[15] Results of the previous paper are referred to, by way of example, by Corollary I.57 in place of Corollary 57 of the previous paper.

Therefore we have

$$c_1 \cdots c_k a_{k+1} \cdots a_n b_{k+1} \cdots b_m \neq 0$$

which implies that

$$\dim \mathfrak{B}(c_1 \cdots c_k a_{k+1} \cdots a_n b_{k+1} \cdots b_m) = n + m - k$$

so that

$$\dim \mathfrak{B}(\mathfrak{R}_1 \mathfrak{R}_2) = n + m - k$$

On the other hand, we have

$$\begin{aligned} \dim \mathfrak{R}_1(c_1 \cdots c_k a_{k+1} \cdots a_n) &= n \\ \dim \mathfrak{R}_2(c_1 \cdots c_k b_{k+1} \cdots b_m) &= m \\ \dim \mathfrak{D}(\mathfrak{R}_1 \mathfrak{R}_2) &= \dim \mathfrak{R}_1(c_1 \cdots c_k) = k \end{aligned}$$

Therefore we can conclude that

$$\dim \mathfrak{D}(\mathfrak{R}_1 \mathfrak{R}_2) + \dim \mathfrak{B}(\mathfrak{R}_1 \mathfrak{R}_2) = \dim \mathfrak{R}_1 + \dim \mathfrak{R}_2$$

□

This is a strengthened version of Proposition I.48, in which it is merely claimed that the right-hand side is greater than or equal to the left-hand side. By dint of Axiom 5 the above equality can now be established.

Corollary 12 *If $a_1 \cdots a_n \neq 0$ and $b_1 \cdots b_m \neq 0$ with* $\dim \mathfrak{R}(a_1 \cdots a_n b_1 \cdots b_m) = s$*, then there exists the intersection of two linear spaces $\mathfrak{R}(a_1 \cdots a_n)$ and $\mathfrak{R}(b_1 \cdots b_m)$, whose dimension is equal to $n + m - s$.*

Corollary 13 *We have*

$$\begin{aligned} \mathfrak{D}(\mathfrak{R}(a_1 \cdots a_n d_1 \cdots d_k), \mathfrak{R}(b_1 \cdots b_m d_1 \cdots d_k)) = \mathfrak{R}(d_1 \cdots d_k) \\ \to a_1 \cdots a_n b_1 \cdots b_m d_1 \cdots d_k \\ \neq 0 \end{aligned}$$

Corollary 14 *We have*

$$\begin{aligned} \mathfrak{D}(\mathfrak{R}(a_1 \cdots a_n), \mathfrak{R}(b_1 \cdots b_m)) = \mathfrak{N} \\ \to a_1 \cdots a_n b_1 \cdots b_m \neq 0 \end{aligned}$$

Circuits

Definition 15 *A cycle $a_1 \cdots a_{n+1}$ (so that $a_1 \cdots a_{n+1} = 0$) with $a_1 \cdots \widehat{a_i} \cdots a_{n+1} \neq 0$ for all $i = 1, \cdots, n+1$ is called a circuit of dimension n, and is denoted by $\overline{a_1 \cdots a_{n+1}}$.*

Proposition 16 *We have*

$$\left.\begin{array}{l} a_1 \cdots a_n b_1 \cdots b_{m-1} \neq 0, \\ \overline{b_1 \cdots b_m} \end{array}\right\} \to \left\{\begin{array}{c} a_1 \cdots a_n b_1 \cdots \widehat{b_i} \cdots b_m \neq 0, \\ (i = 1, \cdots, m). \end{array}\right.$$

Proof. Assume, for the sake of contradiction, that $a_1 \cdots a_n b_1 \cdots \widehat{b_i} \cdots b_m = 0$; then we have by Proposition I.52 that

$$\left.\begin{array}{c} a_1 \cdots a_n b_1 \cdots \widehat{b_i} \cdots b_m, \\ b_1 \cdots b_m \end{array}\right\} \to a_1 \cdots a_n b_1 \cdots b_{m-1} \vee b_1 \cdots \widehat{b_i} \cdots b_m$$

which contradicts our assumptions. On the other hand, we have

$$a_1 \cdots a_n b_1 \cdots b_{m-1}, a_1 \cdots a_n b_1 \cdots b_{m-1} \neq 0 \to \bot$$
$$b_1 \cdots \widehat{b_i} \cdots b_m, \overline{b_1 \cdots b_m} \to \bot$$

which implies that

$$a_1 \cdots a_n b_1 \cdots b_{m-1} \vee b_1 \cdots \widehat{b_i} \cdots b_m \to \bot$$

Therefore we can conclude that

$$a_1 \cdots a_n b_1 \cdots \widehat{b_i} \cdots b_m \neq 0 \ (i = 1, \cdots, m)$$

□

Proposition 17 *We have*

$$\left.\begin{array}{l} \overline{d_1 \cdots d_s a_1 \cdots a_n}, \\ \overline{d_1 \cdots d_s b_1 \cdots b_m}, \\ d_1 \cdots d_s a_1 \cdots a_{n-1} b_1 \cdots b_{m-1} \neq 0 \end{array}\right\}$$
$$\to \left\{\begin{array}{l} d_1 \cdots \widehat{d_i} \cdots d_s a_1 \cdots \widehat{a_j} \cdots a_n b_1 \cdots b_m \neq 0, \\ d_1 \cdots d_s a_1 \cdots \widehat{a_j} \cdots a_n b_1 \cdots \widehat{b_k} \cdots b_m \neq 0, \\ d_1 \cdots \widehat{d_i} \cdots d_s a_1 \cdots a_n b_1 \cdots \widehat{b_k} \cdots b_m \neq 0, \\ (i = 1, \cdots, s;\ j = 1, \cdots, n;\ k = 1, \cdots, m). \end{array}\right.$$

Proof. We have

$$\left.\begin{array}{l} \overline{d_1 \cdots d_s a_1 \cdots a_n}, \\ d_1 \cdots d_s a_1 \cdots a_{n-1} b_1 \cdots b_{m-1} \neq 0 \end{array}\right\}$$
$$\xrightarrow{\text{(Proposition 16)}} d_1 \cdots d_s a_1 \cdots \widehat{a_j} \cdots a_n b_1 \cdots b_{m-1} \neq 0$$

$$\left.\begin{array}{l} \overline{d_1 \cdots d_s b_1 \cdots b_m}, \\ d_1 \cdots d_s a_1 \cdots \widehat{a_j} \cdots a_n b_1 \cdots b_{m-1} \neq 0 \end{array}\right\}$$

$$\xrightarrow{\text{(Proposition 16)}} \begin{cases} d_1 \cdots \widehat{d_i} \cdots d_s a_1 \cdots \widehat{a_j} \cdots a_n b_1 \cdots b_m \neq 0, \\ d_1 \cdots d_s a_1 \cdots \widehat{a_j} \cdots a_n b_1 \cdots \widehat{b_k} \cdots b_m \neq 0 \end{cases}$$

$$\left.\begin{array}{l} \overline{d_1 \cdots d_s a_1 \cdots a_n}, \\ d_1 \cdots d_s a_1 \cdots \widehat{a_j} \cdots a_n b_1 \cdots \widehat{b_k} \cdots b_m \neq 0 \end{array}\right\}$$

$$\xrightarrow{\text{(Proposition 16)}} d_1 \cdots \widehat{d_i} \cdots d_s a_1 \cdots a_n b_1 \cdots \widehat{b_k} \cdots b_m \neq 0$$

Therefore we can conclude that

$$\left.\begin{array}{l} \overline{d_1 \cdots d_s a_1 \cdots a_n}, \\ \overline{d_1 \cdots d_s b_1 \cdots b_m}, \\ d_1 \cdots d_s a_1 \cdots a_{n-1} b_1 \cdots b_{m-1} \neq 0 \end{array}\right\}$$

$$\to \begin{cases} d_1 \cdots \widehat{d_i} \cdots d_s a_1 \cdots \widehat{a_j} \cdots a_n b_1 \cdots b_m \neq 0, \\ d_1 \cdots d_s a_1 \cdots \widehat{a_j} \cdots a_n b_1 \cdots \widehat{b_k} \cdots b_m \neq 0, \\ d_1 \cdots \widehat{d_i} \cdots d_s a_1 \cdots a_n b_1 \cdots \widehat{b_k} \cdots b_m \neq 0 \end{cases}$$

□

Proposition 18 *We have*

$$\overline{a_1 \cdots a_n x} \to (\exists y)(\overline{a_1 \cdots a_{n-1} y})$$

Proof. We have

$$a_1 \cdots a_n x \xrightarrow{\text{(Axiom 5)}} (\exists y)(a_1 \cdots a_{n-1} y, a_n x y)$$

Now if $a_1 \cdots \widehat{a_i} \cdots a_{n-1} y = 0$ were to obtain, then

$$a_1 \cdots \widehat{a_i} \cdots a_{n-1} y,\ a_n x y \xrightarrow{\text{(Corollary I.56)}} a_1 \cdots \widehat{a_i} \cdots a_n x$$

would follow. Therefore we have

$$a_1 \cdots \widehat{a_i} \cdots a_n x, \overline{a_1 \cdots a_n x} \to \bot$$

which confirms our assertion that $a_1 \cdots \widehat{a_i} \cdots a_{n-1} y \neq 0$ $(i = 1, \cdots, n-1)$. □

Using the above proposition repeatedly, we get

Proposition 19 *We have*

$$n \geqq m, \overline{a_1 \cdots a_n x} \to (\exists z)(\overline{a_1 \cdots a_m z})$$

Proposition 20 *We have*

$$\overline{a_1 \cdots a_n b_1 \cdots b_m} \to (\exists z)(\overline{a_1 \cdots a_n z}, \overline{b_1 \cdots b_m z})$$

Proof. We have

$$a_1 \cdots a_n b_1 \cdots b_m \xrightarrow{\text{(Proposition 10)}} (\exists z)(a_1 \cdots a_n z, b_1 \cdots b_m z)$$

If $a_1 \cdots \widehat{a_i} \cdots a_n z = 0$ were to obtain, then

$$a_1 \cdots \widehat{a_i} \cdots a_n z, b_1 \cdots b_m z \xrightarrow{\text{(Corollary 57)}} a_1 \cdots \widehat{a_i} \cdots a_n b_1 \cdots b_m$$

would follow, which implies that

$$a_1 \cdots \widehat{a_i} \cdots a_n b_1 \cdots b_m, \overline{a_1 \cdots a_n b_1 \cdots b_m} \to \perp$$

Therefore we are sure that $a_1 \cdots \widehat{a_i} \cdots a_n z \neq 0$ $(i = 1, \cdots, n)$, which implies that $\overline{a_1 \cdots a_n z}$. Analogously we have $\overline{b_1 \cdots b_m z}$. Therefore

$$\overline{a_1 \cdots a_n b_1 \cdots b_m} \to (\exists z)(\overline{a_1 \cdots a_n z}, \overline{b_1 \cdots b_m z})$$

follows. This completes the proof.

□

Proposition 21 *We have*

$$\left.\begin{array}{l} \overline{a_1 \cdots a_n z}, \overline{b_1 \cdots b_m z}, \\ a_1 \cdots a_{n-1} b_1 \cdots b_{m-1} z \neq 0 \end{array}\right\} \to \overline{a_1 \cdots a_n b_1 \cdots b_m}$$

Proof. On the one hand we have

$$a_1 \cdots a_n z, b_1 \cdots b_m z \xrightarrow{\text{(Corollary I.57)}} a_1 \cdots a_n b_1 \cdots b_m$$

On the other hand we have

$$\left.\begin{array}{l} \overline{z a_1 \cdots a_n}, \overline{z b_1 \cdots b_m}, \\ z a_1 \cdots a_{n-1} b_1 \cdots b_{m-1} \neq 0 \end{array}\right\}$$

$$\xrightarrow{\text{(Proposition 11)}} \left\{\begin{array}{l} a_1 \cdots a_i \cdots a_n b_1 \cdots b_m \neq 0, \\ a_1 \cdots a_n b_1 \cdots b_j \cdots b_m \neq 0. \end{array}\right.$$

Therefore we are sure that $\overline{a_1 \cdots a_n b_1 \cdots b_m}$.

□

Proposition 22 *We have*

$$\overline{d a_1 \cdots a_n x}, dby, d a_1 \cdots a_n b \neq 0 \to (\exists z)(\overline{d a_1 \cdots a_n b z})$$

Proof. We have

$$\left.\begin{array}{l}\overline{da_1\cdots a_nx}, \overline{dby},\\ da_1\cdots a_nb\neq 0\end{array}\right\}\xrightarrow{\text{(Proposition 19)}}\overline{a_1\cdots a_nbxy}$$

$$\xrightarrow{\text{(Proposition 19)}}(\exists z)(\overline{a_1\cdots a_nyz}), a_1\cdots a_nby\neq 0$$

$$\left.\begin{array}{l}\overline{a_1\cdots a_nyz}, \overline{dby},\\ a_1\cdots a_nby\neq 0\end{array}\right\}\xrightarrow{\text{(Proposition 19)}}\overline{da_1\cdots a_nbz}$$

Therefore we have

$$\overline{da_1\cdots a_nx}, dby, da_1\cdots a_nb\neq 0\to\overline{da_1\cdots a_nbz}$$

□

Proposition 23 *We have*

$$\left.\begin{array}{l}\overline{d_1\cdots d_sa_1\cdots a_nx},\\ \overline{d_1\cdots d_sb_1\cdots b_my},\\ d_1\cdots d_sa_1\cdots a_nb_1\cdots b_m\neq 0\end{array}\right\}\to(\exists z)(\overline{d_1\cdots d_sa_1\cdots a_nb_1\cdots b_mz})$$

Proof. We have

$$\overline{d_1\cdots d_sb_1\cdots b_my}\xrightarrow{\text{(Proposition 19)}}\left\{\begin{array}{l}(\exists y_1)(\overline{d_1b_1y_1}),\\ (\exists y_2)(\overline{d_1b_2y_2}),\\ \cdots\cdots\cdots\cdots\cdots,\\ (\exists y_m)(\overline{d_1b_my_m}).\end{array}\right.$$

$$\left.\begin{array}{l}\overline{d_1\cdots d_sa_1\cdots a_nx}, \overline{d_1b_1y_1},\\ d_1\cdots d_sa_1\cdots a_nb_1\neq 0\end{array}\right\}$$

$$\xrightarrow{\text{(Proposition 22)}}(\exists z_1)(\overline{d_1\cdots d_sa_1\cdots a_nb_1z_1}).$$

$$\left.\begin{array}{l}\overline{d_1\cdots d_sa_1\cdots a_nb_1z_1}, \overline{d_1b_2y_2},\\ d_1\cdots d_sa_1\cdots a_nb_1b_2\neq 0\end{array}\right\}$$

$$\xrightarrow{\text{(Proposition 22)}}(\exists z_2)(\overline{d_1\cdots d_sa_1\cdots a_nb_1b_2z_2}).$$

$$\cdots\quad\cdots\quad\cdots\quad\cdots\quad\cdots$$

$$\cdots\quad\cdots\quad\cdots\quad\cdots\quad\cdots$$

$$\left.\begin{array}{l}\overline{d_1\cdots d_sa_1\cdots a_nb_1\cdots b_{m-1}z_{m-1}}, \overline{d_1b_my_m},\\ d_1\cdots d_sa_1\cdots a_nb_1\cdots b_m\neq 0\end{array}\right\}$$

$$\xrightarrow{\text{(Proposition 22)}}(\exists z)(\overline{d_1\cdots d_s a_1\cdots a_n b_1\cdots b_m z}).$$

Therefore we have

$$\left.\begin{array}{l}\overline{d_1\cdots d_s a_1\cdots a_n x},\\ \overline{d_1\cdots d_s b_1\cdots b_m y},\\ d_1\cdots d_s a_1\cdots a_n b_1\cdots b_m \neq 0\end{array}\right\} \to (\exists z)(\overline{d_1\cdots d_s a_1\cdots a_n b_1\cdots b_m z})$$

$\square$

Proposition 24 *We have*

$$\left.\begin{array}{l}\overline{a_1\cdots a_n b_1 x},\\ a_1\cdots a_n b_1 b_2,\\ a_1\cdots a_n b_2 \neq 0\end{array}\right\} \to (\exists z)(\overline{a_1\cdots a_n b_2 z}).$$

Proof. We proceed by induction on n.

1. If n is equal to 0, then we have

$$\overline{b_1 x}, b_1 b_2 \to (\exists b_1)(\overline{b_2 b_1})$$

2. If n is equal to 1, then we have

$$\left.\begin{array}{l}\overline{a_1 b_1 x},\\ a_1 b_1 b_2,\\ a_1 b_2 \neq 0\end{array}\right\} \to \left\{\begin{array}{l} b_1 b_2 \neq 0 \to \overline{a_1 b_2 b_1},\\ b_1 b_2 = 0 \to \overline{a_1 b_2 x}.\end{array}\right.$$

3. Now we would like to show that the case $[n]$ follows from the case $[n-1]$. If $a_1\cdots \widehat{a_i}\cdots a_n b_1 b_2 \neq 0$ $(i = 1, \cdots, n)$, we have

$$\left.\begin{array}{l}a_1\cdots \widehat{a_i}\cdots a_n b_1 b_2 \neq 0\ (i = 1, \cdots, n),\\ a_1\cdots a_n b_2 \neq 0,\\ a_1\cdots a_n b_1 x \to a_1\cdots a_n b_1 \neq 0\end{array}\right\} \to \overline{a_1\cdots a_n b_2 b_1}.$$

If $a_1\cdots \widehat{a_i}\cdots a_n b_1 b_2 = 0$ for some i, say, $i = n$, then we have

$$\left.\begin{array}{r}\overline{a_1\cdots a_n b_1 x} \xrightarrow{\text{(Proposition 18)}} (\exists x_1)(\overline{a_1\cdots a_{n-1} b_1 x_1}),\\ a_1\cdots a_{n-1} b_1 b_2,\\ a_1\cdots a_n b_2 \neq 0 \to a_1\cdots a_{n-1} b_2 \neq 0\end{array}\right\}$$

$$\xrightarrow{[n-1]}(\exists y)(\overline{a_1\cdots a_{n-1} b_2 y}).$$

$$\left.\begin{array}{r}\overline{a\cdots a_n b_1 x} \xrightarrow{\text{(Proposition 18)}} (\exists x_2)\overline{a_1\cdots a_{n-1} a_n x_2},\\ \overline{a_1\cdots a_{n-1} b_2 y},\\ a_1\cdots a_n b_2 \neq 0\end{array}\right\}$$

$$\xrightarrow{\text{(Proposition 23)}}(\exists z)(\overline{a_1 \cdots a_n b_2 z}).$$

Therefore we have

$$\left.\begin{array}{l}\overline{a_1 \cdots a_n b_1 x},\\ a_1 \cdots a_n b_1 b_2,\\ a_1 \cdots a_n b_2 \neq 0\end{array}\right\} \to (\exists z)(\overline{a_1 \cdots a_n b_2 z}).$$

□

Definition 25 *A linear space of dimension n, which contains a circuit of dimension n in itself, is called a principal linear space of dimension n.*

Proposition 26 *We have*

$$\overline{b_1 \cdots b_{n+1}}, \mathfrak{R}(a_1 \cdots a_n) \ni b_1, \cdots, b_{n+1}$$
$$\to (\exists a_{n+1})(\overline{a_1 \cdots a_{n+1}}, \mathfrak{R}(a_1 \cdots a_n) \ni a_{n+1}).$$

Proof. We have $b_1 \cdots b_{n-1} a_i \neq 0$ for some $1 \leq i \leq n$, say, $i = 1$. Then we have

$$\left.\begin{array}{l}\overline{b_1 \cdots b_{n-1} b_n b_{n+1}},\\ b_1 \cdots b_{n-1} b_n a_1,\\ b_1 \cdots b_{n-1} a_1 \neq 0\end{array}\right\}$$

$$\xrightarrow{\text{(Proposition 24)}}(\exists z_1)(\overline{b_1 \cdots b_{n-1} a_1 z_1}).$$

We should have $b_1 \cdots b_{n-2} a_1 a_j \neq 0$ for some $2 \leq j \leq n$, say, $j = 2$. Then we have

$$\left.\begin{array}{l}\overline{b_1 \cdots b_{n-2} a_1 b_{n-1} z_1},\\ b_1 \cdots b_{n-2} a_1 b_{n-1} a_2,\\ b_1 \cdots b_{n-2} a_1 a_2 \neq 0\end{array}\right\}$$

$$\xrightarrow{\text{(Proposition 24)}}(\exists z_2)(\overline{b_1 \cdots b_{n-2} a_1 a_2 z_2}).$$

$$\cdots \quad \cdots \quad \cdots \quad \cdots \quad \cdots$$
$$\cdots \quad \cdots \quad \cdots \quad \cdots \quad \cdots$$

Finally We have $b_1 a_1 \cdots a_{n-2} a_k \neq 0$ for some $n-1 \leq k \leq n$, say, $k = n-1$. Therefore we have

$$\left.\begin{array}{l}\overline{b_1 a_1 \cdots a_{n-2} b_2 a_{n-2}},\\ b_1 a_1 \cdots a_{n-2} b_2 a_{n-1},\\ b_1 a_1 \cdots a_{n-2} a_{n-1} \neq 0\end{array}\right\}$$

$$\xrightarrow{\text{(Proposition 24)}}(\exists z_{n-1})(\overline{b_1 a_1 \cdots a_{n-1} z_{n-1}}).$$

By assumption we have $a_1 \cdots a_n \neq 0$, so that we have

$$\left.\begin{array}{l}\overline{a_1 \cdots a_{n-1} b_1 z_{n-1}},\\ a_1 \cdots a_{n-1} b_1 a_n,\\ a_1 \cdots a_{n-1} a_n \neq 0\end{array}\right\}$$

$$\xrightarrow{\text{(Proposition 24)}} (\exists a_{n+1})(\overline{a_1 \cdots a_n a_{n+1}}).$$

Therefore we are sure that

$$\overline{b_1 \cdots b_{n+1}}, \mathfrak{R}(a_1 \cdots a_n) \ni b_1, \cdots, b_{n+1}$$

$$\rightarrow (\exists a_{n+1})(\overline{a_1 \cdots a_{n+1}}, \mathfrak{R}(a_1 \cdots a_n) \ni a_{n+1}).$$

□

Proposition 27 *Any linear space contained in a principal linear space is always a principal linear space.*

Proof. Let the basis of the contained linear space or containing linear space be $a_1 \cdots a_m$ or $a_1 \cdots a_m \cdots a_n$ respectively; then $\mathfrak{R}(a_1 \cdots a_m \cdots a_n)$ is a principal linear prime space by our assumption. Thus we have

$$(\exists x)(\overline{a_1 \cdots a_m \cdots a_n x}, \mathfrak{R}(a_1 \cdots a_m \cdots a_n) \ni x).$$

$$\overline{a_1 \cdots a_m \cdots a_n x} \xrightarrow{\text{(Proposition 19)}} (\exists y)(\overline{a_1 \cdots a_m y}).$$

Therefore we are sure by Proposition 26 that

$$(\exists y)(\overline{a_1 \cdots a_m y}, \mathfrak{R}(a_1 \cdots a_m) \ni y)$$

which implies that $\mathfrak{R}(a_1 \cdots a_m)$ is also a principal linear space. □

Proposition 28 *A linear space such that all linear subspaces of dimension* 2 *are principal linear spaces is also a principal linear space.*

Proof. Let $a_1 \cdots a_n$ be a basis of the given linear space. Then we have by our assumption that

$$(\exists x_1)(\overline{a_1 a_2 x_1}), (\exists x_2)(\overline{a_2 a_3 x_2}), \ldots, (\exists x_{n-1})(\overline{a_{n-1} a_n x_{n-1}})$$

Therefore we have

$$\overline{a_1 a_2 x_1}, \overline{a_2 a_3 x_2}, a_1 a_2 a_3 \neq 0 \xrightarrow{\text{(Proposition 23)}} (\exists y_3)(\overline{a_1 a_2 a_3 y_3}).$$

$$\overline{a_1 a_2 a_3 y_3}, \overline{a_3 a_4 x_3}, a_1 a_2 a_3 a_4 \neq 0 \xrightarrow{\text{(Proposition 23)}} (\exists y_4)(\overline{a_1 a_2 a_3 a_4 y_4}).$$

$$\cdots \quad \cdots \quad \cdots \quad \cdots \quad \cdots$$

$$\cdots \quad \cdots \quad \cdots \quad \cdots \quad \cdots$$

$$\overline{a_1 \cdots a_{n-1} y_{n-1}}, \overline{a_{n-1} a_n x_{n-1}}, a_1 \cdots a_n \neq 0 \xrightarrow{\text{(Proposition 23)}} (\exists y_n)(\overline{a_1 \cdots a_n y_n}).$$

Therefore

$$(\exists y_n)(\overline{a_1 \cdots a_n y_n}, \mathfrak{R}(a_1 \cdots a_n) \ni y_n)$$

Now we are sure that $\mathfrak{R}(a_1 \cdots a_n)$ is a principal linear space.

□

The Decomposition of a Linear Space

Definition 29 *We say that a linear space $\mathfrak{R}$ can be divided into the direct sum of k linear spaces $\mathfrak{R}_1, \cdots, \mathfrak{R}_k$, and it is denoted by*

$$\mathfrak{R} = \mathfrak{R}_1 \overset{*}{+} \cdots \overset{*}{+} \mathfrak{R}_k,$$

provided that the following two conditions are satisfied:

$$\mathfrak{R} = \mathfrak{R}_1 \cup \cdots \cup \mathfrak{R}_k,$$
$$dim\, \mathfrak{R} = dim\, \mathfrak{R}_1 + \cdots + dim\, \mathfrak{R}_k.$$

Further, in this case, the k linear spaces $\mathfrak{R}_1, \cdots, \mathfrak{R}_k$, are said to be disjoint from each other, and it is denoted by

$$\mathfrak{R}_1 * \cdots * \mathfrak{R}_k.$$

Now it is easy to see that

Claim 30 *We have*

$$\mathfrak{R}_1 * \cdots * \mathfrak{R}_i * \cdots * \mathfrak{R}_k \to \mathfrak{R}_i * \cdots * \mathfrak{R}_1 * \cdots * \mathfrak{R}_k.$$

$$\mathfrak{R}_1 \overset{*}{+} \cdots \overset{*}{+} \mathfrak{R}_i \overset{*}{+} \cdots \overset{*}{+} \mathfrak{R}_k = \mathfrak{R}_i \overset{*}{+} \cdots \overset{*}{+} \mathfrak{R}_1 \overset{*}{+} \cdots \overset{*}{+} \mathfrak{R}_k$$

Proposition 31 *We have*

$$\mathfrak{R} = \mathfrak{R}_1(a_1 \cdots a_n) \overset{*}{+} \cdots \overset{*}{+} \mathfrak{R}_k(l_1 \cdots l_m)$$
$$\to \mathfrak{R} = \mathfrak{R}(a_1 \cdots a_n \cdots l_1 \cdots l_m).$$

Proof. We have

$$\mathfrak{R} = \mathfrak{R}_1 \overset{*}{+} \cdots \overset{*}{+} \mathfrak{R}_k \to \mathfrak{R} = \mathfrak{B}(\mathfrak{R}_1, \cdots, \mathfrak{R}_k)$$

$$\xrightarrow{\text{(Proposition I.47)}} \mathfrak{R} = \mathfrak{B}(a_1 \cdots a_n \cdots l_1 \cdots l_m),$$

so that we have

$$\dim \mathfrak{R} = \dim \mathfrak{R}_1 + \cdots + \dim \mathfrak{R}_k = n + \cdots + m.$$

Therefore we have

$$\dim \mathfrak{B}(a_1 \cdots a_n \cdots l_1 \cdots l_m) = n + \cdots + m$$

so that

$$a_1 \cdots a_n \cdots l_1 \cdots l_m \neq 0.$$

Now we are sure that

$$\mathfrak{R} = \mathfrak{R}(a_1 \cdots a_n \cdots l_1 \cdots l_m).$$

□

Corollary 32 *We have*

$$\mathfrak{R}(a_1 \cdots a_n) * \cdots * \mathfrak{R}(l_1 \cdots l_m) \to a_1 \cdots a_n \cdots l_1 \cdots l_m \neq 0.$$

Proposition 33 *We have*

$$\left.\begin{array}{l} \mathfrak{R}_1(a_1 \cdots a_n) * \cdots * \mathfrak{R}_k(l_1 \cdots l_m), \\ a_1 \cdots a_n x \neq 0, \ \cdots, \ l_1 \cdots l_m x \neq 0 \end{array}\right\} \to a_1 \cdots a_n \cdots l_1 \cdots l_m x \neq 0.$$

Proof. By Proposition 31 we have

$$\mathfrak{R}(a_1 \cdots a_n \cdots l_1 \cdots l_m) = \mathfrak{R}_1(a_1 \cdots a_n) \overset{*}{+} \cdots \overset{*}{+} \mathfrak{R}_k(l_1 \cdots l_m).$$

Therefore we have

$$\left.\begin{array}{l} \mathfrak{R}(a_1 \cdots a_n \cdots l_1 \cdots l_m) = \mathfrak{R}_1 \cup \cdots \cup \mathfrak{R}_k, \\ a_1 \cdots a_n x \neq 0 \to x \notin \mathfrak{R}_1(a_1 \cdots a_n), \\ \cdots \qquad \cdots \qquad \cdots, \\ l_1 \cdots l_m x \neq 0 \to x \notin \mathfrak{R}_k(l_1 \cdots l_m) \end{array}\right\} \to$$

$$x \notin \mathfrak{R}(a_1 \cdots a_n \cdots l_1 \cdots l_m) \to a_1 \cdots a_n \cdots l_1 \cdots l_m x \neq 0.$$

□

Corollary 34 *We have*

$$\left.\begin{array}{c}\mathfrak{R}(a_1\cdots a_n)*\cdots*\mathfrak{R}(l_1\cdots l_m),\\ a_1\cdots a_n\cdots l_1\cdots l_m x\end{array}\right\}\rightarrow$$

exactly one of the statements $a_1\cdots a_n x,\cdots,l_1\cdots l_m x$ *is valid.*

Proposition 35 *We have*

$$\mathfrak{R}_1*\cdots*\mathfrak{R}_i*\cdots*\mathfrak{R}_k\rightarrow\mathfrak{B}(\mathfrak{R}_1\cdots\mathfrak{R}_i)=\mathfrak{R}_1\cup\cdots\cup\mathfrak{R}_i.$$

Proof. Let the bases of $\mathfrak{R}_1,\cdots,\mathfrak{R}_i\cdots,\mathfrak{R}_k$ be $a_1\cdots a_n,\cdots,b_1\cdots b_m,\cdots,l_1\cdots l_s$ respectively. Then we have

$$\mathfrak{R}_1(a_1\cdots a_n)*\cdots*\mathfrak{R}_i(b_1\cdots b_m)*\cdots*\mathfrak{R}_k(l_1\cdots l_s)$$

$$\xrightarrow{\text{(Corollary 32)}}$$

$$a_1\cdots a_n\cdots b_1\cdots b_m\cdots l_1\cdots l_s\neq 0\rightarrow a_1\cdots a_n\cdots b_1\cdots b_m\neq 0$$
$$\rightarrow\mathfrak{B}(\mathfrak{R}_1\cdots\mathfrak{R}_i)=\mathfrak{R}(a_1\cdots a_n\cdots b_1\cdots b_m)$$

Therefore we have

$$\mathfrak{B}(\mathfrak{R}_1\cdots\mathfrak{R}_i)\ni x\rightarrow\mathfrak{R}(a_1\cdots a_n\cdots b_1\cdots b_m)\ni x\rightarrow a_1\cdots a_n\cdots b_1\cdots b_m x$$
$$\xrightarrow{\text{(Corollary 34)}} a_1\cdots a_n x\vee\cdots\vee b_1\cdots b_m x$$
$$\rightarrow\mathfrak{R}_1\cup\cdots\cup\mathfrak{R}_i\ni x$$

so that

$$\mathfrak{B}(\mathfrak{R}_1\cdots\mathfrak{R}_i)\subseteqq\mathfrak{R}_1\cup\cdots\cup\mathfrak{R}_i.$$

On the other hand we have

$$\mathfrak{B}(\mathfrak{R}_1\cdots\mathfrak{R}_i)\supseteqq\mathfrak{R}_1\cup\cdots\cup\mathfrak{R}_i$$

so that we are sure that

$$\mathfrak{B}(\mathfrak{R}_1\cdots\mathfrak{R}_i)=\mathfrak{R}_1\cup\cdots\cup\mathfrak{R}_i.$$

□

Corollary 36 *We have*

$$\mathfrak{R}_1(a_1\cdots a_n)*\cdots*\mathfrak{R}_i(b_1\cdots b_m)*\cdots*\mathfrak{R}_k(l_1\cdots l_s)$$
$$\rightarrow\mathfrak{B}\{\mathfrak{R}_1(a_1\cdots a_n),\cdots,\mathfrak{R}_i(b_1\cdots b_m)\}=\mathfrak{R}(a_1\cdots a_n\cdots b_1\cdots b_m).$$

Proposition 37 *We have*

$$\begin{aligned}&\mathfrak{R}_1 * \cdots * \mathfrak{R}_i * \mathfrak{R}_{i+1} * \cdots * \mathfrak{R}_k \\ &\to \mathfrak{D}\{\mathfrak{B}(\mathfrak{R}_1 \cdots \mathfrak{R}_i), \mathfrak{B}(\mathfrak{R}_{i+1} \cdots \mathfrak{R}_k)\} = \mathfrak{N}.\end{aligned}$$

Proof. Let the bases of $\mathfrak{R}_1, \cdots, \mathfrak{R}_i, \mathfrak{R}_{i+1}, \cdots, \mathfrak{R}_k$ be $a_1 \cdots a_n, \cdots, b_1 \cdots b_m, c_1 \cdots c_r, \cdots, d_1 \cdots d_s$ respectively. Then, by Corollary 36, we have

$$\begin{aligned}\mathfrak{B}(\mathfrak{R}_1 \cdots \mathfrak{R}_i) &= \mathfrak{R}(a_1 \cdots a_n \cdots b_1 \cdots b_m), \\ \mathfrak{B}(\mathfrak{R}_{i+1} \cdots \mathfrak{R}_k) &= \mathfrak{R}(c_1 \cdots c_r \cdots d_1 \cdots d_s).\end{aligned}$$

Therefore we have

$$\mathfrak{D}\left\{\begin{array}{l}\mathfrak{B}(\mathfrak{R}_1 \cdots \mathfrak{R}_i), \\ \mathfrak{B}(\mathfrak{R}_{i+1} \cdots \mathfrak{R}_k)\end{array}\right\} \ni x \to \mathfrak{D}\left\{\begin{array}{l}\mathfrak{R}(a_1 \cdots a_n \cdots b_1 \cdots b_m), \\ \mathfrak{R}(c_1 \cdots c_r \cdots d_1 \cdots d_s)\end{array}\right\} \ni x$$

$$\to \begin{array}{l}a_1 \cdots a_n \cdots b_1 \cdots b_m x, \\ c_1 \cdots c_r \cdots d_1 \cdots d_s x\end{array}\Big\} \to a_1 \cdots a_n \cdots b_1 \cdots b_m c_1 \cdots c_r \cdots d_1 \cdots d_s \xrightarrow{\text{(Corollary 32)}\perp}$$

so that we are sure that

$$\mathfrak{D}\{\mathfrak{B}(\mathfrak{R}_1 \cdots \mathfrak{R}_i), \mathfrak{B}(\mathfrak{R}_{i+1} \cdots \mathfrak{R}_k)\} = \mathfrak{N}.$$

□

Proposition 38 *We have*

$$\mathfrak{R}_1 * \cdots * \mathfrak{R}_i * \cdots * \mathfrak{R}_k \to \mathfrak{R}_1 * \cdots * \mathfrak{R}_i.$$

Proof. We have

$$\mathfrak{R}_1 * \cdots * \mathfrak{R}_i * \cdots * \mathfrak{R}_k \xrightarrow{\text{(Proposition 36)}} \mathfrak{B}(\mathfrak{R}_1 \cdots \mathfrak{R}_i) = \mathfrak{R}_1 \cup \cdots \cup \mathfrak{R}_i$$

so that

$$\begin{aligned}\dim \mathfrak{R}_1 + \cdots + \dim \mathfrak{R}_k &= \dim (\mathfrak{R}_1 \cup \cdots \cup \mathfrak{R}_k) \\ &\leqq \dim (\mathfrak{R}_1 \cup \cdots \cup \mathfrak{R}_i) + \dim \mathfrak{R}_{i+1} \\ &\quad + \cdots + \dim \mathfrak{R}_k \\ &= \dim \mathfrak{B}(\mathfrak{R}_1 \cdots \mathfrak{R}_i) + \dim \mathfrak{R}_{i+1} + \cdots + \dim \mathfrak{R}_k \\ &\leqq \dim \mathfrak{R}_1 + \cdots + \dim \mathfrak{R}_k.\end{aligned}$$

Therefore we are sure that

$$\dim \mathfrak{B}(\mathfrak{R}_1 \cdots \mathfrak{R}_i) = \dim \mathfrak{R}_1 + \cdots + \dim \mathfrak{R}_i$$

so that

$$\mathfrak{R}_1 * \cdots * \mathfrak{R}_i.$$

□

Proposition 39 *We have*

$$\mathfrak{R} = \mathfrak{R}_1 \overset{*}{+} \cdots \overset{*}{+} \mathfrak{R}_i \overset{*}{+} \cdots \overset{*}{+} \mathfrak{R}_k, \mathfrak{R}_i = \mathfrak{R}_{i1} \overset{*}{+} \cdots \overset{*}{+} \mathfrak{R}_{is}$$
$$\rightarrow \mathfrak{R} = \mathfrak{R}_1 \overset{*}{+} \cdots \overset{*}{+} \mathfrak{R}_{i1} \overset{*}{+} \cdots \overset{*}{+} \mathfrak{R}_{is} \overset{*}{+} \cdots \overset{*}{+} \mathfrak{R}_k.$$

Proof. We have

$$\left.\begin{array}{l} \mathfrak{R} = \mathfrak{R}_1 \cup \cdots \cup \mathfrak{R}_i \cup \cdots \cup \mathfrak{R}_k, \\ \mathfrak{R}_i = \mathfrak{R}_{i1} \cup \cdots \cup \mathfrak{R}_{is} \end{array}\right\}$$
$$\rightarrow \mathfrak{R} = \mathfrak{R}_1 \cup \cdots \cup \mathfrak{R}_{i1} \cup \cdots \cup \mathfrak{R}_{is} \cup \cdots \cup \mathfrak{R}_k.$$
$$\left.\begin{array}{l} \dim \mathfrak{R} = \dim \mathfrak{R}_1 + \cdots + \dim \mathfrak{R}_i + \cdots + \dim \mathfrak{R}_k, \\ \dim \mathfrak{R}_i = \dim \mathfrak{R}_{i1} + \cdots + \dim \mathfrak{R}_{is} \end{array}\right\}$$
$$\rightarrow \dim \mathfrak{R} = \dim\mathfrak{R}_1 + \cdots + \dim\mathfrak{R}_{i1} + \cdots + \dim\mathfrak{R}_{is} + \cdots + \dim\mathfrak{R}_k$$

Therefore we are sure that

$$\mathfrak{R} = \mathfrak{R}_1 \overset{*}{+} \cdots \overset{*}{+} \mathfrak{R}_{i1} \overset{*}{+} \cdots \overset{*}{+} \mathfrak{R}_{is} \overset{*}{+} \cdots \overset{*}{+} \mathfrak{R}_k.$$

□

Proposition 40 *We have*

$$\mathfrak{R} = \mathfrak{R}_1 \overset{*}{+} \cdots \overset{*}{+} \mathfrak{R}_{i1} \overset{*}{+} \cdots \overset{*}{+} \mathfrak{R}_{is} \overset{*}{+} \cdots \overset{*}{+} \mathfrak{R}_k$$
$$\rightarrow \mathfrak{R} = \mathfrak{R}_1 \overset{*}{+} \cdots \overset{*}{+} (\mathfrak{R}_{i1} \overset{*}{+} \cdots \overset{*}{+} \mathfrak{R}_{is}) \overset{*}{+} \cdots \overset{*}{+} \mathfrak{R}_k.$$

Proof. We have

$$\mathfrak{R}=\mathfrak{R}_1 \cup \cdots \cup \mathfrak{R}_{i1} \cup \cdots \cup \mathfrak{R}_{is} \cup \cdots \cup \mathfrak{R}_k$$
$$\rightarrow \mathfrak{R} = \mathfrak{R}_1 \cup \cdots \cup (\mathfrak{R}_{i1} \cup \cdots \cup \mathfrak{R}_{is}) \cup \cdots \cup \mathfrak{R}_k.$$
$$\mathfrak{R}_1 * \cdots * \mathfrak{R}_{i1} * \cdots * \mathfrak{R}_{is} * \cdots * \mathfrak{R}_k \xrightarrow{\text{(Proposition 38)}} \mathfrak{R}_{i1} * \cdots * \mathfrak{R}_{is}$$
$$\rightarrow \mathfrak{R}_{i1} \cup \cdots \cup \mathfrak{R}_{is} = \mathfrak{R}_{i1} \overset{*}{+} \cdots \overset{*}{+} \mathfrak{R}_{is} \rightarrow$$
$$\dim(\mathfrak{R}_{i1} \cup \cdots \cup \mathfrak{R}_{is}) = \dim(\mathfrak{R}_{i1} \overset{*}{+} \cdots \overset{*}{+} \mathfrak{R}_{is}) = \dim\mathfrak{R}_{i1} + \cdots + \dim \mathfrak{R}_{is}.$$

Therefore we have

$$\begin{aligned} \dim \mathfrak{R} &= \dim \mathfrak{R}_1 + \cdots + \dim \mathfrak{R}_{i1} + \cdots + \dim \mathfrak{R}_{is} + \cdots + \dim \mathfrak{R}_k \\ &= \dim \mathfrak{R}_1 + \cdots + \dim (\mathfrak{R}_{i1} \cup \cdots \cup \mathfrak{R}_{is}) + \cdots + \dim \mathfrak{R}_k \end{aligned}$$

so that

$$\begin{aligned} \mathfrak{R} &= \mathfrak{R}_1 \overset{*}{+} \cdots \overset{*}{+} (\mathfrak{R}_{i1} \cup \cdots \cup \mathfrak{R}_{is}) \overset{*}{+} \cdots \overset{*}{+} \mathfrak{R}_k \\ &= \mathfrak{R}_1 \overset{*}{+} \cdots \overset{*}{+} (\mathfrak{R}_{i1} \overset{*}{+} \cdots \overset{*}{+} \mathfrak{R}_{is}) \overset{*}{+} \cdots \overset{*}{+} \mathfrak{R}_k. \end{aligned}$$

□

Proposition 41 *We have*

$$\left.\begin{array}{l}\mathfrak{R} = \mathfrak{R}_1 \cup \cdots \cup \mathfrak{R}_k, \\ \mathfrak{D}\{\mathfrak{R}(\mathfrak{R}_1 \cdots \mathfrak{R}_{i-1}), \mathfrak{R}_i\} = \mathfrak{R}\ (i = 2, \cdots, k)\end{array}\right\} \to \mathfrak{R} = \mathfrak{R}_1 \overset{*}{+} \cdots \overset{*}{+} \mathfrak{R}_k.$$

Proof. We have

$$\mathfrak{D}\{\mathfrak{B}(\mathfrak{R}_1 \cdots \mathfrak{R}_{i-1}), \mathfrak{R}_i\} = \mathfrak{R}\underrightarrow{\text{(Proposition 11)}}$$

$$\dim \mathfrak{B}\{\mathfrak{B}(\mathfrak{R}_1 \cdots \mathfrak{R}_{i-1}), \mathfrak{R}_i\} = \dim \mathfrak{B}(\mathfrak{R}_1 \cdots \mathfrak{R}_{i-1}) + \dim \mathfrak{R}_i$$

so that

$$\dim (\mathfrak{R}_1 \cup \cdots \cup \mathfrak{R}_i) = \dim (\mathfrak{R}_1 \cup \cdots \cup \mathfrak{R}_{i-1}) + \dim \mathfrak{R}_i.$$

Therefore we have

$$\dim (\mathfrak{R}_1 \cup \cdots \cup \mathfrak{R}_k) = \dim (\mathfrak{R}_1 \cup \cdots \cup \mathfrak{R}_{k-1}) + \dim \mathfrak{R}_k,$$

$$\dim (\mathfrak{R}_1 \cup \cdots \cup \mathfrak{R}_{k-1}) = \dim (\mathfrak{R}_1 \cup \cdots \cup \mathfrak{R}_{k-2}) + \dim \mathfrak{R}_{k-1},$$

$$\cdots \quad \cdots \quad \cdots \quad \cdots \quad \cdots \quad \cdots \quad \cdots$$

$$\dim (\mathfrak{R}_1 \cup \mathfrak{R}_2) = \dim \mathfrak{R}_1 + \dim \mathfrak{R}_2.$$

Therefore we have

$$\dim (\mathfrak{R}_1 \cup \cdots \cup \mathfrak{R}_k) = \dim \mathfrak{R}_1 + \cdots + \dim \mathfrak{R}_k$$

so that

$$\mathfrak{R} = \mathfrak{R}_1 \overset{*}{+} \cdots \overset{*}{+} \mathfrak{R}_k.$$

□

Definition 42 *A linear space $\mathfrak{R}(a_1 \cdots a_n)$ which can not be divided into the direct sum of two linear spaces is called an indecomposable linear space and is denoted by $\overline{\mathfrak{R}}(a_1 \cdots a_n)$.*

Proposition 43 *We have*

$$\left.\begin{array}{l} a_1 \cdots a_n b_1 \cdots b_m c_1 \cdots c_k z, \\ a_1 \cdots a_n b_1 \cdots b_m z \neq 0, \\ b_1 \cdots b_m c_1 \cdots c_k z \neq 0, \\ c_1 \cdots c_k a_1 \cdots a_n z \neq 0 \end{array}\right\}$$

$$\to \left\{\begin{array}{l} (\exists z_1)(a_1 \cdots a_n b_1 \cdots b_m z_1, a_1 \cdots a_n z_1 \neq 0, b_1 \cdots b_m z_1 \neq 0), \\ (\exists z_2)(b_1 \cdots b_m c_1 \cdots c_k z_2, b_1 \cdots b_m z_2 \neq 0, c_1 \cdots c_k z_2 \neq 0), \\ (\exists z_3)(c_1 \cdots c_k a_1 \cdots a_n z_3, c_1 \cdots c_k z_3 \neq 0, a_1 \cdots a_n z_3 \neq 0). \end{array}\right.$$

Proof. We have

$$a_1 \cdots a_n b_1 \cdots b_m c_1 \cdots c_k z \to (\exists z_1)(a_1 \cdots a_n b_1 \cdots b_m z_1, c_1 \cdots c_k z z_1).$$

$$c_1 \cdots c_k a_1 \cdots a_n z \neq 0, c_1 \cdots c_k z z_1 \to a_1 \cdots a_n z_1 \neq 0.$$

$$b_1 \cdots b_m c_1 \cdots c_k z \neq 0, c_1 \cdots c_k z z_1 \to b_1 \cdots b_m z_1 \neq 0.$$

Therefore we have

$$(\exists z_1)(a_1 \cdots a_n b_1 \cdots b_m z_1, a_1 \cdots a_n z_1 \neq 0, b_1 \cdots b_m z_1 \neq 0)$$

By the same token we have

$$(\exists z_2)(b_1 \cdots b_m c_1 \cdots c_k z_2, b_1 \cdots b_m z_2 \neq 0, c_1 \cdots c_k z_2 \neq 0),$$

$$(\exists z_3)(c_1 \cdots c_k a_1 \cdots a_n z_3, c_1 \cdots c_k z_3 \neq 0, a_1 \cdots a_n z_3 \neq 0).$$

□

Proposition 44 *We have*

$$\left.\begin{array}{l} ab_1 \cdots b_m c_1 \cdots c_k \neq 0, \\ ab_1 \cdots b_m z_1,\ az_1 \neq 0, b_1 \cdots b_m z_1 \neq 0, \\ ac_1 \cdots c_k z_2,\ az_2 \neq 0, c_1 \cdots c_k z_2 \neq 0 \end{array}\right\}$$

$$\to (\exists z)(ab_1 \cdots b_m c_1 \ldots c_k z, ab_1 \cdots b_m z \neq 0,$$
$$ac_1 \cdots c_k z \neq 0, b_1 \cdots b_m c_1 \cdots c_k z \neq 0).$$

Proof. We have

$$ab_1 \cdots b_m z_1, ac_1 \cdots c_k z_2 \to b_1 \cdots b_m z_1 c_1 \cdots c_k z_2$$
$$\to (\exists z)(b_1 \cdots b_m z_2 z, c_1 \cdots c_k z_1 z).$$

$$ab_1 \cdots b_m z_1, c_1 \cdots c_k z_1 z \to ab_1 \cdots b_m c_1 \cdots c_k z.$$

$$ab_1 \cdots b_m c_1 \cdots c_k \neq 0, b_1 \cdots b_m z_1 \neq 0, ab_1 \cdots b_m z_1 \to b_m c_1 \cdots c_k z_1 \neq 0.$$

$$b_1 \cdots b_m c_1 \cdots c_k z_1 \neq 0, c_1 \cdots c_k z_1 z \to b_1 \cdots b_m z \neq 0.$$

$$ab_1 \cdots b_m c_1 \cdots c_k \neq 0, ac_1 \cdots c_k z_2, az_2$$
$$\neq 0 \to ab_1 \cdots b_m z_2 \neq 0.$$

$$b_1 \cdots b_m z \neq 0, ab_1 \cdots b_m z_2 \neq 0, b_1 \cdots b_m z_2 z$$
$$\to ab_1 \cdots b_m z \neq 0.$$

$$ab_1 \cdots b_m c_1 \cdots c_k \neq 0, c_1 \cdots c_k z_2 \neq 0, ac_1 \cdots c_k z_2$$
$$\to b_1 \cdots b_m c_1 \cdots c_k z_2 \neq 0.$$

$$b_1 \cdots b_m c_1 \cdots c_k z_2 \neq 0, b_1 \cdots b_m z_2 z \to c_1 \cdots c_k z \neq 0.$$

$$ab_1 \cdots b_m c_1 \cdots c_k \neq 0, ab_1 \cdots b_m z_1, az_1 \neq 0$$

$$\begin{aligned}
&\to ac_1\cdots c_k z_1 \neq 0.\\
c_1,\cdots c_k z \neq 0, ac_1\cdots c_k z_1 \neq 0, c_1,\cdots c_k z_1 z&\\
&\to ac_1\cdots c_k z \neq 0.\\
b_1\cdots b_m c_1\cdots c_k z_2 \neq 0, b_1\cdots b_m z \neq 0, b_1\cdots b_m z_2 z&\\
&\to b_1\cdots b_m c_1\cdots c_k z \neq 0.
\end{aligned}$$

Therefore we are sure that

$$(\exists z)(ab_1\cdots b_m c_1\ldots c_k z, ab_1\cdots b_m z \neq 0,\\ ac_1\cdots c_k z \neq 0, b_1\cdots b_m c_1\cdots c_k z \neq 0).$$

□

Proposition 45 *We have*

$$\overline{\mathfrak{R}^s} \supset \mathfrak{R}^{s-1} \to \overline{\mathfrak{R}}^{s-1}.$$

Proof. Let

$$\begin{aligned}
\overline{\mathfrak{R}^s} &= \bar{\mathfrak{R}}(ab_1\cdots b_m c_1\cdots c_k),\\
\mathfrak{R}^{s-1} &= \mathfrak{R}(b_1\cdots b_m c_1\cdots c_k).
\end{aligned}$$

Suppose, for the sake of contradiction, that

$$\mathfrak{R}(b_1\cdots b_m c_1\cdots c_k) = \mathfrak{R}(b_1\cdots b_m) \overset{*}{+} \mathfrak{R}(c_1\cdots c_k).$$

Then we have

$$\begin{aligned}
&\bar{\mathfrak{R}}(ab_1\cdots b_m c_1\cdots c_k) \to\\
&(\exists x)(\mathfrak{R}(ab_1\cdots b_m c_1\cdots c_k) \ni x, ab_1\cdots b_m x \neq 0, c_1\cdots c_k x \neq 0),\\
&(\exists y)(\mathfrak{R}(ab_1\cdots b_m c_1\cdots c_k) \ni y, ac_1\cdots c_k y \neq 0, b_1\cdots b_m y \neq 0).
\end{aligned}$$

Furthermore we have

$$\left.\begin{array}{l}
\mathfrak{R}(b_1\cdots b_m) * \mathfrak{R}(c_1\cdots c_k),\\
b_1\cdots b_m x \neq 0, c_1\cdots c_k x \neq 0,\\
b_1\cdots b_m y \neq 0, c_1\cdots c_k y \neq 0
\end{array}\right\}$$

$$\xrightarrow{\text{(Proposition 33)}} \left\{\begin{array}{l}
b_1\cdots b_m c_1\cdots c_k x \neq 0,\\
b_1\cdots b_m c_1\cdots c_k y \neq 0.
\end{array}\right.$$

If $ac_1\cdots c_k x \neq 0$, then we have

$$\left.\begin{array}{l}
ab_1\cdots b_m c_1\cdots c_k x,\\
ab_1\cdots b_m x \neq 0,\\
b_1\cdots b_m c_1\cdots c_k x \neq 0,\\
c_1\cdots c_k ax \neq 0
\end{array}\right\} \xrightarrow{\text{(Proposition 43)}}$$

$$(\exists z)(b_1 \cdots b_m c_1 \cdots c_k z, b_1 \cdots b_m z \neq 0, c_1 \cdots c_k z \neq 0).$$

$$\left.\begin{array}{l} \mathfrak{R}(b_1 \cdots b_m) * \mathfrak{R}(c_1 \cdots c_k), \\ b_1 \cdots b_m z \neq 0, c_1 \cdots c_k z \neq 0, \\ b_1 \cdots b_m c_1 \cdots c_k z \end{array}\right\} \xrightarrow{\text{(Proposition 33)}} \perp$$

so that we are sure that $ac_1 \cdots c_k z = 0$ must hold. Similarly we are sure that $ab_1 \cdots b_m y = 0$. Therefore we have

$$\left.\begin{array}{l} ab_1 \cdots b_m c_1 \cdots c_k \neq 0, \\ ab_1 \cdots b_m y,\ ay \neq 0, b_1 \cdots b_m y \neq 0, \\ ac_1 \cdots c_k x,\ ax \neq 0, c_1 \cdots c_k x \neq 0 \end{array}\right\} \xrightarrow{\text{(Proposition 44)}}$$

$$\left\{\begin{array}{l} Ez,\ ab_1 \cdots b_m c_1 \cdots c_k z, \\ ab_1 \cdots b_m z \neq 0,\ ac_1 \cdots c_k z \neq 0, \\ b_1 \cdots b_m c_1 \cdots c_k z \neq 0 \end{array}\right. \xrightarrow{\text{(Proposition 43)}}$$

$$(\exists w)(b_1 \cdots b_m c_1 \cdots c_k w, b_1 \cdots b_m w \neq 0, c_1 \cdots c_k w \neq 0).$$

However we have

$$\left.\begin{array}{l} \mathfrak{R}(b_1 \cdots b_m) * \mathfrak{R}(c_1 \cdots c_k), \\ b_1 \cdots b_m w \neq 0, c_1 \cdots c_k w \neq 0, \\ b_1 \cdots b_m c_1 \cdots c_k w \end{array}\right\} \xrightarrow{\text{(Proposition 33)}} \perp.$$

Therefore we are sure that

$$\bar{\mathfrak{R}}(ab_1 \cdots b_m c_1 \cdots c_k), \mathfrak{R}(b_1 \cdots b_m) * \mathfrak{R}(c_1 \cdots c_k) \to \perp$$

so that

$$\bar{\mathfrak{R}}(ab_1 \cdots b_m c_1 \cdots c_k) \to \bar{\mathfrak{R}}(b_1 \cdots b_m c_1 \cdots c_k).$$

□

Proposition 46 *Any linear space contained in an indecomposable linear space is always an indecomposable linear space.*

Proof. If $\overline{\mathfrak{R}}(a_1 \cdots a_m \cdots a_n) \supseteqq \mathfrak{R}(a_1 \cdots a_m)$, then we have

$$\bar{\mathfrak{R}}(a_1 \cdots a_m \cdots a_n) \to \bar{\mathfrak{R}}(a_1 \cdots a_m \cdots a_{n-1}) \to \cdots \to \bar{\mathfrak{R}}(a_1 \cdots a_m)$$

so that

$$\bar{\mathfrak{R}}(a_1 \cdots a_m \cdots a_n) \xrightarrow{\text{(Proposition 45)}} \bar{\mathfrak{R}}(a_1 \cdots a_m).$$

□

Proposition 47 *Any indecomposable linear space of dimension 2 is a principal linear space.*

Proof. Let the given indecomposable linear space of dimension 2 be $\mathfrak{R}(a_1a_2)$. Then we have

$$\begin{aligned}\bar{\mathfrak{R}}(a_1a_2) &\to (\exists z)(\mathfrak{R}(a_1a_2) \ni z, a_1z \neq 0, a_2z \neq 0)\\ &\to (\exists z)(\mathfrak{R}(a_1a_2) \ni z, \overline{a_1a_2z})\end{aligned}$$

so that we are sure that $\mathfrak{R}(a_1a_2)$ is a principal linear space. □

Proposition 48 *Any indecomposable linear space is a principal linear space.*

Proof. In case that the dimension of the given indecomposable linear space is equal to 1, there is nothing to prove. Therefore it suffices to deal with the case that the dimension is equal to or more than 2. It follows from Propositions 46 and 47 that any linear space of dimension 2 contained in an indecomposable linear space is a principal linear space. Thus we are sure by Proposition 28 that any indecomposable linear space is a principal linear space. □

Proposition 49 *Any principal linear space is an indecomposable linear space.*

Proof. Suppose, for the sake of contradiction, that

$$\mathfrak{R}(a_1 \cdots a_n b_1 \cdots b_m) = \mathfrak{R}_1(a_1 \cdots a_n) \overset{*}{+} \mathfrak{R}_2(b \cdots b_m),$$

though $\mathfrak{R}(a_1 \cdots a_n b_1 \cdots b_m)$ is assumed to be a principal linear space. We have by Definition 25 and Proposition 26 that

$$(\exists z)(\overline{a_1 \cdots a_n b_1 \cdots b_m z}, \mathfrak{R}(a_1 \cdots a_n b_1 \cdots b_m) \ni z).$$

$$\overline{a_1 \cdots a_n b_1 \cdots b_m z} \to a_1 \cdots a_n z \neq 0, b_1 \cdots b_m z \neq 0.$$

$$\left.\begin{array}{l}\mathfrak{R}(a_1 \cdots a_n) * \mathfrak{R}(b_1 \cdots b_m),\\ a_1 \cdots a_n z \neq 0, b_1 \cdots b_m z \neq 0\end{array}\right\} \xrightarrow{\text{(Proposition 33)}} a_1 \cdots a_n b_1 \cdots b_m z \neq 0.$$

$$\overline{a_1 \cdots a_n b_1 \cdots b_m z}, a_1 \cdots a_n b_1 \cdots b_m z \neq 0 \to \perp$$

which is the desired contradiction. This completes the proof. □

It follows from Propositions 48 and 49 that

Proposition 50 *The notion of a principal linear space and that of an indecomposable linear space are equivalent.*

Proposition 51 *A linear space such that all of the linear spaces of dimension 2 lying in it are indecomposable linear spaces is also an indecomposable linear space.*

Proof. This follows readily from Propositions 28 and 50. □

Proposition 52 *We have*

$$\left.\begin{array}{l}\bar{\mathfrak{R}}(d_1\cdots d_s a_1\cdots a_n),\\ \bar{\mathfrak{R}}(d_1\cdots d_s b_1\cdots b_m),\\ d_1\cdots d_s a_1\cdots a_n b_1\cdots b_m\neq 0\end{array}\right\}\rightarrow \bar{\mathfrak{R}}(d_1\cdots d_s a_1\cdots a_n b_1\cdots b_m).$$

Proof. We have

$$\bar{\mathfrak{R}}(d_1\cdots d_s a_1\cdots a_n)\xrightarrow{\text{(Proposition 50, Proposition 26)}}$$

$$(\exists x)(\overline{d_1\cdots d_s a_1\cdots a_n x}).$$

$$\bar{\mathfrak{R}}(d_1\cdots d_s b_1\cdots b_m)\xrightarrow{\text{(Proposition 50, Proposition 26)}}$$

$$(\exists y)(d_1\cdots d_s b_1\cdots b_m y).$$

$$\left.\begin{array}{l}\overline{d_1\cdots d_s a_1\cdots a_n x},\\ \overline{d_1\cdots d_s b_1\cdots b_m y},\\ d_1\cdots d_s a_1\cdots a_n b_1\cdots b_m\neq 0\end{array}\right\}\xrightarrow{\text{(Proposition 23)}}$$

$$(\exists z)(\overline{d_1\cdots d_s a_1\cdots a_n b_1\cdots b_m z}),$$

so that

$$(\exists z)(\mathfrak{R}(d_1\cdots d_s a_1\cdots a_n b_1\cdots b_m)\ni z, d_1\cdots d_s a_1\cdots a_n b_1\cdots b_m z),$$

which implies by Proposition 50 that

$$\bar{\mathfrak{R}}(d_1\cdots d_s a_1\cdots a_n b_1\cdots b_m).$$

□

Proposition 53 *We have*

$$\mathfrak{R}_1 * \cdots * \mathfrak{R}_k, \mathfrak{R}_1 \supseteqq \mathfrak{R}'_1, \cdots, \mathfrak{R}_k \supseteqq \mathfrak{R}'_k \rightarrow \mathfrak{R}'_1 * \cdots * \mathfrak{R}'_k.$$

Proof. Let the bases of $\mathfrak{R}'_1, \mathfrak{R}_1, \cdots, \mathfrak{R}'_k, \mathfrak{R}_k$ be $a_1\cdots a_m, a_1\cdots a_m\cdots a_n, \cdots, l_1\cdots l_r, l_1\cdots l_r\cdots l_s$, respectively. Then, by Corollary 32, we have

$$\begin{aligned}\mathfrak{R}_1 * \cdots * \mathfrak{R}_k &\rightarrow a_1\cdots a_m\cdots a_n\cdots l_1\cdots l_r\cdots l_s\neq 0\\ &\rightarrow a_1\cdots a_m\cdots l_1\cdots l_r\neq 0\end{aligned}$$

so that

$$\dim\mathfrak{R}'_1(a_1\cdots a_m)+\cdots+\dim\mathfrak{R}'_k(l_1\cdots l_r)=\dim\mathfrak{R}(a_1\cdots a_m\cdots l_1\cdots l_r).$$

$$\mathfrak{R}(a_1 \cdots a_m \cdots l_1 \cdots l_r) \ni x \to a_1 \cdots a_m \cdots l_1 \cdots l_r x$$
$$\to a_1 \cdots a_m \cdots a_n \cdots l_1 \cdots l_r \cdots l_s x.$$

$$\left.\begin{array}{l} a_1 \cdots a_m \cdots a_n \cdots l_1 \cdots l_r \cdots l_s x, \\ \mathfrak{R}_1(a_1 \cdots a_n) * \cdots * \mathfrak{R}_k(l_1 \cdots l_s) \end{array}\right\} \xrightarrow{\text{(Corollary 34)}}$$
$$a_1 \cdots a_m \cdots a_n x \vee \ldots \vee l_1 \cdots l_r \cdots l_s x.$$

$$\left.\begin{array}{l} a_1 \cdots a_m \cdots a_n x \vee \cdots \vee l_1 \cdots l_r \cdots l_s x, \\ a_1 \cdots a_m \cdots l_1 \cdots l_r x, a_1 \cdots a_m \cdots a_n \cdots l_1 \cdots l_r \cdots l_s \neq 0 \end{array}\right\} \to$$
$$a_1 \cdots a_m x \vee \ldots \vee l_1 \cdots l_r x.$$

$$a_1 \cdots a_m x \vee \ldots \vee l_1 \cdots l_r x \to \mathfrak{R}'_1 \cup \cdots \cup \mathfrak{R}'_k \ni x.$$

Therefore we have

$$\mathfrak{R}(a_1 \cdots a_m \cdots l_1 \cdots l_r) \ni x \to \mathfrak{R}'_1 \cup \cdots \cup \mathfrak{R}'_k \ni x$$

so that

$$\mathfrak{R}(a_1 \cdots a_m \cdots l_1 \cdots l_r) \subseteqq \mathfrak{R}'_1 \cup \cdots \cup \mathfrak{R}'_k.$$

On the other hand we have

$$\mathfrak{R}(a_1 \cdots a_m \cdots l_1 \cdots l_r) \supseteqq \mathfrak{R}'_1 \cup \cdots \cup \mathfrak{R}'_k$$

so that

$$\mathfrak{R}(a_1 \cdots a_m \cdots l_1 \cdots l_r) = \mathfrak{R}'_1 \cup \cdots \cup \mathfrak{R}'_k,$$

which implies that

$$\mathfrak{R}(a_1 \cdots a_m \cdots l_1 \cdots l_r) = \mathfrak{R}'_1 \overset{*}{+} \cdots \overset{*}{+} \mathfrak{R}'_k.$$

Therefore we have

$$\mathfrak{R}'_1 * \cdots * \mathfrak{R}'_k$$

so that

$$\mathfrak{R}_1 * \cdots * \mathfrak{R}_k, \mathfrak{R}_1 \supseteqq \mathfrak{R}'_1, \cdots, \mathfrak{R}_k \supseteqq \mathfrak{R}'_k \to \mathfrak{R}'_1 * \cdots * \mathfrak{R}'_k.$$

□

Proposition 54 *We have*

$$\mathfrak{R} = \mathfrak{R}_1 \overset{*}{+} \cdots \overset{*}{+} \mathfrak{R}_k, \mathfrak{R} \supseteqq \overline{\mathfrak{R}}' \to$$

exactly one of the statements $\mathfrak{R}_1 \supseteqq \overline{\mathfrak{R}}', \cdots, \mathfrak{R}_k, \supseteqq \overline{\mathfrak{R}}'$ *holds*

Proof. Let $\mathfrak{D}(\overline{\mathfrak{R}}', \mathfrak{R}_1) = \mathfrak{R}'_1, \cdots, \mathfrak{D}(\overline{\mathfrak{R}}', \mathfrak{R}_k) = \mathfrak{R}'_k$. Then we have

$$\mathfrak{R}_1 \cup \cdots \cup \mathfrak{R}_k \supseteqq \overline{\mathfrak{R}}' \to \mathfrak{R}'_1 \cup \cdots \cup \mathfrak{R}'_k = \overline{\mathfrak{R}}'$$

We have

$$\mathfrak{R}_1 * \cdots * \mathfrak{R}_k, \mathfrak{R}_1 \supseteqq \mathfrak{R}'_1, \cdots, \mathfrak{R}_k \supseteqq \mathfrak{R}'_k$$

$$\xrightarrow{\text{(Proposition 53)}} \mathfrak{R}'_1 * \cdots * \mathfrak{R}'_k$$

so that

$$\overline{\mathfrak{R}}' = \mathfrak{R}'_1 \overset{*}{+} \cdots \overset{*}{+} \mathfrak{R}'_k.$$

Therefore we are sure, by way of example, that

$$\overline{\mathfrak{R}}' = \mathfrak{R}'_1$$

$$\mathfrak{R}'_2 = \cdots = \mathfrak{R}'_k = \mathfrak{N}$$

Since

$$\overline{\mathfrak{R}}' = \mathfrak{R}'_1 \to \mathfrak{R}_1 \supseteqq \overline{\mathfrak{R}}',$$

our proof is now complete. □

Corollary 55 *We have*

$$\overline{a_1 \cdots a_{n+1}}, \mathfrak{R}_1 \overset{*}{+} \cdots \overset{*}{+} \mathfrak{R}_k \ni a_1, \cdots, a_{n+1} \to$$

all of $a_1, \cdots, a_{n+1}$ are contained in exacly one of $\mathfrak{R}_1, \cdots, \mathfrak{R}_k$.

Proposition 56 *Any linear space can be decomposed into the direct sum of indecomposable linear spaces in a unique way up to the order of factors.*

Remark 57 *(Translator's) The following proof is more refined than Nakasawa's original one. Nakasawa's original proof appears mathematically porous.*

Proof. Let

$$\mathfrak{R} = \bar{\mathfrak{R}}_1 \overset{*}{+} \cdots \overset{*}{+} \bar{\mathfrak{R}}_k = \bar{\mathfrak{R}}'_1 \overset{*}{+} \cdots \overset{*}{+} \bar{\mathfrak{R}}'_k.$$

Then we have

$$\mathfrak{R} = \bar{\mathfrak{R}}_1 \overset{*}{+} \cdots \overset{*}{+} \bar{\mathfrak{R}}_k, \mathfrak{R} \supseteqq \bar{\mathfrak{R}}'_1 \xrightarrow{\text{(Proposition 54)}}$$

$$\bar{\mathfrak{R}}_1 \supseteqq \bar{\mathfrak{R}}'_1 \vee \ldots \vee \bar{\mathfrak{R}}_k \supseteqq \bar{\mathfrak{R}}'_1$$

so that we can assume, by way of example, that $\overline{\mathfrak{R}}_1 \supseteqq \overline{\mathfrak{R}}'_1$. On the other hand we have

$$\mathfrak{R} = \overline{\overline{\mathfrak{R}'_1}} \overset{*}{+} \cdots \overset{*}{+} \overline{\mathfrak{R}'_k}, \mathfrak{R} \supseteqq \overline{\mathfrak{R}}_1 \xrightarrow{\text{(Proposition 54)}}$$

$$\overline{\overline{\mathfrak{R}'_1}} \supseteqq \overline{\mathfrak{R}}_1 \vee \ldots \vee \overline{\mathfrak{R}'_k} \supseteqq \overline{\mathfrak{R}}_1$$

so that we have $\overline{\mathfrak{R}}'_i \supseteqq \overline{\mathfrak{R}}_1$ for some $\overline{\mathfrak{R}}'_i$. However we have

$$\overline{\mathfrak{R}}'_i \supseteqq \overline{\mathfrak{R}}_1, \overline{\mathfrak{R}}_1 \supseteqq \overline{\mathfrak{R}}'_1 \to \overline{\mathfrak{R}}'_i \supseteqq \overline{\mathfrak{R}}'_1 \to \overline{\mathfrak{R}}'_i = \overline{\mathfrak{R}}'_1$$

so that $\overline{\mathfrak{R}}_1 = \overline{\mathfrak{R}}'_1$. By using the same token again and again, we can arrive at the desired conclusion. □

The Decomposition of a $\mathfrak{B}_2$-space

Definition 58 *In case that ab or $(Ez)(\overline{abz})$ for two elements a, b of $\mathfrak{B}_2$, then it is said that a doesn't separate itself from b, and it is denoted by $a \sim b$. Otherwise, it is said that a separates itself from b, and it is denoted by $a \nsim b$.*

Proposition 59 *The binary relation "$\sim$" is an equivalence relation.*

Proof.

1. It is reflexive, because we have

$$aa \to a \sim a.$$

2. It is symmetric, for we have

$$a \sim b \to ab \vee (\exists z)(\overline{abz})$$
$$ab \to ba \to b \sim a$$
$$(\exists z)(\overline{abz}) \to (\exists z)(\overline{baz}) \to b \sim a$$

so that

$$a \sim b \to b \sim a.$$

3. To see that it is transitive, we have to show that

$$a \sim b, b \sim c \to a \sim c.$$

By Definition 58, $a \sim b$ means $ab \vee (Ex)(\overline{abx})$, while $b \sim c$ means that $bc \vee (Ey)(\overline{bcy})$, so that we have to consider the following four cases:

$$ab, bc$$
$$ab, (\exists y)(\overline{bcy})$$
$$(\exists x)(\overline{abx}), bc$$
$$(\exists x)(\overline{abx}), (\exists y)(\overline{bcy})$$

Our treatments of the first three cases go as follows:

$$ab, bc \to ac \to a \sim c.$$
$$ab, (\exists y)(\overline{bcy}) \to (\exists y)(\overline{acy}) \to a \sim c.$$
$$(\exists x)(\overline{abx}), bc \to (\exists x)(\overline{acx}) \to a \sim c.$$

Now we deal with the last case. We have

$$\overline{abx}, \overline{bcy} \to \overline{acxy} \vee (abc, ab \neq 0, bc \neq 0).$$
$$\overline{acxy} \to (\exists z)(\overline{acz}) \to a \sim c.$$
$$abc, ab \neq 0, bc \neq 0 \to \overline{abc} \vee ac.$$
$$\overline{abc} \to (\exists z)(\overline{acz}) \to a \sim c.$$
$$ac \to a \sim c.$$

Therefore we are finally sure that

$$a \sim c, b \sim c \to a \sim c.$$

□

Corollary 60 *The above equivalence relation $\sim$ decomposes elements of $\mathfrak{B}_2$ into disjoint classes.*

Definition 61 *Such a class of elements of $\mathfrak{B}_2$ is called a principal space associated with $\mathfrak{B}_2$, in short, a $\mathfrak{B}$-space, of the $\mathfrak{B}_2$-space, and is denoted by $\mathscr{B}$ or the like. Further, it is said that the $\mathfrak{B}_2$-space decomposes into the direct sum of principal spaces $\mathfrak{B}_1, \mathfrak{B}_2, \mathfrak{B}_3, \cdots$ associated with $\mathfrak{B}_2$, and it is denoted by*

$$\mathfrak{B}_2 = \mathscr{B}_1 \overset{*}{+} \mathscr{B}_2 \overset{*}{+} \mathscr{B}_3 \overset{*}{+} \cdots .$$

Proposition 62 *We have*

$$\mathfrak{B} \ni a_1, \cdots, a_n \text{ and } a_1 \cdots a_n \neq 0 \to \mathfrak{B} \supseteqq \mathfrak{R}(a_1 \cdots a_n).$$

Proof of Proposition 62. We have

$$
\begin{aligned}
&\mathfrak{N}(a_1\cdots a_n)\ni x\to a_1\cdots a_n\neq 0, a_1\cdots a_n x.\\
&a_1\cdots a_n\neq 0, a_1\cdots a_n x\to\overline{a_1\cdots a_n x}\vee\\
&\quad\widehat{a_1}a_2\cdots a_n x\vee a_1\widehat{a_2}\cdots a_n x\vee\cdots\vee a_1\cdots a_{n-1}\widehat{a_n}x\\
&\qquad\qquad\to\overline{a_1\cdots a_n x}\vee(\text{e.g.})a_2\cdots a_n x.\\
&\therefore\left.\begin{array}{l} a_1\cdots a_n\neq 0,\\ a_1\cdots a_n x\end{array}\right\}\to\overline{a_1\cdots a_n x}\vee(\text{e.g.})a_2\cdots a_n x.\\
&\quad\left.\begin{array}{l} a_2\cdots a_n\neq 0,\\ a_2\cdots a_n x\end{array}\right\}\to\overline{a_2\cdots a_n x}\vee(\text{e.g.})a_3\cdots a_n x.\\
&\qquad\cdots\quad\cdots\quad\cdots\quad\cdots\quad\cdots\quad\cdots\quad\cdots\quad\cdots\\
&\qquad\cdots\quad\cdots\quad\cdots\quad\cdots\quad\cdots\quad\cdots\quad\cdots\quad\cdots\\
&\quad\left.\begin{array}{l} a_{n-1}a_n\neq 0,\\ a_{n-1}a_n x\end{array}\right\}\to\overline{a_{n-1}\cdots a_n x}\vee(\text{e.g.})\overline{a_n x}.\\
&\therefore\left.\begin{array}{l} a_1\cdots a_n\neq 0,\\ a_1\cdots a_n x\end{array}\right\}\to\overline{a_1\cdots a_n x}\vee(\text{e.g.})\overline{a_2\cdots a_n x}\\
&\qquad\qquad\qquad\qquad\vee(\text{e.g.})\cdots\vee(\text{e.g.})\overline{a_n x}.\\
&\therefore\left.\begin{array}{l} a_1\cdots a_n\neq 0,\\ a_1\cdots a_n x\end{array}\right\}\to(\text{e.g.})\overline{a_m\cdots a_n x}.\\
&\therefore\mathfrak{N}(a_1\cdots a_n)\ni x\to(\text{e.g.})\overline{a_m\cdots a_n x}.\\
&\overline{a_m\cdots a_n x}\underset{\longrightarrow}{(\text{Proposition 19})(\exists z)}(\overline{a_n x z})\vee\overline{a_n x}\to a_n\sim x.\\
&\therefore\mathfrak{N}(a_1\cdots a_n)\ni x\to(\text{e.g.})a_n\sim x.\\
&\quad\mathfrak{B}\ni a_n,\ a_n\sim x\to\mathfrak{B}\ni x.\\
&\therefore\mathfrak{N}(a_1\cdots a_n)\ni x\to\mathfrak{B}\ni x.\\
&\therefore\mathfrak{N}(a_1\cdots a_n)\subseteqq\mathfrak{B}.
\end{aligned}
$$

□

Corollary 63 *We have*

$$\mathfrak{B}\supseteqq M\to\mathfrak{B}\supseteqq\mathfrak{B}(M).$$

Proposition 64 $\mathfrak{B}\ni\mathfrak{N}\to\bar{\mathfrak{N}}$.

Proof. In case that the dimension of $\mathfrak{N}$ is equal to 1, there is nothing to prove. Therefore it suffices to deal with the case that the dimension of $\mathfrak{N}$ is equal to or more than 2. We have

$$
\begin{aligned}
&\mathfrak{N}\supseteqq\mathfrak{N}^2(a_1a_2).\\
&\mathfrak{B}\ni a_1,a_2\to a_1\sim a_2.\\
&a_1\sim a_2, a_1a_2\neq 0\to(\exists z)(\overline{a_1a_2z}).\\
&(\exists z)(\mathfrak{N}^2(a_1a_2)\ni z,\overline{a_1a_2z})\to\bar{\mathfrak{N}}^2(a_1a_2).\\
&\therefore(\text{Proposition 51})\to\bar{\mathfrak{N}}.\\
&\therefore\mathfrak{B}\supseteqq\mathfrak{N}\to\bar{\mathfrak{N}}.
\end{aligned}
$$

□

Corollary 65 $(\mathfrak{B} \ni a_1, \cdots, a_n) \wedge (a_1 \cdots a_n \neq 0) \rightarrow (\exists z)(\mathfrak{B} \ni z, \overline{a_1 \cdots a_n z})$.

Corollary 66 $\mathfrak{B} \ni \mathfrak{R}_1, \mathfrak{R}_2 \rightarrow$*it is not the case that* $\mathfrak{R}_1 * \mathfrak{R}_2$.

By Propositions 62 and 64, we have

Proposition 67 *If the dimension of a $\mathfrak{B}$-space is finite, then the $\mathfrak{B}$-space is a linear space.*

Proposition 68 *We have*

$$\left.\begin{array}{l}(\mathfrak{B}_1 \ni a_1, \cdots, a_n) \wedge (a_1 \cdots a_n \neq 0),\\ (\mathfrak{B}_2 \ni b_1, \cdots, b_m) \wedge (b_1 \cdots b_m \neq 0)\end{array}\right\} \rightarrow a_1 \cdots a_n b_1 \cdots b_m \neq 0.$$

Proof of Proposition 68. Assume, for the sake of contradiction, that $a_1 \cdots a_n b_1 \cdots b_m = 0$ so that

$$a_1 \cdots a_n b_1 \cdots b_m \rightarrow (\exists x)(a_1 \cdots a_n x, b_1 \cdots b_m x)$$

$(\mathfrak{B}_1 \ni a_1, \cdots, a_n) \wedge (a_1 \cdots a_n \neq 0) \wedge (a_1 \cdots a_n x) \xrightarrow{\text{(Proposition 62)}} \mathfrak{B}_1 \ni x.$
$(\mathfrak{B}_2 \ni b_1, \cdots, b_m) \wedge (b_1 \cdots b_m \neq 0) \wedge (b_1 \cdots b_m x) \xrightarrow{\text{(Proposition 62)}} \mathfrak{B}_2 \ni x.$
$(\exists x)(\mathfrak{B}_1 \ni x, \mathfrak{B}_2 \ni x) \rightarrow \perp.$
$\therefore a_1 \cdots a_n b_1 \cdots b_m \rightarrow \perp.$
$\therefore a_1 \cdots a_n b_1 \cdots b_m \neq 0.$

□

Proposition 69 *We have*

$$\mathfrak{B}_1 \ni a_1, \cdots, \mathfrak{B}_k \ni a_k \rightarrow a_1 \cdots a_k \neq 0.$$

Proof. Assume, for the sake of contradiction, that $a_1 \cdots a_k = 0$ so that we have

$$a_1 \cdots a_k \rightarrow \overline{a_1 \cdots a_k} \vee \widehat{a_1} a_2 \cdots a_k \vee a_1 \widehat{a_2} \cdots a_k \vee \cdots \vee a_1 \cdots a_{k-1} \widehat{a_k}.$$

$\therefore a_1 \cdots a_k \rightarrow \overline{a_1 \cdots a_k} \vee \text{(e.g.)} a_1 \cdots a_{k-1}.$
$a_1 \cdots a_{k-1} \rightarrow \overline{a_1 \cdots a_{k-1}} \vee \text{(e.g.)} a_1 \cdots a_{k-2}.$
$\cdots \quad \cdots \quad \cdots \quad \cdots \quad \cdots \quad \cdots$
$a_1 a_2 a_3 \rightarrow \overline{a_1 a_2 a_3} \vee \text{(e.g.)} \overline{a_1 a_2}.$
$\therefore a_1 \cdots a_k \rightarrow \overline{a_1 \cdots a_k} \vee \text{(e.g.)} \overline{a_1 \cdots a_{k-1}} \vee \text{(e.g.)} \cdots \vee \text{(e.g.)} \overline{a_1 a_2}.$
$\therefore a_1 \cdots a_k \rightarrow \text{(e.g.)} \overline{a_1 \cdots a_m},\ 2 \leqq m \leqq k.$ □
$\overline{a_1 \cdots a_m} \rightarrow Ez,\ \overline{a_1 a_2 z}$ or $\overline{a_1 a_2} \rightarrow a_1 \sim a_2.$
$\therefore a_1 \cdots a_k \rightarrow$ for example $a_1 \sim a_2.$
$\mathfrak{B}_1 \ni a_1, \mathfrak{B}_2 \ni a_2, a_1 \sim a_2 \rightarrow \perp.$
$\therefore a_1 \cdots a_k \rightarrow \perp.$
$\therefore a_1 \cdots a_k \neq 0.$

Proposition 70 *We have*

$$\left.\begin{array}{l}\mathfrak{B}_1 \ni a_1, \cdots, a_n \wedge a_1 \cdots a_n \neq 0,\\ \mathfrak{B}_2 \ni b_1, \cdots, b_m \wedge b_1 \cdots b_m \neq 0,\\ \quad \cdots \quad \cdots \quad \cdots \quad \cdots\\ \mathfrak{B}_k \ni l_1, \cdots, l_s \wedge l_1 \cdots l_s \neq 0\end{array}\right\} \to a_1 \cdots a_n b_1 \cdots b_m \cdots l_1 \cdots l_s \neq 0.$$

Proof. Assume, for the sake of contradiction, that $a_1 \cdots a_n b_1 \cdots b_m \cdots l_1 \cdots l_s = 0$ so that we have

$$a_1 \cdots a_n b_1 \cdots b_m \cdots l_1 \cdots l_s \to (\exists a)(a_1 \cdots a_n a, ab_1 \cdots b_m \cdots l_1 \cdots l_s).$$

$$\mathfrak{B}_1 \ni a_1, \cdots, a_n \wedge a_1 \cdots a_n \neq 0 \wedge a_1 \cdots a_n a \xrightarrow{\text{(Proposition 62)}} \mathfrak{B}_1 \ni a.$$

$$\begin{array}{l}\therefore a_1 \cdots a_n b_1 \cdots b_m \cdots l_1 \cdots l_s \to (\exists a)(\mathfrak{B}_1 \ni a_1), ab_1 \cdots b_m \cdots l_1 \cdots l_s.\\ \quad ab_1 \cdots b_m \cdots l_1 \cdots l_s \to (\exists b)(\mathfrak{B}_2 \ni b), ab \cdots l_1 \cdots l_s.\\ \quad \cdots \quad \cdots \quad \cdots \quad \cdots \quad \cdots \quad \cdots\\ \quad ab \cdots l_1 \cdots l_s \to (\exists l)(\mathfrak{B}_k \ni l), ab \cdots l.\\ \therefore a_1 \cdots a_n b_1 \cdots b_m \cdots l_1 \cdots l_s \to \left\{\begin{array}{l}(\exists a)(\exists b) \cdots (\exists l)(ab \cdots l,\\ \mathfrak{B}_1 \ni a, \mathfrak{B}_2 \ni b, \cdots, \mathfrak{B}_k \ni l).\end{array}\right.\\ \quad \mathfrak{B}_1 \ni a, \mathfrak{B}_2 \ni b, \cdots, \mathfrak{B}_k \ni l \xrightarrow{\text{(Proposition 69)}} ab \cdots l \neq 0.\\ \quad ab \cdots l, ab \cdots l \neq 0 \to \bot.\\ \therefore a_1 \cdots a_n b_1 \cdots b_m \cdots l_1 \cdots l_s \to \bot.\\ \therefore a_1 \cdots a_n b_1 \cdots b_m \cdots l_1 \cdots l_s \neq 0.\end{array}$$

□

Corollary 71 *We have*

$$\left.\begin{array}{l}\mathfrak{B}_1 \ni a_1, \cdots, a_n,\\ \cdots \quad \cdots \quad \cdots,\\ \mathfrak{B}_k \ni l_1, \cdots, l_s,\\ a_1 \cdots a_n \cdots l_1 \cdots l_s\end{array}\right\} \to a_1 \cdots a_n \vee \cdots \vee l_1 \cdots l_s.$$

Proposition 72 *We have*

$$\left.\begin{array}{l}\mathfrak{B}_1 \ni a_1, \cdots, a_n \wedge a_1 \cdots a_n x \neq 0,\\ \mathfrak{B}_2 \ni b_1, \cdots, b_m \wedge b_1 \cdots b_m x \neq 0,\\ \cdots \quad \cdots \quad \cdots \quad \cdots \quad \cdots,\\ \mathfrak{B}_k \ni l_1, \cdots, l_s \wedge l_1 \cdots l_s x \neq 0\end{array}\right\} \to a_1 \cdots a_n b_1 \cdots b_m \cdots l_1 \cdots l_s x \neq 0.$$

Proof. Proof: If $\mathfrak{B}_1 \ni x$ by way of example, we have

$$\left.\begin{array}{l}\mathfrak{B}_1 \ni a_1, \cdots, a_n, x \wedge a_1 \cdots a_n x \neq 0,\\ \mathfrak{B}_2 \ni b_1, \cdots, b_m \wedge b_1 \cdots b_m \neq 0,\\ \cdots \quad \cdots \quad \cdots \quad \cdots \quad \cdots,\\ \mathfrak{B}_k \ni l_1, \cdots, l_s \wedge l_1 \cdots l_s \neq 0\end{array}\right\} \xrightarrow{\text{(Proposition 70)}}$$

$$a_1 \cdots a_n b_1 \cdots b_m \cdots l_1 \cdots l_s x \neq 0.$$

If $x \notin \mathfrak{B}_i$ $(i = 1, \cdots, k)$, then we have

$$\left.\begin{array}{l}\mathfrak{B}_1 \ni a_1, \cdots, a_n \wedge a_1 \cdots a_n \neq 0,\\ \mathfrak{B}_2 \ni b_1, \cdots, b_m \wedge b_1 \cdots b_m \neq 0,\\ \cdots\ \cdots\ \cdots\ \cdots\ \cdots,\\ \mathfrak{B}_k \ni l_1, \cdots, l_s \wedge l_1 \cdots l_s \neq 0,\\ \mathfrak{B} \ni x \wedge x \neq 0\end{array}\right\} \xrightarrow{\text{(Proposition 70)}}$$

$$a_1 \cdots a_n b_1 \cdots b_m \cdots l_1 \cdots l_s x \neq 0.$$

□

Corollary 73 *We have*

$$\left.\begin{array}{l}\mathfrak{B}_1 \ni a_1, \cdots, a_n,\\ \cdots\ \cdots\ \cdots\ \cdots,\\ \mathfrak{B}_k \ni l_1, \cdots, l_s,\\ a_1 \cdots a_n \cdots l_1 \cdots l_s x\end{array}\right\} \rightarrow a_1 \cdots a_n x \vee \cdots \vee l_1 \cdots l_s x.$$

Corollary 74 *We have*

$$\left.\begin{array}{l}\mathfrak{B}_1 \ni a_1, \cdots, a_n \wedge a_1 \cdots a_n \neq 0,\\ \cdots\ \cdots\ \cdots\ \cdots\ \cdots,\\ \mathfrak{B}_k \ni l_1, \cdots, l_s, \wedge l_1 \cdots l_s \neq 0,\\ a_1 \cdots a_n \cdots l_1 \cdots l_s x\end{array}\right\} \rightarrow$$

Exactly one of the statements $a_1 \cdots a_n x, \cdots$, and $l_1 \cdots l_s x$ holds.

Proposition 75 *We have*

$$\mathfrak{B}_1 \supseteqq \mathfrak{R}_1, \cdots, \mathfrak{B}_k \supseteqq \mathfrak{R}_k, \rightarrow \mathfrak{R}_1 * \cdots * \mathfrak{R}_k.$$

Proof. If the bases of $\mathfrak{R}_1, \cdots, \mathfrak{R}_k$ are $a_1 \cdots a_n, \cdots, l_1 \cdots l_s$ respectively, then we have

$$\left.\begin{array}{l}\mathfrak{B}_1 \ni a_1, \cdots, a_n \wedge a_1 \cdots a_n \neq 0,\\ \cdots\ \cdots\ \cdots\ \cdots\ \cdots,\\ \mathfrak{B}_k \ni l_1, \cdots, l_s \wedge l_1 \cdots l_s \neq 0\end{array}\right\} \xrightarrow{\text{(Proposition 70)}}$$

$$a_1 \cdots a_n \cdots l_1 \cdots l_s \neq 0.$$

Therefore we have

$$\dim \mathfrak{R}(a_1 \cdots a_m \cdots l_1 \cdots l_s) = \dim \mathfrak{R}_1(a_1 \cdots a_n) + \cdots + \dim \mathfrak{R}_k(l_1 \cdots l_s).$$

On one hand we have

$$\left.\begin{array}{l}\mathfrak{B}_1 \ni a_1, \cdots, a_n,\\ \cdots\ \cdots\ \cdots\ \cdots,\\ \mathfrak{B}_k \ni l_1, \cdots, l_s,\\ a_1 \cdots a_n \cdots l_1 \cdots l_s x\end{array}\right\} \xrightarrow{\text{(Corollary 73)}}$$

$$a_1 \cdots a_n x \vee \cdots \vee l_1 \cdots l_s x$$

$$\therefore \mathfrak{R}(a_1 \cdots a_n \cdots l_1 \cdots l_s) \ni x \to \mathfrak{R}_1(a_1 \cdots a_n) \ni x \vee \cdots \vee \mathfrak{R}_k(l_1 \cdots l_s) \ni x.$$
$$\therefore \mathfrak{R}(a_1 \cdots a_n \cdots l_1 \cdots l_s) \subseteqq \mathfrak{R}_1(a_1 \cdots a_n) + \cdots + \mathfrak{R}_k(l_1 \cdots l_s).$$

On the other hand we have

$$\mathfrak{R}(a_1 \cdots a_n \cdots l_1 \cdots l_s) \supseteqq \mathfrak{R}_1(a_1 \cdots a_n) \cup \cdots \cup \mathfrak{R}_k(l_1 \cdots l_s).$$
$$\therefore \mathfrak{R}(a_1 \cdots a_n \cdots l_1 \cdots l_s) = \mathfrak{R}_1(a_1 \cdots a_n) \cup \cdots \cup \mathfrak{R}_k(l_1 \cdots l_s).$$
$$\therefore \mathfrak{R}(a_1 \cdots a_n \cdots l_1 \cdots l_s) = \mathfrak{R}_1(a_1 \cdots a_n) \overset{*}{+} \cdots \overset{*}{+} \mathfrak{R}_k(l_1 \cdots l_s).$$
$$\therefore \mathfrak{R}_1(a_1 \cdots a_n) * \cdots * \mathfrak{R}_k(l_1 \cdots l_s).$$

□

Corollary 76 *We have*

$$\mathfrak{B}_1 \ni \mathfrak{R}_1, \cdots, \mathfrak{B}_k \ni \mathfrak{R}_k \to \mathfrak{D}\{\mathfrak{B}(\mathfrak{R}_1 \cdots \mathfrak{R}_i), \mathfrak{B}(\mathfrak{R}_{i+1} \cdots \mathfrak{R}_k)\} = \mathfrak{N}$$
$$(1 \leqq i \leqq k-1).$$

Proposition 77 *We have*

$$\mathfrak{R} = \overline{\mathfrak{D}}(\mathfrak{B}_1, \mathfrak{R}) \overset{*}{+} \overline{\mathfrak{D}}(\mathfrak{B}_2, \mathfrak{R}) \overset{*}{+} \overline{\mathfrak{D}}(\mathfrak{B}_3, \mathfrak{R}) \overset{*}{+} \cdots .$$

Proof. We have

$$\left.\begin{array}{l} \mathfrak{B}_2 = \mathfrak{B}_1 \cup \mathfrak{B}_2 \cup \mathfrak{B}_3 \cup \cdots, \\ \mathfrak{B}_2 \supseteqq \mathfrak{R} \end{array}\right\} \to$$

$$\mathfrak{R} = \mathfrak{D}(\mathfrak{B}_1, \mathfrak{R}) \cup \mathfrak{D}(\mathfrak{B}_2, \mathfrak{R}) \cup \mathfrak{D}(\mathfrak{B}_3, \mathfrak{R}) + \cdots .$$

Furthermore $\mathfrak{D}(\mathfrak{B}_1, \mathfrak{R})$ $(i = 1, 2, 3, \cdots)$ is a linear space according to Corollary to Proposition 62.

$$\left.\begin{array}{l} \mathfrak{B}_1 \supseteqq \mathfrak{D}(\mathfrak{B}_1, \mathfrak{R}), \\ \mathfrak{B}_2 \supseteqq \mathfrak{D}(\mathfrak{B}_2, \mathfrak{R}), \\ \mathfrak{B}_3 \supseteqq \mathfrak{D}(\mathfrak{B}_3, \mathfrak{R}), \\ \cdots \quad \cdots \quad \cdots \end{array}\right\} \xrightarrow{\text{(Proposition 75)}}$$

$$\mathfrak{D}(\mathfrak{B}_1, \mathfrak{R}) * \mathfrak{D}(\mathfrak{B}_2, \mathfrak{R}) * \mathfrak{D}(\mathfrak{B}_3, \mathfrak{R}) * \cdots .$$

Furthermore we have

$$\mathfrak{B}_1 \supseteqq \mathfrak{D}(\mathfrak{B}_i, \mathfrak{R}) \ (i = 1, 2, 3, \cdots) \xrightarrow{\text{(Proposition 64)}} \overline{\mathfrak{D}}(\mathfrak{B}_i, \mathfrak{R})$$

Therefore we have

$$\mathfrak{R} = \overline{\mathfrak{D}}(\mathfrak{B}_1, \mathfrak{R}) \overset{*}{+} \overline{\mathfrak{D}}(\mathfrak{B}_2, \mathfrak{R}) \overset{*}{+} \overline{\mathfrak{D}}(\mathfrak{B}_3, \mathfrak{R}) \overset{*}{+} \cdots .$$

□

Corollary 78 *For each principal linear space* $\bar{\mathfrak{R}}$, *exactly one of the statements* $\mathfrak{B}_1 \supseteqq \bar{\mathfrak{R}}, \mathfrak{B}_2 \supseteqq \bar{\mathfrak{R}}, \mathfrak{B}_3 \supseteqq \bar{\mathfrak{R}}, \cdots$ *holds.*

Corollary 79 *For each circuit* $\overline{a_1 \cdots a_{n+1}}$, *exactly one of the statements* $\mathfrak{B}_1 \ni a_1, \cdots, a_{n+1}$, $\mathfrak{B}_2 \ni a_1, \cdots, a_{n+1}$, $\mathfrak{B}_3 \ni a_1, \cdots, a_{n+1}$, $\cdots$ *holds.*

Proposition 80 *Any sum of principal linear spaces or any sum of principal spaces associated with* $\mathfrak{B}_2$ *is also a* $\mathfrak{B}_2$*-space.*

Proof. The proof is clear. □

Remark 81 *It remains to show that if the finiteness of the dimension is added as an axiom, our principal linear space is reduced to a projective space of finite dimension, which is built on the basis of the principality and dimension axiom*[16]. *Thus, at this time, the* $\mathfrak{B}_2$*-space is divided into the finite direct sum of projective spaces of finite dimension.*

[16] For example, cf. Veblen-Young : Projective Geometry!

On Axiomatics of Linear Dependence III

Takeo Nakasawa
Received July, 20, 1936

Introduction

In my earlier work[1], I gave a method which can be used for investigations of linear dependence in the projective space. The method is an algebraic symbol calculation, in which certain sequences of elements are called linearly independent and the other sequences of elements are called linearly dependent. Therefore, I called this calculation *cycle calculation*[2], following G. Thomsen.

The present paper is concerned with

1. the complete structure of projective geometry in which calculation principles listed at the beginning[3] are interpreted as spatial axioms of a geometrical space;
2. some applications of the general linear space and its comparison with the related works by G. Birkhoff and H. Whitney;

---

[1] T. Nakasawa, "*Zur Axiomatik der linearen Abhängigkeit. I*", Science Reports of the Tokyo Bunrika Daigaku, Sec. A, Vol. **2**, No. 43, (233–255); *II*, ibid., Sec. A, Vol. 3, No. 51, (45–69). For the sake of convenience, we describe these two works shortly as A1 and A2 respectively.

[2] The word and the idea of the cycle calculation probably go back to the book by G. Thomsen, "*Grundlagen der Elementargeometrie*" (Leipzig 1933). Influenced by his idea, I began my axiomatic investigation of the linear dependence.

[3] Cf. p. 236 in A1, and p. 46 in A2!

3. a proof of the independence of our axiom system.

Now, if we write $a_1 \cdots a_s = 0$ (usually abbreviated to $a_1 \cdots a_s$) or $a_1 \cdots a_s \neq 0$ for s points $a_1, a_2, \cdots, a_s$ of the classical n-dimensional projective space[4] depending on whether they are linearly dependent or linearly independent, then it is well known that the following axioms hold[5]:

Axiom 1 $a \neq 0, aa$

Axiom 2 $a_1 \cdots a_s \to a_1 \cdots a_s x \ (s \geqq 2)$

Axiom 3 $a_1 \cdots a_i \cdots a_s \to a_i a_1 \cdots \widehat{a_i} \cdots a_s \ (s \geqq i \geqq 2)$

Axiom 4 $x a_1 \cdots a_s,\ a_1 \cdots a_s y \to (a_1 \cdots a_s) \vee (x a_1 \cdots a_{s-1} y) \ (s \geqq 1)$

Axiom 5 $a_1 \cdots a_s x y \to (\exists z)((a_1 \cdots a_s z) \wedge (xyz)) \ (s \geqq 2)$

Conversely, if one takes the first four axioms only as calculation principles for elements $a_1, a_2, \cdots, a_s \cdots$ of an abstract set $\mathfrak{B}$, then almost all properties of linear dependence which contain no existence statement can be derived. Furthermore, if one takes the last axiom into consideration, then *the dimension axiom* of linear spaces

$$\dim \mathfrak{D}(AB) + \dim \mathfrak{B}(AB) = \dim A + \dim B^6$$

is provable, and almost all properties of the projective space are provable. I have already thoroughly demonstrated the details about it in my works A1 and A2.

In the present paper, I will try to make it clear which propositions of projective geometry can be derived from the above five axioms. To this end, it seems appropriate to compare the above axiom system with that by Veblen and Young[7], which is considered to be one of the most complete and exemplary systems of projective geometry.

The Comparison with the System of Veblen and Young

Definition 6 *Let $a_1, a_2, \cdots, a_n$ be n linear independent elements, i.e., $a_1 \cdots a_n \neq 0$. We call the set of all elements x of $\mathfrak{B}$ with $a_1 \cdots a_n x = 0$ the linear space generated by $a_1, a_2, \cdots, a_n$ in $\mathfrak{B}$ of the dimension n, in characters $\mathfrak{R}^n(a_1 \cdots a_n)$, and we call the cycle $a_1 \cdots a_n$ the basis of $\mathfrak{R}^n(a_1 \cdots a_n)$. Especially, we call a linear space of dimension* 1 *or* 2 *a point or a line accordingly.*

4 Cf. for example Eugenio Bertini, "*Einführung in die projektive Geometrie mehrdimensionaler Räume*", Wien, 1906, (1–23)!

5 Concerning the indications used here cf. p. 235 in A1 or p. 45 in A2!

6 Proposition 29 on p. 48 in A2.

7 O. Veblen and J. W. Young, "Projective geometry, I", Boston, 1910, (1–30).

We have the following simple propositions.

Proposition 7 *Two points $\mathfrak{R}(a)$, $\mathfrak{R}(b)$ agree only if $ab = 0$.*

Proposition 8 *Three points $\mathfrak{R}(a)$, $\mathfrak{R}(b)$, and $\mathfrak{R}(c)$ lie in the same line only if $abc = 0$.*

Proposition 9 *Two distinct points $\mathfrak{R}(a)$, $\mathfrak{R}(b)$ determine a line $\mathfrak{R}(ab)$.*

Proposition 10 *Two distinct points $\mathfrak{R}(x)$, $\mathfrak{R}(y)$ lying on a line $\mathfrak{R}(ab)$ determine the same line. In other words, we have $\mathfrak{R}(ab) = \mathfrak{R}(xy)$.*

For proofs of the above results the reader is referred to $A1$.

Proposition 11 *If $\mathfrak{R}(a)$, $\mathfrak{R}(b)$, and $\mathfrak{R}(c)$ do not lie on a line, and $\mathfrak{R}(d)$ and $\mathfrak{R}(e)$ lie on lines $\mathfrak{R}(bc)$ and $\mathfrak{R}(ac)$ respectively with $\mathfrak{R}(d) \neq \mathfrak{R}(e)$, then there is a point $\mathfrak{R}(f)$ lying on lines $\mathfrak{R}(ab)$ and $\mathfrak{R}(de)$ simultaneously.*

Proof. It follows from the assumption and Propositions 7 and 8 that $abc \neq 0$, $bcd = 0$, $ace = 0$, and $de \neq 0$. Therefore we have

$$\left.\begin{matrix} abc \neq 0, \\ bcd, \\ ace \end{matrix}\right\} \to \left.\begin{matrix} abc \neq 0, \\ abcd, \\ abce \end{matrix}\right\} \to abde \to (\exists f)(abf \wedge def)$$

Thus we can conclude that the point $\mathfrak{R}(f)$ lies on lines $\mathfrak{R}(ab)$ and $\mathfrak{R}(de)$ simultaneously. □

Proposition 12 *Let $\mathfrak{R}(a_1 \cdots a_k)$ be a linear space of dimension k and $\mathfrak{R}(a_{k+1})$ a point not lying therein. Then the linear space $\mathfrak{R}(a_1 \cdots a_{k+1})$ of dimension $k+1$ agrees with the set $\sum_x \mathfrak{R}(a_{k+1}x)$ which consists of all lines $\mathfrak{R}(a_{k+}x)$ arising from any point $\mathfrak{R}(x)$ in $\mathfrak{R}(a_1 \cdots a_k)$.*

Proof. Let $\mathfrak{R}(y)$ be any point lying in $\sum_x \mathfrak{R}(a_{k+1}x)$. It follows that

$$a_1 \cdots a_k x,\ x a_{k+1} y \to {}^8 a_1 \cdots a_k + 1y \to \mathfrak{R}(a_1 \cdots a_k + 1) \ni \mathfrak{R}(y)$$

Therefore we have

$$\sum_x \mathfrak{R}(a_{k+1}x) \subseteqq \mathfrak{R}(a_1 \cdots a_{k+1}) \tag{1}$$

---

[8] Corollary 2 to the proposition 24 on p. 252 in A1.

Let $\mathfrak{R}(y)$ be a point lying in $\mathfrak{R}(a_1 \cdots a_{k+1})$. Then we have

$$a_1 \cdots a_{k+1} y \to (\exists x)(a_1 \cdots a_k x \wedge a_{k+1} y x)$$

Thus we have

$$(\exists x)(\mathfrak{R}(a_1 \cdots a_k) \ni \mathfrak{R}(x) \wedge \mathfrak{R}(a_{k+1} x) \ni \mathfrak{R}(y))$$

which implies that

$$\sum_x \mathfrak{R}(a_{k+1} x) \supseteqq \mathfrak{R}(a_1 \cdots a_{k+1}) \tag{2}$$

It follows from (1) and (2) that

$$\sum_x \mathfrak{R}(a_{k+1} x) = \mathfrak{R}(a_1 \cdots a_{k+1})$$

□

Proposition 13 *If two points lying on a line of at least three points are considered to belong to the same class, then $\mathfrak{B}$ decomposes into a number of classes of elements*[9] *which contain no common elements among them. Let $\mathfrak{B}_v$ be such a class of elements, then $\mathfrak{B}_v$ is also a $\mathfrak{B}_2$-space*[10]*, and $\mathfrak{B}$ decomposes into the direct sum of $\mathfrak{B}_1$, $\mathfrak{B}_2$, $\mathfrak{B}_3$, $\cdots$*[11] *. Then, at least three points exist on every line of $\mathfrak{B}_v$*[12].

This Proposition is shown already in A2, §5, pp. 63–69.

Veblen and Young's axiomatics for projective geometry goes as follows[13]: By a *projective geometry* we mean a set of elements which are called points and are subject to the following three axioms:

Axiom 14 *If A and B are distinct points, there is one and only one line that contains A and B.*

Axiom 15 *If A, B, C are non-collinear points and if a line l contains a point D of the line (BC) and a point E of the line (AC), where D and E are distinct points, then the line l contains a point F of the line (AB).*

Axiom 16 *There are at least three points on every line.*

---

9 Proposition 63 and corollary on p. 63 in A2.

10 Definition VI on p. 47 and proposition 73 on p. 69 in A2.

11 Proposition 71, corollary, proposition 72, and corollary 1 on pp. 67–68 in A2.

12 $\mathfrak{B}_v$ is called the $\mathfrak{B}$-space. Cf. definition XII on p. 63 in A2!

13 O. Veblen and W. H. Bussy, "*Finite projective geometries*", Trans. Amer. Math. Soc., Vol. 7, 1906, (241–242).

In projective geometry of Veblen and Young's style a k-space is defined by the following inductive definition. A point is a 1-space. If $A_1, A_2, \cdots, A_{k+1}$ are points not all in the same k-space, the set of all points collinear with the line determined by the point A_{k+1} and any point of the k-space $(A_1 A_2 \cdots A_k)$ is the $(k+1)$-space $(A_1 A_2 \cdots A_{k+1})$. Thus a line is a 2-space, a plane is a 3-space, and so on.

By comparison of their axiomatic system with our propositions proved above, we see obviously that our Propositions 9 and 10 correspond to Axiom 14, while Proposition 11 corresponds to Axiom 15. Veblen and Young's definition of k-space is justifiable by our Proposition 12. However Veblen and Young's Axiom 16 is not always valid in a $\mathfrak{B}_2$-space, since an n-simplex[14] can be regarded as a $\mathfrak{B}_2$-space. Nevertheless, this holds exactly in a $\mathfrak{B}$-space, so that it does also in a prime space[15].

Therefore, one can claim the following by dint of Proposition 13.

Proposition 17 *Any $\mathfrak{B}_2$-space can be divided into the direct sum of projective spaces in a unique way up to the order of factors. Any $\mathfrak{B}$-space or any linear prime space is a projective space.*

The following axiom on finite dimensionality is usually assumed in Veblen and Young's axiomatic system.

Axiom 18 *There is a finite upper bound to the dimensions of spaces.*

Let n be the upper bound, then our set of points is called an *n-dimensional projective space.*

Now we have the following proposition.

Proposition 19 *Any linear prime space of dimension n is an n-dimensional projective space. Any linear space can be divided into the finite direct sum of projective spaces of finite dimensions in a unique way up to the order of factors*[16].

---

[14] We regard an n-simplex as the set of n elements in which the *be valid*—relation is defined as follows, i.e., the cycle which contains more than two same elements, we set $= 0$, otherwise $\neq 0$.

[15] Definition VIII on p. 52 and corollary 1 to the proposition 72 on p. 68 in A2.

[16] Proposition 62 on p. 62 and proposition 57 on p. 60 in A2.

Application and Related Works

The Application to General Linear Spaces[17] and Hilbert Spaces[18]

Let $\mathfrak{R}$ be a general linear space[19] over the real numbers (more generally, over an abstract field if the reader wants). Now, let $x_1, \cdots, x_s$ be s elements of $\mathfrak{R}$ different from zero for each. If there are s real numbers $a_1, \cdots, a_s$, not all of which are equal to zero and for which $a_1x_1 + \cdots + a_sx_s = 0$ (i.e., it by $x_1, \cdots, x_s$ are linearly dependent!) holds, we denote it by $x_1 \cdots x_s = 0$, while we denote $x_1 \cdots x_s \neq 0$ otherwise. Then, the five axioms of a $\mathfrak{B}_2$-space are obviously satisfied. Consequently we have

Proposition 20 *The set of all non-zero elements of a general linear space is a $\mathfrak{B}_2$-space with respect to linear dependence explained above.*

If, further, one additional condition that $\sum_i a_i = 0$ is required in the above definition of linear dependence (i.e., $x_1, \cdots, x_s$ are dependent in position!), then the four axioms of the $\mathfrak{B}_1$-space are obviously satisfied. Thus, one can claim

Proposition 21 *A general linear space is a $\mathfrak{B}_1$-space with respect to dependence in position.*

Since there are at least three points on every line with respect to both definitions of dependence, any general linear space is a $\mathfrak{B}$-space in both senses. Hilbert spaces are, as is well known, an exceptionally important example of general linear spaces, and so our cycle calculation can be applied to this.

---

[17] Cf. for example S. Banach, "*Theorie des operations lineaires*, Warszawa, 1932, Ch. II!

[18] D. Hilbert, "*Grundzüge einer allgemeinen Theorie der linearen Integralgleichungen*", 4. Mitteilung, Göttingen Nachrichten, 1906, (pp. 157–227).

[19] S. Banach, "*Sur les operations dans les ensembles abtraits et leur application aus equations integrals*", Fund. Math., **3**, (1922), p. 135.
A. Tychonoff, "Ein Fixpunktssatz", Math. Ann., 111, (1935), p. 767.

The Application to *Lattice* Theory[20], *Boolean* Algebras[21], and Association Theory[22]

Remark 22 *(Translator's) The following description of lattice theory is a streamlined one of Nakasawa's from our modern point, so that it should be more readable than Nakasawa's original one. You should note that lattice theory was a very fresh subject in the* 1930*s.*

A lattice C is a set with two binaray operations $\curlyvee$ and $\curlywedge$ satisfying the following conditions:

1. The associative laws hold. I.e.,
$$(a \curlyvee b) \curlyvee c = a \curlyvee (b \curlyvee c)$$
$$(a \curlywedge b) \curlywedge c = a \curlywedge (b \curlywedge c)$$
for any $a, b, c \in C$.
2. The commutative laws hold. I.e.,
$$a \curlyvee b = b \curlyvee a$$
$$a \curlywedge b = b \curlywedge a$$
for any $a, b \in C$.
3. The absorptive laws hold[23]. I.e.,
$$(a \curlyvee b) \curlywedge a = a$$
$$(a \curlywedge b) \curlyvee a = a$$
for any $a, b \in C$.

It is well known that there is naturally a partial order on a lattice C, in which a$\leq$ b iff $a \curlyvee b = b$. A lattice C is said to satisfy the *finite chain*

[20] G. Birkhoff, "*On the combination of subalgebras*", Proc. of the Cam. Phil. Soc., 29, 1933, (441–464), and "*Applications of lattice algebra*", ibid., **30**, 1934, (115–122).

[21] E.V. Huntington, "*Set of independent postulates for the algebra of logic*", Trans. of Amer. Math. Soc., Vol. **5**, 1904, (288–290).
G. Bergnn, "*Zur Axiomatik der Elementargeometrie*", Monat. für Math. und Phys., **36**, 1929, (269–290).

[22] Fritz Klein, "*Zur Theorie der abstrakten Verknüpfungen*", Math. Ann., **105**, 1931, (308–323) and *Über einen Zerlegungssatz in der Theorie der abstrakten Verknüpfungen*", Math. Ann., **106**, 1932, (114–130).

[23] i.e., $a \curlywedge (a \curlyvee b) = a \curlyvee (a \curlywedge b) = a$ for any a, b. L4 on p. 743 in G. Birkhoff, "Combinatorial relations in projective geometries", Ann. of Math., Vol. 36, 1935, (743–749). For the sake of convenience, we describe this work briefly as "G. Birkhoff C".

condition if there exists no infinite chain $a_1 < a_2 < a_3 < \cdots$ in C[24]. A lattice C is called *weakly distributive*[25] if it satisfies the following weakly distributive law:

$$a \leq c \to a \curlyvee (b \curlywedge c) = (a \curlyvee b) \curlywedge c$$

Weakly distributive lattices are called modular lattices by G. Birkhoff. We deal with a modular lattice satisfying the finite chain condition[26]. We denote the set of all one-dimensional elements, namely, of all points of C by L, while we write $a_1 \cdots a_s \neq 0$ or $a_1 \cdots a_s = 0$ for s points $a_1, \cdots, a_s$, depending on dim $(a_1 \curlyvee \cdots \curlyvee a_s) = s$ or dim $(a_1 \curlyvee \cdots \curlyvee a_s) < s$. Then the five axioms of a $\mathfrak{B}_2$-space obviously hold in L. Therefore we have

Proposition 23 *The point set of a modular lattice which satisfies the multiple chain proposition is a $\mathfrak{B}_2$-space of finite dimension.*

It should be remarked here, that $a_1 \neq a_2$ implies $a_1 a_2 \neq 0$ in L, so that the point $\mathfrak{R}(a)$ consists solely of a. If one assumes that the lattice C is complemented, then every n-dimensional element a can be expressed[27] as $a = a_1 \curlyvee \cdots \curlyvee a_n$ for some n points $a_1, \cdots, a_n$. Therefore one can claim the following:

Proposition 24 *The point set of every "complemented modular Lattice" which satisfies the finite chain condition is a $\mathfrak{B}_2$-space of finite dimension, and every element of the lattice clearly corresponds to a linear space of the $\mathfrak{B}_2$-space.*

If one sets the distributive rule

$$a \curlywedge (b \curlyvee c) = (a \curlywedge b) \curlyvee (a \curlywedge c)^{28}$$

instead of the weak distributive rule in a complemented modular lattice, then a Boolean algebra arises; thus, every Boolean algebra which satisfies the finite chain condition is also a $\mathfrak{B}_2$-space naturally. Since no three points can lie on a line, the Boolean algebra decomposes into the direct sum of finite points so that it is finite as a set. It is easy to see that the so-called association discussed by F. Klein can also be interpreted as a lattice.

---

[24] i.e., some two paragraphs in every sequence of the products a_1, $a_1 \curlywedge a_2$, $a_1 \curlywedge a_2 \curlywedge a_3$, $\cdots$ are equal respectively. L_4 on p. 801 in G. Birkhoff, "*Abstract linear dependence and lattices*", Amer. Journ. of Math., Vol. **57**, 1935, (800–804).

[25] i.e., $(a \curlyvee c) \curlywedge \{b \curlyvee (a \curlywedge c)\} = \{(a \curlyvee c) \curlywedge b\} \curlyvee (a \curlywedge e)$ for any a, b, c. L5 on p. 442 in G. Birkhoff, "On the structure of abstract algebras", Proc. of the Cam. Phil. Soc., **31**, 1935, (433–454).

[26] p. 745 in G. Birkhoff C.

[27] p. 746 in G. Birkhoff C.

[28] p. 705 in M. H. Stone, "*Postulates for Boolean algebras and generalized Boolean algebras*", Amer. Joun. of Math., Vol. **57**, 1935, (703–732).

The Relation to H. Whitney's Matroid Theory[29]

Let M be a $\mathfrak{B}_1$-space[30] of finitely many elements $e_1, e_2, \cdots, e_n$, and N a subset of it. Then we have the following propositions.

Proposition 25 *dim* $\mathfrak{N} = 0$[31].

Proposition 26 *We have*

$$\dim(N + e) = \dim N + k$$

where $k = 1$ *or* $k = 0$.

Proposition 27 *We have*

$$\dim(N + e_1) = \dim(N + e_2) = \dim N \longrightarrow \dim(N + e_1 + e_2) = \dim N$$

Conversely, if one starts with an abstract set M endowed with dim satisfying the above three propositions, it naturally gives rise to a $\mathfrak{B}_1$-space by defining $e_1 \cdots e_s = 0$ or $e_1 \cdots e_s \neq 0$ according to dim $(e_1 \cdots e_s) < s$ or dim $(e_1 \cdots e_s) = s$. The above three propositions 25, 26 and 27 are actually the axioms of H. Whitney's so-called matroid[32], and so one can state the following proposition.

Proposition 28 *Any finite* $\mathfrak{B}_1$*-space is a matroid, and vice versa.*

The Independence of the Axiomatic System

In this section we will present a new axiomatic system, which is finally to be shown to be equivalent to our old axiomatic system Axioms 1-5. Our new axiomatic system goes as follows:

Axiom 29 $a \neq 0$

Axiom 30 aba

Axiom 31 $ac, bc \rightarrow ab$

---

29 H. Whitney, "*On the abstract properties of linear dependence*", Amer. Journ. of Math., Vol. **57**, 1935, (509–533).
S. Maclane, "*Some interpretations of abstract linear dependence in terms of projective geometry*", Amer. Journ. of Math., Vol. **58**, 1936, (236–240).

30 p. 236 in A1.

31 $\mathfrak{N}$ means the empty set.

32 p. 510 in the work by H. Whitney above.

Axiom 32 $abc, abd \to ab \vee bcd$

Axiom 33 $abcd \to acbd$

Axiom 34 $a_1 \cdots a_s z, zb_1 b_2 \to a_1 \cdots a_{s1} b_1 b_2$ $(s \geqq 1)$

Axiom 35 $a_1 \cdots a_s b_1 b_2 \to (\exists z)(a_1 \cdots a_s z \wedge zb_1 b_2)$ $(s \geqq 1)$

It is easy to see that Axioms 29–35 are derivable from Axioms 1-5. In order to show the converse, we need to establish the following chain of lemmas:

Lemma 36 *We have*

$$aa$$

for every a.

Proof. We have

$$\begin{array}{c} aaa \\ az, zaa \\ az, az \to aa \end{array}$$

□

Lemma 37 *We have*

$$ab \to ba$$

Proof. We have

$$bb, ab \to ba$$

□

Lemma 38 *We have*

$$baa$$

for any a, b.

Proof. We have

$$\begin{array}{c} \left.\begin{array}{r} aba, \\ aba \end{array}\right\} \to baa \\ \downarrow \\ ab \\ ba, aaa \to baa \end{array}$$

□

Lemma 39 *We have*

$$aab$$

for any a, b.

Proof. We have

$$\left.\begin{array}{r} baa, \\ bab \end{array}\right\} \to aab$$
$$\downarrow$$
$$ba$$
$$ab,\ bab \to aab$$

□

Lemma 40 *We have*

$$abc \to bac$$

Proof. We have

$$\left.\begin{array}{r} aba, \\ abc \end{array}\right\} \to bac$$
$$\downarrow$$
$$ab$$
$$ba, aac \to bac$$

□

Lemma 41 *We have*

$$abc \to bca$$

Proof. We have

$$\left.\begin{array}{r} abc, \\ aba \end{array}\right\} \to bca$$
$$\downarrow$$
$$ab$$
$$ba,\ aca \to bca$$

□

Lemma 42 *We have*

$$a_1 \cdots a_s \to a_1 \cdots a_s x$$

for any $x, s \geqq 2$.

Proof. We have

$$
\begin{aligned}
&a_1 \cdots a_s \\
&a_1 \cdots a_s, a_s a_s x \\
&a_1 \cdots a_s x
\end{aligned}
$$

□

Lemma 43 *We have*

$$a_1 \cdots a_m z, z b_1 \cdots b_n \to a_1 \cdots a_m b_1 \cdots b_n$$

for $m \geqq 1,\ n \geqq 2$.

Proof. We proceed by induction on n. We have

$$
\begin{aligned}
&a_1 \cdots a_m z, z b_1 \cdots b_n \\
&a_1 \cdots a_m z, z b_1 \cdots b_{n-2} z_1, z_1 b_{n-1} b_n \\
&a_1 \cdots a_m b_1 \cdots b_{n-2} z_1, z_1 b_{n-1} b_n \\
&a_1 \cdots a_m b_1 \cdots b_n
\end{aligned}
$$

□

Lemma 44 *We have*

$$a_1 \cdots a_m b_1 \cdots b_n \to (\exists z)(a_1 \cdots a_m z \wedge z b_1 \cdots b_n)$$

for $m \geqq 1,\ n \geqq 2$.

Proof. We proceed by induction on n.

$$
\begin{aligned}
&a_1 \cdots a_m b_1 \cdots b_n \\
&a_1 \cdots a_m b_1 \cdots b_{n-2} z_1, z_1 b_{n-1} b_n \\
&a_1 \cdots a_m z, z b_1 \cdots b_{n-2} z_1, z_1 b_{n-1} b_n \\
&a_1 \cdots a_m z, z b_1 \cdots b_n
\end{aligned}
$$

□

Lemma 45 *We have*

$$a_1 \cdots a_i \cdots a_s \to a_i a_1 \cdots \widehat{a_i} \cdots a_s$$

for $s = 4, 4 \geqq i \geqq 2$.

Proof. We have

$$\begin{array}{lll} abcd & abcd & abcd \\ abz,\ zcd & bacd & badc \\ baz,\ zcd & bcad & bdac \\ bacd & cbad & dbca \end{array}$$

□

Lemma 46 *We have*

$$a_1 \cdots a_i \cdots a_s \to a_i a_1 \cdots \widehat{a_i} \cdots a_s$$

for $s \geqq 5, s \geqq i \geqq 2$.

Proof. We proceed by induction on s.

1. In case of $i \leqq s - 2$:We have

$$\begin{array}{l} a_1 \cdots a_i \cdots a_s \\ a_1 \cdots a_i \cdots a_{s-2} z,\ z a_{s-1} a_s \\ a_i \cdots a_1 \cdots a_{s-2} z,\ z a_{s-1} a_s \\ a_i \cdots a_1 \cdots a_s \end{array}$$

2. In case of $i = s - 1$:We have

$$\begin{array}{l} a_1 \cdots a_{s-3} a_{s-2} a_{s-1} a_s \\ a_{s-2} \cdots a_{s-3} a_1 a_{s-1} a_s \\ a_{s-2} \cdots a_{s-3} z,\ z a_1 a_{s-1} a_s \\ a_{s-2} \cdots a_{s-3} z,\ z a_{s-1} a_1 a_s \\ a_{s-2} \cdots a_{s-3} a_{s-1} a_1 a_s \\ a_{s-1} \cdots a_{s-3} a_{s-2} a_1 a_s \end{array}$$

3. In case of $i = s$:We have

$$\begin{array}{l} a_1 \cdots a_{s-3} a_{s-2} a_{s-1} a_s \\ a_{s-2} \cdots a_{s-3} a_1 a_{s-1} a_s \\ a_{s-2} \cdots a_{s-3} z,\ z a_1 a_{s-1} a_s \\ a_{s-2} \cdots a_{s-3} z,\ z a_s a_{s-1} a_1 \\ a_{s-2} \cdots a_{s-3} a_{s-1} a_1 \\ a_s \cdots a_{s-3} a_{s-1} a_1 \end{array}$$

□

Lemma 47 *We have*

$$\left.\begin{array}{l} abx_1\cdots x_m, \\ aby \end{array}\right\} \longrightarrow ab \vee bx_1\cdots x_m y$$

for $m \geqq 1$.

Proof. We have

$$\left.\begin{array}{l} abx_1\cdots x_m, \\ aby \end{array}\right\}$$
$$\left.\begin{array}{l} abz,\ zx_1,\ x_m, \\ aby \end{array}\right\}$$
$$(ab \vee bzy), zx_1\cdots x_m$$
$$ab \vee (byz, zx_1\cdots x_m)$$
$$ab \vee bx_1\cdots x_m y$$

□

Lemma 48 *We have*

$$\left.\begin{array}{l} a_1\cdots a_s x_1\cdots x_m, \\ a_1\cdots a_s y \end{array}\right\} \to a_1\cdots a_s \vee a_2\cdots a_s x_1\cdots x_m y$$

for $s \geqq 2,\ m \geqq 1$.

Proof. We proceed by induction on s. We have

$$\left.\begin{array}{l} a_1\cdots a_s x_1\cdots x_m, \\ a_1\cdots a_s y \end{array}\right\}$$
$$\left.\begin{array}{l} a_1\cdots a_s x_1\cdots x_m, \\ a_1\cdots a_{s-1} z, za_s y \end{array}\right\}$$
$$(a_1\cdots a_{s-1} \vee a_2\cdots a_s x_1\cdots x_m z), za_s y$$
$$a_1\cdots a_{s-1} \vee (a_2\cdots a_s x_1\cdots x_m z, za_s y)$$
$$a_1\cdots a_{s-1} a_1\cdots a_{s-1} \vee a_s z \vee a_2\cdots a_s x_1\cdots x_m y$$
$$a_1\cdots a_{s-1} \vee (a_1\cdots a_{s-1} z, za_s) \vee a_2\cdots a_s x_1\cdots x_m y$$
$$a_1\cdots a_{s-1} \vee a_1\cdots a_s \vee a_2\cdots a_s x_1\cdots x_m y$$
$$a_1\cdots a_s \vee a_2\cdots a_s x_1\cdots x_m y$$

□

Now we recapitulate.

Theorem 49 *The two axiomatic systems Axioms* 1–5 *and Axioms* 29–35 *are equivalent.*

Now we want to show the independence of each axiom of Axioms 29–35 from the others.

Proposition 50 *Axiom* 29 *is independent from the other axioms.*

Proof. We have only to imagine a set consisting of a single element, in which all cycles are decreed as "$= 0$". □

Proposition 51 *Axiom* 30 *is independent from the others.*

Proof. We need only to imagine a set consisting of only one element, in which all cycles are decreed as "$\neq 0$". □

Proposition 52 *Axiom* 31 *is independent from the others.*

Proof. We imagine a set consisting of two elements a, b and decree the dependency of cycles as follows:

number of elements of the cycle	1	2	$3 \leqq$
the dependency of the cycle	$\neq 0$	$ab \neq 0$ $= 0$ otherwise	$= 0$

□

Proposition 53 *Axiom* 32 *is independent from the others.*

Proof. We imagine a set consisting of three elements a, b, c, for which the dependency of cycles is decreed as follows:

number of elements of the cycle	1	2	3	$4 \leqq$
the dependency of the cycle	$\neq 0$	$A \neq 0$ $B = 0$	$abc \neq 0$ $= 0$ otherwise	$= 0$

where A is any cycle consisting of three distinct elements, and B is any cycle of length 3 in which an element occurs at least twice. □

Proposition 54 *Axiom* 33 *is independent from the others.*

Proof. We imagine a set consisting of five elements a, b, c, d, e and decree the dependency of cycles as follows:

number of elements of the cycle	1	2	3	4	$5 \leqq$
the dependency of the cycle	$\neq 0$	$A \neq 0$ $B = 0$	$B = 0$ $(abe = 0)$ $(cde) = 0$ $\neq 0$ otherwise	$B = 0$ $(e) = 0$ $((ab)(cd)) = 0$ $\neq 0$ otherwise	$= 0$

where

1. A is an arbitrary cycle consisting of distinct elements;
2. B is an arbitrary cycle in which the same element occurs at least twice;
3. (abe) is an arbitrary cycle consisting of a, b, e in any order, and the same for (cde);
4. (e) is any cycle containing e at least once;
5. $((ab)(cd))$ is one of $abcd$, $bacd$, $abdc$ and $bade$.

□

Proposition 55 *Axiom* 34 *in case of* $s = 1$ *is independent from the others.*

Proof. We imagine a set consisting of three elements a, b, c, for which the dependency of cycles are decreed as follows:

number of elements of the cycle	1	2	3	$4 \leqq$
the dependency of the cycle	$\neq 0$	$= 0$	$aab \neq 0$ $= 0$ otherwise	$= 0$

□

Proposition 56 *Axiom* 34 *in case of* $s \geqq 2$ *is independent from the others.*

Proof. We consider a set consisting of only one element, in which the dependency of cycles is defined as follows:

number of elements of the cycle	1	$2 \leqq, \leqq s+1$	$s+2 \leqq$
the dependency of the cycle	$\neq 0$	$= 0$	$\neq 0$

□

Proposition 57 *Axiom* 35 *in case of* $s = 1$ *is independent from the others.*

Proof. We consider a set consisting of only one element, in which the dependency of cycles is defined as follows:

number of elements of the cycle	$1 \leqq, \leqq 2$	$3 \leqq$
the dependency of the cycle	$\neq 0$	$= 0$

□

Proposition 58 *Axiom* 35 *in case of* $s \geqq 2$ *is independent from the others.*

Proof. We imagine a set consisting of $s+2$ elements which has the following dependency of cycles:

number of elements of the cycle	$1 \leqq, \leqq s+1$	$s+2 \leqq$
the dependency of the cycle	$A \neq 0$ $= 0$ otherwise	$= 0$

where A is any cycle consisting of distinct elements. □

Finally we recapitulate:

Theorem 59 *Each axiom of Axioms* 29–35 *is independent from the others.*

I do not insist that the axiomatic system considered here is in final form.

Acknowledgement In preparing this work, I am indebted to Professor K. Nakamura, who spent much time in reading several versions of the paper until it got its final form. He supported me through critical remarks, valuable advice and kind suggestions. I would like to express my hearty thanks to him on this occasion.

On Mapping Sequences of a Projective Spectrum

Takeo Nakasawa
received on Dec. 13, 1937

Introduction

In the present work we will introduce two new notions, i.e., *mapping sequence* and *projective point sequence*, and by these notions, we study what one may call *Alexandroff's Homeomorphism Theorem*[1]. Namely by means of mapping sequences, we will divide that theorem into several homeomorphism theorems which clearly express the topological character of the mapping sequences, and by the use of projective point sequences, we simplify the proofs of these theorems. Then, we will give a proof of Borsuk's[2] theorem by making use of the notion of mapping sequences. Finally, we consider an extension of the notion of a mapping sequence, and a few results are derived from this extension.

Notation

First, we present detailed definitions of frequently occurring letters or symbols: in this way, we avoid laborious repetition of the same definitions:

1. A or B means a compactum, i.e., a compact metrizable space.

---

[1] P. Alexandroff, *Gestalt und Lage*, Ann. of Math. **30** (1929), p. 134.

[2] K. Borsuk, *Sur les rétracts*, Fund. Math. **17** (1931), p. 165.

2. $\{A_m\}$ and $\{B_m\}$ mean geometrically realized spectrum developments of A and of B respectively, i.e., a projective spectrum constructed by A or B and geometrically realized in A or B.

3. $\{\mathfrak{A}_m\}$ and $\{\mathfrak{B}_m\}$ mean partitions of A and B respectively , of which the nerve sequences are $\{A_m\}$ and $\{B_m\}$ resp, i.e., the nerve of $\mathfrak{A}_m = A_m$ and the nerve of $\mathfrak{B}_m = B_m$.

4. α_m and β_m mean the diameters of the covering $\mathfrak{A}_m$ resp. of the covering $\mathfrak{B}_m$, i.e., $\mathfrak{A}_m = \alpha_m$-covering and $\mathfrak{B}_m = \beta_m$-covering.

5. $d(\mathfrak{A}_m)$ and $d(\mathfrak{B}_m)$ mean *Lebesgue's constants*, which are fixed with respect to the covering $\mathfrak{A}_m$ and with respect to the covering $\mathfrak{B}_m$ respectively (cf. Gestalt und Lage, p. 115, Lemma).

6. a_m (or $a_{i_1 \cdots i_m}$) means a vertex of A_m, and b_m (or $b_{j_1 \cdots j_m}$) a vertex of B_m.

7. A sequence which is a projective sequence and at the same time a sequence of vertices will be called a *projective point sequence*.

For the meaning of other notions, see the cited paper[1] by Alexandroff!

Remark 1 *(Translator's Note) This means the paper in the footnote* 1.

1.

Proposition 2 *If a simplicial mapping f_{ν_m} from A_{ν_m} **into**[3] B_m exists for every m in the way that the relation*

$$(1) \quad f_{\nu_p} \pi_p^q (a_{\nu_q}) \cdot - \cdot \pi_p^q f_{\nu_q}(a_{\nu_q})$$

*takes place in B_p for any p and q with $p \leqq q$, and any vertex $a_{\nu q}$ of A_{ν_q}, then the sequence $\{f_{\nu_m}\}$ of the simplicial mappings f_{ν_m} defines a mapping from A **into** B.*

Remark 3 *(Translator's Note) For vertices x, y of B_p, "$x \cdot - \cdot y$" means that x and y are connected by an edge of B_m This is an Alexandroff notation (c.f. footnote* 1 *and Definition 18 of the present paper)*

[3] See footnote 4

Proof. Let $\{a_{i_1 \cdots i_m}\}$ be a projective point sequence defining a point x of A. Because of the inequality

$$\rho(f_{\nu_p}(a_{i_1 \cdots i_{\nu_p}}), f_{\nu q}(a_{i_1 \cdots i_{\nu_q}})) \leqq 2\beta_p,$$

the sequence $\{f_{\nu_m}(a_{i_1 \cdots i_{\nu_m}})\}$ converges. We denote the accumulation point by y, and we define the mapping f of A into B by $f(x) = y$. As projective point sequences defining x are close to each other, their images by f_{ν_m} are also close to each other. Hence, f is a well-defined mapping of A into B. Now, let $\{x^n\} \to x$ be in A, and $f(x^n) = y^n$ in B. Let $\{a^n_{i_1 \cdots i_m}\}$ be a projective point sequence defining x^n, then for any p a sufficiently large n can be chosen so that the relation $a_{i_1 \cdots i_{\nu_p}} \cdot - \cdot a^n_{i_1 \cdots i_{\nu_p}}$ takes place; therefore we have

$$\rho(f_{\nu_p}(a_{i_1 \cdots i_{\nu_p}}), f_{\nu_p}(a^n_{i_1 \cdots i_{\nu_p}})) < 2\beta_p.$$

On the other hand, from

$$\rho(f_{\nu_p}(a_{i_1 \cdots i_{\nu_q}}), f_{\nu q}(a_{i_1 \cdots i_{\nu q}})) < 2\beta,$$

we have

$$\rho(f_{\nu_p}(a_{i_1 \cdots i_{\nu_q}}), y) \leqq 2\beta_p.$$

Similarly we have

$$\rho(f_{\nu_p}(a^n_{i_1 \cdots i_{\nu_p}}), y^n) \leqq 2\beta_p.$$

Therefore we have

$$\rho(y, y^n) < 6\beta_p.$$

Consequently, f is continuous. □

Proposition 4 *If a simplicial mapping f_{ν_m} from A_{ν_m} **onto**[4] B_m exists for every m in such a way that the relation*

$$(1) \quad f_{\nu_p} \pi^{a}_{p}(a_{\nu q}) \cdot - \cdot \pi^{a}_{p} f_{\nu_q}(a_{\nu q})$$

*takes place in B_p for any p and q, $p \leqq q$, and any vertex a_{ν_q} of A_{ν_q}, then the sequence $\{f_{\nu_m}\}$ of the simplicial mappings f_{ν_m} defines a mapping from A **onto** B.*

---

[4] Here "*into*" means "*onto a proper or the whole of complex (or subset)* " and similarly, "*onto*" means "*onto the whole of the complex (or subset)* ", and a subcomplex Q of a given complex P shall be called the *whole of the complex*, if Q contains all vertices of P.

Proof. According to Proposition 2, it suffices to prove that for any point y of B, at least one point of A is mapped onto y. Let $\{a^m_{i_1 \cdots i_{\nu m}}\}$ be a sequence of points such that points $a^m_{i_1 \cdots i_{\nu m}}$ are mapped by $\{f_{\nu_m}\}$ onto a projective point sequence $\{b_{j_1 \cdots j_m}\}$ defining y, and further let x^m be the accumulation point of the projective point sequence $\{a^m_{i_1 \cdots i_s}\}$ of which the ν_m-th vertex is $a^m_{i_1 \cdots i_{\nu m}}$. Since A is compact, an appropriate chosen subsequence $\{x^{m_\mu}\}$ of the sequence $\{x^m\}$ converges to a point x of A. Let further $f(x^m) = y^m$, $f(x) = y^*$, so that

$$\lim \rho(y, b_{j_1 \ldots j_{m_\mu}}) = 0,$$
$$\lim \rho(b_{j_1 \ldots j_{m_\mu}}, y^{m_\mu}) \leqq \lim 2\beta_{m_\mu} = 0,$$
$$\lim \rho(y^{m_\mu}, y^*) = 0.$$

Therefore $\rho(y, y^*) = 0$. Thus $f(x) = y$. □

Proposition 5 *If a continuous mapping f of A **into** B exists, then a subsequence $\{A_{\nu_m}\}$ of $\{A_m\}$ and a simplicial mapping f_{ν_m} of A_{ν_m} **into** B_m can be found in such a way that for every m the relation*

$$(1) \qquad f_{\nu_p} \pi^q_p (a_{\nu_q}) \cdot - \cdot \pi^a_p f_{\nu_q}(a_{\nu_q}) \qquad (3)$$

takes place in B_p for any p and q, $p \leqq q$ and any vertex a_{ν_q} of A_{ν_q}, and in such a way that the sequence $\{f_{\nu_m}\}$ defines the mapping f.

Proof. If one sets $f(A) = A'$, then the sequence $\{f(\mathfrak{A}_m)\}$ is a sequence of A'. Then a sufficiently large ν_m can be found for any m in such a way that the diameter of $f(\mathfrak{A}_{\nu_m})$ is smaller than the Lebesgue constant $d(\mathfrak{B}_m)$ of $\mathfrak{B}_m$. Then the complex $f(A_{\nu_m})$ in B can be mapped via a simplicial mapping φ_{ν_m} into B_m. Because of the inequality

$$\rho(f(a_{i_1 \cdots i_{\nu_p}}), f(a_{i_1 \cdots i_{\nu_q}})) < d(\mathfrak{B}_p),$$

we have

$$f_{\nu_p}(a_{i_1 \ldots i_{\nu_p}}) \cdot - \cdot \pi^q_p f_{\nu_q}(a_{i_1 \ldots i_{\nu_q}})$$

in B_p, which is exactly condition (1). Because of the inequality

$$\rho(f(a_{i_1 \cdots i_{\nu m}}), f_{\nu m}(a_{i_1 \cdots i_{\nu m}})) < \beta_m,$$

we have also

$$\lim f_{\nu_m}(a_{i_1 \ldots i_{\nu m}}) = \lim f(a_{i_1 \ldots i_{\nu m}}) = f(x).$$

Hence the sequence $\{f_{\nu_m}\}$ defines the mapping f. □

Proposition 6 *If a continuous mapping f of A onto B exists, then a subsequence $\{A_{\nu m}\}$ of $\{A_m\}$ and a simplicial mapping $f_{\nu m}$ of $A_{\nu m}$* **onto** *B_m can be found in such a way that for every m, the relation*

$$(1) \quad f_{\nu_p}\pi^{q}_{p}(a_{\nu_q}) \cdot - \cdot \pi^{q}_{p} f_{\nu_q}(a_{\nu_q})$$

takes place in B_p for any p and q with $p \leqq q$ and any vertex a_{ν_q} of A_{ν_q}, and that the sequence $\{f_{\nu m}\}$ defines the mapping f.

Proof. If we take a sufficiently small Lebesgue constant in the proof of Proposition 5, then, because of the equality $f(A) = B$, $f(A_{\nu m})$ is mapped onto B_m via a simplicial mapping. What remains to be proved goes very analogously to that of Proposition 5. □

Now, we will give a definition of the mapping sequence in detail:

Definition 7 *A sequence $\{f_{\nu m}\}$ of simplicial mappings which satisfies the condition* (1) *shall be called a mapping sequence of $\{A_m\}$ into (or onto) $\{B_m\}$.*

With this notion, we can express the propositions established above in the following form:

Theorem 8 *Continuous mappings of A into B can be completely characterized by the mapping sequences of $\{A_m\}$ into $\{B_m\}$.*

Theorem 9 *Continuous mappings of A onto B can be completely characterized by the mapping sequences of $\{A_m\}$ onto $\{B_m\}$.*

2.

Theorem 10 [5]*If A and B are homeomorphic to each other if and only if a subsequence $\{A_{\lambda_{2m-1}}\}$ of $\{A_m\}$, a subsequence $\{B_{\lambda_{2m}}\}$of $\{B_m\}$, and, for every m, a simplicial mapping $f_{\lambda_{2m+1}}$of $A_{\lambda_{2m+1}}$ into $B_{\lambda_{2m}}$ and a simplicial mapping $g_{\lambda_{2m}}$ of $B_{\lambda_{2m}}$into $A_{\lambda_{2m-1}}$can be found in such a way that the relations*

$$\begin{aligned}(A) \quad & f_{\nu_{2p+1}}\pi^{2q+1}_{2p+1}(a_{\lambda_{2q+1}}) \cdot - \cdot \pi^{2q}_{2p} f_{\lambda_{2q+1}}(a_{\lambda_{2q+1}}),\\ & g_{\lambda_{2p}}\pi^{2q}_{2p}(b_{\lambda_{2q}}) \cdot - \cdot \pi^{2q-1}_{2p-1} g_{\lambda_{2q}}(b_{\lambda_{2q}}),\end{aligned}$$

$$\begin{aligned}(B) \quad & g_{\lambda_{2q}} f_{\lambda_{2q+1}}(a_{\lambda_{2q+1}}) \cdot - \cdot \pi^{2q+1}_{2q-1}(a_{\lambda_{2q+1}}),\\ & f_{\lambda_{2q-1}} g_{\lambda_{2q}}(b_{\lambda_{2q}}) \cdot - \cdot \pi^{2q}_{2q-2}(b_{\lambda_{2q}})\end{aligned}$$

---

[5] This is literally the same as , so-called, Alexandroff's *Homeomorphism theorem* (Ann. of Math., **30** (1929), p. 134).

in $B_{\lambda_{2p}}$, in $A_{\lambda_{2p-1}}$, in $A_{\lambda_{2q-1}}$, and in $B_{\lambda_{2q-2}}$ respectively, take place for any p and q with $p \leq q$, and any vertex $a_{\lambda_{2q+1}}$ of $A_{\lambda_{2q+1}}$ and any vertex $b_{\lambda_{2q}}$ of $B_{\lambda_{2q}}$ resp.

Proof. Let A and B be homeomorphic to each other. Then, according to Proposition 5, a subsequence $\{A_{\nu_m}\}$ of $\{A_m\}$, a subsequence $\{B_{\mu_m}\}$ of $\{B_m\}$, and for every m a simplicial mapping f_{ν_m} of A_{ν_m} into B_m and a simplicial mapping g_{μ_m} of B_{μ_m} into A_m can be found in such way that the two relations of (A) are fulfilled. If we choose further subsequences $\{A_{\lambda_{2m-1}}\}$ resp. $\{B_{\lambda_{2m}}\}$ of $\{A_{\nu_m}\}$ resp. $\{B_{\mu_m}\}$ following the rule illustrated below:

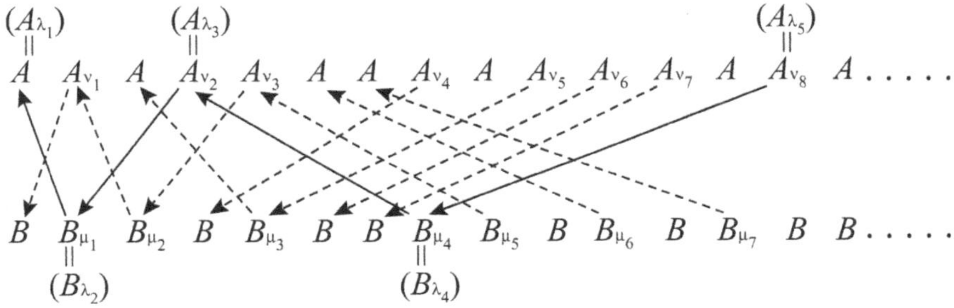

Then we have

$$\rho(gf_{\lambda_{2q+1}}(a_{i_1 \cdots i_{\lambda_{2q+1}}}), gf(a_{i_1 \cdots i_{\lambda_{2q+1}}})) < d(\mathfrak{U}_{\lambda_{2q-1}})^6$$

so that we have

$$g_{\lambda_{2q}} f_{\lambda_{2q+1}}(a_{i_1 \cdots i_{\lambda_{2q+1}}}) \cdot - \cdot a_{i_1 \cdots i_{\lambda_{2q-1}}}.$$

This is the first relation of (B), and the second is proved in a similar way. Consequently, the condition is necessary. In order to prove the sufficiency, it is enough, using the notation of Proposition 2 to show that $gf(x) = x$ in A and $fg(y) = y$ in B. Let $\{a_{i_1 \cdots i_m}\}$ be the projective point sequence defining a point x of A; then the sequence $\{f_{\lambda_{2m+1}}(a_{i_1 \cdots i_{\lambda_{2m+1}}})\}$, which is written as $\{b^{2m+1}_{j_1 \cdots j_{\lambda_{2m}}}\}$, converges to the point $f(x)$ in B which is also called y. Further, let $\{b_{j_1 \cdots j_{\lambda_{2m}}}\}$be a projective point sequence defining y. Then for any k, a sufficiently large n exists so that the relation $b_{j_1 \cdots j_{\lambda_{2k}}} \cdot - \cdot b^{2n+1}_{j_1 \cdots j_{\lambda_{2k}}}$ is valid. Therefore we have

$$\begin{aligned} &g_{\lambda_{2k}}(b_{j_1 \cdots j_{\lambda_{2k}}}) \cdot - \cdot g_{\lambda_{2k}} \pi^{2n}_{2k}(b^{2n+1}_{j_1 \cdots j_{\lambda_{2n}}}) \cdot - \cdot \pi^{2n-1}_{2k-1} g_{\lambda_{2n}}(b^{2n+1}_{j_1 \cdots j_{\lambda_{2n}}}) \\ &= \pi^{2n-1}_{2k-1} g_{\lambda_{2n}} f_{\lambda_{2n+1}}(a_{i_1 \cdots i_{\lambda_{2n+1}}}) \cdot - \cdot \pi^{2n+1}_{2k-1}(a_{i_1 \cdots i_{\lambda_{2n+1}}}) = a_{i_1 \cdots i_{\lambda_{2k-1}}} \end{aligned}$$

6 Here g just means f^{-1}.

Thus we have

$$g_{\lambda_{2k}}(b_{j_1\cdots j_{\lambda_{2k}}})\cdot - \cdot - \cdot - \cdot a_{i_1\cdots i_{\lambda_{2k-1}}}$$

so that

$$gf(x) = \lim g_{\lambda_{2k}}(b_{j_1\cdots j_{\lambda_{2k}}}) = \lim a_{i_1\cdots i_{\lambda_{2k-1}}} = x.$$

Therefore, $gf(x) = x$. The relation $fg(y) = y$ is to be proved in a similar way. □

One can modify the condition of Theorem 10 a little by replacing the condition of being mapped *into* with a stronger one, i.e., $A_{\lambda_{2m+1}}$ being mapped *onto* the whole complex $B_{\lambda_{2m}}$, and removing the second relation of (B). The condition being so changed, it is again necessary and sufficient, because the modified condition is actually necessary according to Proposition 6, and the necessity proof of the Theorem 10 remains valid. On the other hand, $f(A) = B$ follows from Proposition 4, and gf = identity is valid according to the sufficiency proof of Theorem 10; then, the second formula of (B) is not used at all, by which the condition is sufficient. Consequently, one can state the following theorem:

Theorem 11 *A and B are homeomorphic to each other if and only if a subsequence $\{A_{\lambda_{2m-1}}\}$ of $\{A_m\}$, a subsequence $\{B_{\lambda_{2m}}\}$ of $\{B_m\}$, and for every m a simplicial mapping $f_{\lambda_{2m+1}}$ of $A_{\lambda_{2m+1}}$ onto $B_{\lambda_{2m}}$ and a simplicial mapping $g_{\lambda_{2m}}$ of $B_{\lambda_{2m}}$ into $A_{\lambda_{2m-1}}$ can be found in such a way that the relations*

$$(A)\quad f_{\lambda_{2p+1}}\pi^{2q+1}_{2p+1}(a_{\lambda_{2p+1}})\cdot - \cdot \pi^{2q}_{2p} f_{\lambda_{2q+1}}(a_{\lambda_{2q+1}}),$$
$$g_{\lambda_{2p}}\pi^{2q}_{2p}(b_{\lambda_{2q}})\cdot - \cdot \pi^{2q-1}_{2p-1} q_{\lambda_{2q}}(b_{\lambda_{2q}}),$$

$$(B)\quad g_{\lambda_{2q}} f_{\lambda_{2q+1}}(a_{\lambda_{2q+1}})\cdot - \cdot \pi^{2q+1}_{2q-1}(a_{\lambda_{2q+1}})$$

in $B_{\lambda_{2p}}$, in $A_{\lambda_{2p-1}}$ and $A_{\lambda_{2q-1}}$ respectively, take place for any p and q with $p \leqq q$, and any vertex $a_{\lambda_{2q+1}}$ of $A_{\lambda_{2q+1}}$ and any vertex of $b_{\lambda_{2q}}$ of $B_{\lambda_{2q}}$.

3.

Proposition 12 *Let $\{f_{\lambda_m}\}$ and $\{g_{\mu_m}\}$ be mapping sequences of $\{A_m\}$ into $\{B_m\}$, and let f and g be the continuous mappings which $\{f_{\lambda_m}\}$ and $\{g_{\mu_m}\}$ define respectively. If for a given ε, a sufficiently large s exists so that the relation*[7]

$$(C)\quad f_{\lambda_s}(a_{i_1\cdots i_{\lambda_s}})\cdot - \cdot - \cdot - \cdot g_{\mu_s}(a_{i_1\cdots i_{\mu_s}})$$

[7] It would be necessary to point out that the indices of $a_{i_1\cdots i_{\lambda_s}}$ and $a_{i_1\cdots i_{\mu_s}}$, i.e., two vertices in a single projective point sequence, are involved (in the condition: translator's supplement).

takes place in B_s, for any vertex $a_{i_1 \cdots i_{\lambda_s}}$ of A_{λ_s} and $a_{i_1 \cdots i_{\mu_s}}$ of A_{μ_s}, then the inequality $\rho(f(x),\ g(x)) < \varepsilon$ is valid in B for any point x of A.

Proof. Take s so large that the inequality $\varepsilon > 8\beta_s$ is valid. Let $\{a_{i_1 \cdots i_m}\}$ be the projective point sequence defining a point x of A. Then we have

$$\rho(f_{\lambda_s}(a_{i_1 \cdots i_{\lambda_s}}), f(x)) \leqq 2\beta_s,$$
$$\rho(g_{\mu_s}(a_{i_1 \cdots i_{\mu_s}}), g(x)) \leqq 2\beta_s.$$

On the other hand, according to (C), we have

$$\rho(f_{\lambda_s}(a_{i_1 \cdots i_{\lambda_s}}), g_{\mu_s}(a_{i_1 \cdots i_{\mu_s}})) \leqq 4\beta_s.$$

Therefore we have

$$\rho(f(x), g(x)) < 8\beta_s < \varepsilon$$

in B. □

Proposition 13 *Let $\{f_{\lambda_m}\}$ and $\{g_{\mu_m}\}$ be mapping sequences of $\{A_m\}$ into $\{B_m\}$, and let f and g be the continuous mappings which $\{f_{\lambda_m}\}$ and $\{g_{\mu_m}\}$ define respectively. If for a given s, a sufficiently small ε exists so that the inequality $\rho(f(x), g(x)) < \varepsilon$ in B is valid for any point x of A, then, the following relation in B_s is valid for any vertex $a_{i_1 \cdots i_{\lambda_s}}$ of A_{λ_s} resp. $a_{i_1 \cdots i_{\mu_s}}$ of A_{μ_s}:*

$$(C) \quad f_{\lambda_s}(a_{i_1 \cdots i_{\lambda_s}}) \cdot - \cdot - \cdot - \cdot g_{\mu_s}(a_{i_1 \cdots i_{\mu_s}}).$$

Proof. Take ε so small that the inequality $d(\mathfrak{B}_s) > \varepsilon$ is valid, and further n so large that the inequality $d(\mathfrak{B}_s) - \varepsilon > 4\beta_n$ is valid. Then we have

$$\rho(f_{\lambda_n}(a_{i_1 \cdots i_{\lambda n}}), f(x)) \leqq 2\beta_n,$$
$$\rho(g_{\mu_n}(a_{i_1 \cdots i_{\lambda n}}), g(x)) \leqq 2\beta_n.$$

On the other hand, $\rho(f(x), g(x)) < \varepsilon$ is valid according to the assumption. So we have

$$\rho(f_{\lambda_n}(a_{i_1 \cdots i_{\lambda n}}), g_{\mu_n}(a_{i_1 \cdots i_{\mu n}})) < 4\beta_n + \varepsilon < d(\mathfrak{B}_s)$$

so that we have

$$\pi_s^n f_{\lambda_n}(a_{i_1 \cdots i_{\lambda n}}) \cdot - \cdot \pi_s^n g_{\mu_n}(a_{i_1 \cdots i_{\mu n}}).$$

On the other hand, we have, by condition (A), that

$$f_{\lambda_s}(a_{i_1 \cdots i_{\lambda s}}) \cdot - \cdot \pi_s^n f_{\lambda_n}(a_{i_1 \cdots i_{\lambda n}}),$$
$$g_{\mu_s}(a_{i_1 \cdots i_{\mu_s}}) \cdot - \cdot \pi_s^n g_{\mu_n}(a_{i_1 \cdots i_{\mu n}}).$$

Therefore we have

$$f_{\lambda_s}(a_{i_1 \cdots i_{\lambda_s}}) \cdot - \cdot - \cdot - \cdot g_{\mu_s}(a_{i_1 \cdots i_{\mu_s}}).$$

□

The following theorems follow from these:

Theorem 14 *Two continuous mappings of A into B have so called "small distance" if and only if their mapping sequences of $\{A_m\}$ into $\{B_m\}$ fulfill the condition (C) for sufficiently large s.*

Theorem 15 *Two mapping sequences $\{f_{\lambda_m}\}$, $\{g_{\mu_m}\}$ of $\{A_m\}$ into $\{B_m\}$ define the same continuous mapping of A into B if and only if the condition (C) is fulfilled for any $\{A_m\}$ and $\{B_m\}$.*

Theorem 16 *That the condition (C) is fulfilled for any m is an equivalence relation on the set of the mapping sequences.*

Using the notion of mapping sequences, I will give a new proof of a theorem of Borsuk[8].

Theorem 17 ***(Borsuk's Theorem)*** *The mapping space[9] is separable.*

Proof. For any fixed number λ_m, we can divide the set of the mapping sequences of which the mth parts map A_{λ_m} to B_m into classes by considering that two mapping sequences are equal if their simplicial mappings of A_{λ_m} into B_m produce the same vertex-assignment. Since the number of simplicial mappings of A_{λ_m} into B_m is obviously finite, the number of the classes is finite, too. From every class, we pick a mapping sequence as a representative of the class; let $f^{\ m}_{1\lambda_m}, f^{\ m}_{2\lambda_m}, \cdots, f^{\ m}_{l(\lambda_m)\lambda_m}$ be these representatives. Then, the set of all representatives $M = \sum_{m-1}^{\infty} \sum_{\lambda_m=m}^{\infty} \sum_{i=1}^{l(\lambda_m)} f^{\ m}_{i\lambda_m}$ is obviously a countable set. Now, we consider any mapping sequence $g = \{g_{\mu_m}\}$, and then, we take a finite subset $f^{\ s}_{1\mu_s}, f^{\ s}_{2\mu_s}, \cdots, f^{\ s}_{l(\mu_s)\mu_s}$ from M with respect to μ_m and for any fixed s, so that g belongs to the same class with some mapping sequence of $\{f^{\ s}_{i\mu_s}, \cdots, f^{\ s}_{l(\mu_s)\mu_s}\}$ according to the above classification of simplicial mappings of A_{μ_s} into B_s. Now, for example, g belongs to the same class with $f^{\ s}_{i\mu_s}$, and let $f^{\ s}_{i\mu_s} = \{f_{\nu_1}, f_{\nu_2}, \cdots, f_{\nu_{s-1}}, f_{\mu_s}, f_{\nu_{s+1}}, \cdots\}$,

---

8 We define a metric on the set of continuous mappings of A into B by $\rho(f, g) = Max\ \rho(f(x), g(x))$ in B. We call the so metrized set of mappings the *mapping space* of A into B. The theorem that the mapping space is bounded, complete, and, separable is proved by *K. Borsuk* (K. Borsuk, Sur les retracts, Fund. Math. 17 (1931), p. 164–165). What I consider here is only the proof of separability.

9 The same as 8

so for any vertex $a_{i_1 \cdots i_{\mu_s}}$ of A_{μ_s} $f_{\mu_s}(a_{i_1 \cdots i_{\mu_s}}) = g_{\mu_s}(a_{i_1 \cdots i_{\mu_s}})$ is valid; this is the condition (C). Therefore, the inequality $\rho(f^{s}_{i}, g) < \varepsilon$ is valid according to Proposition 12; Thus, the countable set M lies in the mapping space everywhere densely. Therefore, the mapping space is separable. □

4.

Definition 18 *In these paragraphs, k, k_1, $\cdots$, k_6 mean given constants, independent of each other. And "$a \cdot - (k) - \cdot b$" means that one can connect a and b with a sequence consisting of at most k edges, and a is said to be k-neighboured by b. Especially, a is neighboured by b in the ordinary sense of Alexandroff if $k = 1$.*

Definition 19 *A vertex-to-vertex mapping f of A_{ν_m} into B_m is called a pseudo simplicial mapping if the relation $f(a) \cdot - (k) - \cdot f(b)$ in B_m follows from $a \cdot - \cdot b$ in A_{ν_m}. The pseudo simplicial mapping is obviously the usual simplicial mapping for $k = 1$.*

Definition 20 *A sequence $\{f_{\nu_m}\}$ of pseudo simplicial mappings f_{ν_m} $(m = 1, 2, \cdots)$ of A_{ν_m} into (onto resp.) B_m is called a mapping sequence in the wider sense of $\{A_m\}$ into (onto resp.) $\{B_m\}$ if the relation*

$$(A') \quad f_{\nu_p} \pi^{q}_{p}(a_{\nu_q}) \cdot - (k_2) - \cdot \pi^{q}_{p} f_{\nu_p}(a_{\nu_q}) \quad in \quad B_p$$

takes place for any p and q with $p \leq q$, and for any vertex a_{ν_q} of A_{ν_q}. And a mapping sequence in the wider sense is a usual mapping sequence for $k_1 = k_2 = 1$.

Remark 21 *(Translator's Note) k_1 refers to the constant required in Definition 19 of the pseudo simplicial mappings $\{f_{\nu_m}\}$.*

Proposition 22 *A mapping sequence in the wider sense of $\{A_m\}$ into (onto resp.) $\{B_m\}$ defines a continuous mapping of A into (onto resp.) B.*

The proof goes quite similarly to the proof of Proposition 2 (Proposition 4 resp.).

Theorem 23 *A and B are homeomorphic to each other if a subsequence $\{A_{\lambda_{2m-1}}\}$ of $\{A_m\}$, a subsequence $\{B_{\lambda_{2m}}\}$ of $\{B_m\}$ and for any m, a pseudo simplicial mapping $f_{\lambda_{2m+1}}$ of $A_{\lambda_{2m+1}}$ into $B_{\lambda_{2m}}$ and a pseudo simplicial mapping $g_{\lambda_{2m}}$ of $B_{\lambda_{2m}}$ into $A_{\lambda_{2m-1}}$ can be found in such a way that the relations*

$$\begin{aligned} (A') \quad & f_{\lambda_{2p+1}} \pi^{2q+1}_{2p+1}(a_{\lambda_{2q+1}}) \cdot - (k_3) - \cdot \pi^{2q}_{2p} f_{\lambda_{2q+1}}(a_{\lambda_{2q+1}}), \\ & g_{\lambda_{2p}} \pi^{2q}_{2p}(b_{\lambda_{2q}}) \cdot - (k_4) - \cdot \pi^{2q-1}_{2p-1} g_{\lambda_{2q}}(b_{\lambda_{2q}}), \end{aligned}$$

$$(B') \quad g_{\lambda_{2q}} f_{\lambda_{2q+1}}(a_{\lambda_{2q+1}}) \cdot - (k_5) - \cdot \pi^{2q+1}_{2q-1}(a_{\lambda_{2q+1}}),$$
$$f_{\lambda_{2q-1}} g_{\lambda_{2q}}(b_{\lambda_{2q}}) \cdot - (k_6) - \cdot \pi^{2q}_{2q-2}(b_{\lambda_{2q}})$$

take place in $B_{\lambda_{2p}}$, in $A_{\lambda_{2p-1}}$, in $A_{\lambda_{2q-2}}$, and in $B_{\lambda_{2q-2}}$ respectively for any p and q with $p \leq q$, and any vertex $a_{\lambda_{2q+1}}$ of $A_{\lambda_{2q+1}}$ and any vertex $b_{\lambda_{2q}}$ of $B_{\lambda_{2q}}$ respectively.

The proof goes quite similarly to the sufficiency proof of Theorem 10.

Theorem 24 *A and B are homeomorphic to each other if for a subsequence $\{f_{\lambda_{2m-1}}\}$ of $\{A_m\}$, a subsequence $\{B_{\lambda_{2m}}\}$ of $\{B_m\}$, and for any m, a pseudo simplicial mapping $f_{\lambda_{2m+1}}$ of $A_{\lambda_{2m+1}}$ onto $B_{\lambda_{2m}}$ and a pseudo simplicial mapping $g_{\lambda_{2m}}$ of $B_{\lambda_{2m}}$ into $A_{\lambda_{2m-1}}$ can be found in such a way that the relations*

$$(A') \quad f_{\lambda_{2p+1}} \pi^{2q+1}_{2p+1}(a_{\lambda_{2q+1}}) \cdot - (k_3) - \cdot \pi^{2q}_{2p} f_{\lambda_{2q+1}}(a_{\lambda_{2q+1}}),$$
$$g_{\lambda_{2p}} \pi^{2q}_{2p}(b_{\lambda_{2q}}) \cdot - (k_4) - \cdot \pi^{2q-1}_{2p-1} g_{\lambda_{2q}}(b_{\lambda_{2q}}),$$

$$(B') \quad g_{\lambda_{2q}} f_{\lambda_{2q+1}}(a_{\lambda_{2q+1}}) \cdot - (k_5) - \cdot \pi^{2q+1}_{2q-1}(a_{\lambda_{2q+1}})$$

take place for any p and q with $p \leq q$, and any vertex $a_{\lambda_{2q+1}}$ of $A_{\lambda_{2q+1}}$ and any vertex $b_{\lambda_{2q}}$ of $B_{\lambda_{2q}}$ respectively, and in $B_{\lambda_{2p}}$, in $A_{\lambda_{2p-1}}$ and in $A_{\lambda_{2q-1}}$ respectively.

The proof goes quite similarly to the sufficiency proof of Theorem 11.

Proposition 25 *Let $\{f_{\lambda_m}\}$ and $\{g_{\mu_m}\}$ be mapping sequences in the wider sense of $\{A_m\}$ into $\{B_m\}$, and let f and g be the continuous mappings which $\{f_{\lambda_m}\}$ and $\{g_{\mu_m}\}$ define respectively. If for given ε, there is a sufficiently large s so that the relation*

$$(C') \quad f_{\lambda_s}(a_{i_1 \cdots i_{\lambda_s}}) \cdot - (k_5) - \cdot g_{\mu_s}(a_{i_1 \cdots i_{\mu_s}})$$

takes place in B_s for any vertex $a_{i_1 \cdots i_{\lambda_s}}$ of A_{λ_s} resp. $a_{i_1 \cdots i_{\mu_s}}$ of A_{μ_s}, then the inequality $\rho(f(x), g(x)) < \varepsilon$ in B is valid for any point x of A.

The proof goes quite similarly to the proof of Proposition 12.

Proposition 26 *Let $\{f_{\lambda_m}\}$ and $\{g_{\mu_m}\}$ be mapping sequences in the wider sense of $\{A_m\}$ into $\{B_m\}$, and let f and g be the continuous mappings which $\{f_{\lambda_m}\}$ and $\{g_{\mu_m}\}$ define respectively. If for a given s, there is a sufficiently small ε so that the inequality $\rho(f(x), g(x)) < \varepsilon$ in B is valid for any point x of A, then the following relation in B_s is valid for any vertex $a_{i_1 \cdots i_{\lambda_s}}$ of A_{λ_s} and any vertex. $a_{i_1 \cdots i_{\mu_s}}$ of A_{μ_s} respectively:*

$$(C') \quad f_{\lambda_s}(a_{i_1 \cdots i_{\lambda_s}}) \cdot - (k_5) - \cdot g_{\mu_s}(a_{i_1 \cdots i_{\mu_s}}).$$

The proof goes quite similarly to the proof of Proposition 13. It should be pointed out here that $k_5 \geqq k_3 + k_4 + 1$ should hold.

Remark 27 *(Translator's Note) k_3 and k_4 refer to the constant required $\{f_{\lambda_m}\}$ of $\{g_{\mu_m}\}$ and respectively in Definition 19. The above remark states that k_5, chosen appropriately so that the relation (C) holds, should satisfy that inequality.*

Theorem 28 *Two mapping sequences in the wider sense $\{f_{\lambda_m}\}$ and $\{g_{\mu_m}\}$ of $\{A_m\}$ into $\{B_m\}$ define the same continuous mapping of A into B if the condition (C') is fulfilled for each $\{A_m\}$ and $\{B_m\}$.*

Theorem 29 *If, for two mapping sequences $\{f_{\lambda_m}\}$ and $\{g_{\mu_m}\}$ of $\{A_m\}$ into $\{B_m\}$, the relation*

$$f_{\lambda_m}(a_{i_1 \cdots i_{\lambda_m}}) \cdot - (k) - \cdot g_{\mu_m}(a_{i_1 \cdots i_{\mu_m}})$$

is valid for any m, then the relation

$$f_{\lambda_m}(a_{i_1 \cdots i_{\lambda_m}}) \cdot - \cdot - \cdot - \cdot g_{\mu_m}(a_{i_1 \cdots i_{\mu_m}})$$

is valid for any m.

Remark 30 *(Translator's Note) No indication on k is given in the paper, which makes the above statement unclear.*

BIRKHÄUSER

Science Networks – Historical Studies (SNHS)

Edited by
Eberhard Knobloch, Technische Universität Berlin, Germany
Helge Kragh, University of Aarhus, Denmark
Erhard Scholz, Bergische Universität Wuppertal, Germany
In cooperation with an international editorial board

The publications in this series are limited to the fields of mathematics, physics, astronomy, and their applications. The publication language is preferentially English. The series is primarily designed to publish monographs. Annotated sources and exceptional biographies might be accepted in rare cases. The series is aimed primarily at historians of science and libraries; it should also appeal to interested specialists, students, and diploma and doctoral candidates. In cooperation with their international editorial board, the editors hope to place a unique publication at the disposal of science historians throughout the world.

SNHS 37: Kragh, H.
The Moon that Wasn't. The Saga of Venus' Spurious Satellite (2008)
ISBN 978-3-7643-8908-6

SNHS 36: Kvasz, L.
Patterns of Change. Linguistic Innovations in the Development of Classical Mathematics (2008)
ISBN 978-3-7643-8839-3

This book offers a reconstruction of linguistic innovations in the history of mathematics; innovations which changed the ways in which mathematics was done, understood and philosophically interpreted. It argues that there are at least three ways in which the language of mathematics has been changed throughout its history, thus determining the lines of development that mathematics has followed.
One of these patterns of change, called a re-coding, generates two developmental lines. The first of them connecting arithmetic, algebra, differential and integral calculus and predicate calculus led to a gradual increase of the power of our calculating tools, turning difficult problems of the past into easy exercises. The second developmental line connecting synthetic geometry, analytic geometry, fractal geometry, and set theory led to a sophistication of the ways we construct geometrical objects, altering our perception of form and increasing our sensitivity to complex visual patterns.
Another important pattern of change, called relativization, is illustrated by the development of synthetic geometry, connecting Euclid's geometry, projective geometry, non-Euclidean geometry, and Klein's Erlanger Programm up to Hilbert's Grundlagen der Geometrie. In this development the notions of space and geometric object underwent deep and radical changes culminating in the liberation of objects from the supremacy of space and so bringing to existence geometric objects which space would never tolerate.

SNHS 35: Caramalho Domingues, J.
Lacroix and the Calculus (2008)
ISBN 978-3-7643-8637-5

Silvestre François Lacroix (Paris, 1765 – ibid., 1843) was a most influential mathematical book author. His most famous work is the three-volume *Traité du calcul différentiel et du calcul intégral* — an encyclopedic appraisal of 18th-century calculus which remained the standard reference on the subject through much of the 19th century, in spite of Cauchy's reform of the subject in the 1820's.
Lacroix and the Calculus is the first major study of Lacroix's large *Traité*. It uses the unique and massive bibliography given by Lacroix to explore late 18th-century calculus, and the way it is reflected in Lacroix's account. Several particular aspects are addressed in detail, including: the foundations of differential calculus, analytic and differential geometry, conceptions of the integral, and types of solutions of differential equations (singular/complete/general integrals, geometrical interpretations, and generality of arbitrary functions).

SNHS 34: Høyrup, J.
Jacopo da Firenze's *Tractatus Algorismi* and Early Italian Abbacus Culture (2007)
ISBN 978-3-7643-8390-9

Science Networks – Historical Studies (SNHS)

BIRKHÄUSER

SNHS 33: De Risi, V.
Geometry and Monadology. Leibniz's *Analysis Situs* and Philosophy of Space (2007)
ISBN 978-3-7643-7985-8

SNHS 32: Krömer, R.
Tool and Object. A History and Philosophy of Category Theory (2007)
ISBN 978-3-7643-7523-2

SNHS 31: Keller, A.
Expounding the Mathematical Seed. Vol. 2: The Supplements (2006). ISBN 978-3-7643-7292-7

SNHS 30: Keller, A.
Expounding the Mathematical Seed. Vol. 1: The Translation (2006). ISBN 978-3-7643-7291-0

SNHS 30/31 Set: ISBN 978-3-7643-7299-6

SNHS 29: Guerraggio, A. / Nastasi, P.
Italian Mathematics Between the Two World Wars (2005). ISBN 3-7643-6555-2

SNHS 28: Hesseling, D.
Gnomes in the Fog. The Reception of Brouwer's Intuitionism in the 1920s (2003)
ISBN 978-3-7643-6536-3

SNHS 27: Dauben, J.W. / Scriba, C.J.
Writing the History of Mathematics – Its Historical Development (2002)
ISBN 978-3-7643-6166-2 (Hardcover)
ISBN 978-3-7643-6167-9 (Softcover)

SNHS 26: Israel, G. / Millán Gasca, A.
The Biology of Numbers. The Correspondence of Vito Volterra on Mathematical Biology (2002)
ISBN 978-3-7643-6514-1

SNHS 25: Siegmund-Schultze, R.
Rockefeller and the Internationalization of Mathematics Between the Two World Wars. (2001). ISBN 978-3-7643-6468-7

SNHS 24: Jensen, C.
Controversy and Consensus: Nuclear Beta Decay 1911–1934. (2000)
ISBN 978-3-7643-5313-1

SNHS 23: Ferreirós, J.
Labyrinth of Thought. A History of Set Theory and its Role in Modern Mathematics (2001)
ISBN 978-3-7643-5749-8

SNHS 22: Marage, P. / Wallenborn, G. (eds.)
The Solvay Councils and the Birth of Modern Physics (1999). ISBN 978-3-7643-5705-4

SNHS 21: Sakarovitch, J.
Épures d'architecture. De la coupe des pierres à la géométrie descriptive XVI–XIX siècles (1998)
ISBN 978-3-7643-5701-6

SNHS 20: Grattan-Guinness, I. / Bornet, G.
George Boole – Selected Manuscripts on Logic and its Philosophy (1997)
ISBN 978-3-7643-5456-5

SNHS 19: Ullmann, D.
Chladni und die Entwicklung der Akustik 1750–1860 (1996). ISBN 978-3-7643-5398-8

SNHS 18: Hentschel, K. (ed.)
Physics and National Socialism. An Anthology of Primary Sources (1996)
ISBN 978-3-7643-5312-4

SNHS 17: Corry, L.
Modern Algebra and the Rise of Mathematical Structures (1996). ISBN 978-3-7643-5311-7

SNHS 16: Yavetz, I.
From Obscurity to Enigma. The Work of Oliver Heaviside, 1872–1889 (1995)
ISBN 978-3-7643-5180-9

SNHS 15: Sasaki, Ch. / Sugiura, M. / Dauben, J.W., The Intersection of History and Mathematics (1994). ISBN 978-3-7643-5029-1

GPSR Compliance
The European Union's (EU) General Product Safety Regulation (GPSR) is a set of rules that requires consumer products to be safe and our obligations to ensure this.

If you have any concerns about our products, you can contact us on

ProductSafety@springernature.com

In case Publisher is established outside the EU, the EU authorized representative is:

Springer Nature Customer Service Center GmbH
Europaplatz 3
69115 Heidelberg, Germany

www.ingramcontent.com/pod-product-compliance
Ingram Content Group UK Ltd.
Pitfield, Milton Keynes, MK11 3LW, UK
UKHW050116300526
471597UK00005BA/209
* 9 7 8 3 7 6 4 3 8 5 7 2 9 *